Martian Geometry Book 2

Preface

This preface refers to twelve new books of Martian anomalies. There are also 12 in eBook form, the corresponding eBook should be free with the purchase of a paperback. The eBook may be useful as some images may be dark or blurry in the paperback, this is the nature of the printing process. The eBook may have lower resolution but some images may be clearer in it. Each book is approximately 250-270 pages in length, they also have the same Introduction and Global Hypothesis section which is about 70 pages long. This is repeated at the end of books 2 to 12. There are about ten more books partially completed to be published, the books cover anomalies all over Mars and have about 3000 images in total. If you like these books, and would like to support this work, then you can buy the books on Amazon. You can search for "Greg Orme" and "Martian Hypotheses" there. You can also support this work at Patreon at this link: https://www.patreon.com/ultor. If you enjoy the books you can also help with reviewing them at Amazon.

The aim is to raise money with these books to fund an institute to study these formations. If these are artificial then they will need to be studied by scientists from many fields such as biology (examining the faces, their bodies, and fish sculptures), geology (analysing the materials used in their construction), anthropology (why repeated faces with crowns were constructed, perhaps gods or rulers), mathematics (for geometric formations), sociology (how these societies worked), economists (working out how the society functioned, for example with farming, fishing, working together for large scale constructions), engineering (how these formations were constructed), and archaeology (examining ruins). How this would be done is not clear, but this institute would try to make a start on understanding these formations. No one really knows how to study an extinct alien civilization, if this is one. Most likely, if they are real, then a more professional organization would take over this work later. The intention then is to bridge the gap between amateur analysis of these formation to a much better funded organization, perhaps at the government level. The evidence gives a reasonable case for artificiality, but much study needs to be done to determine how plausible this is.

It may be valuable to read the Introduction section more than once, to see how the images you see are connecting into these classifications. Often the images have a lot of details, each time they are examined more of these can be seen. They might also inspire you to see other connections, for example one image might be similar to another in a different part of Mars. This is likely to happen, even with so many images the surface of this hypothesis is barely being scratched. Mars has an area similar to the land area of Earth, this is because much of Earth is covered in oceans. For this much land then 3000 images is likely to have missed many important discoveries.

You can also use the indexes in each book, they refer to many similar formations throughout them. For example, if you are looking at hypothetical road formations then roads in many different areas can be found in the indexes. It would be possible then to quickly see all the different kinds of hypothetical roads in all 10 books. The last 2 books are not indexed the same way, they have a collection of published papers. Some of the formations are similar to those in the first 10 books and can also be compared to them. For example the paper on Free Standing Walls has many hollow hill formations, these also occur in the Protonilus region. The idea behind the global hypothesis is to suggest gow these different formations connect together into a hypothetical Martian civilization. It's important then to get an intuition of how these formations connect together globally. Some areas for example might have hypothetical roads for transport, other might have hypothetical tubes like a covered road. Different terrain, available materials, and climate might have led to one being used over the other. It may be as Mars cooled it became necessary to travel under cover because of the cold. Another possibility is predators or meteors made traveling on roads too dangerous. Also there are many hypothetical dam formations, but the construction techniques vary between areas. Some are formed with dam walls attached to the crater, when they break some show a cavity under them and others do not. This would indicate the dam wall was dug into this cavity to keep it from sliding down the crater wall. In other areas this was not necessary, it may be that there the crater wall was harder rock which the dam wall could be cemented to. Some show columns and layers in them but others have evenly spaced vertical grooves on the dam walls. Some dams are excavated out of the crater wall or the material at the bottom of the crater, these may depend on the rock type in the crater. For example, if the crater wall is too easily broken then an excavated dam might have been the best engineering solution.

Some areas have hollow hills, these are where a hollow habitat may have been built on an existing hill or the whole hill was constructed. In some areas these have layers similar to a Cobler Dome, this is where bricks form the dome in decreasing circles as the dome is built up. These are called amphitheatres as a friendly name, the first amphitheatre formation looked more like seating around an amphitheatre. Other hypothetical buildings have no layers in their roofs. This may have depended on the materials available. Many appear to have a smooth skin like cement which has broken up in some parts of the roof, and is intact in others. In many areas this is more intact on the southern side, as the skin breaks off the softer inner parts of the roof appear to have eroded faster and collapse. The one sided erosion may imply a prevailing wind, or as the oceans and air froze at the pole this created the erosion.

There are also large areas of walls and room like shapes, these are hypothetical cities. Other areas connect these hollow hills together with tubes or roads as another kind of hypothetical city. Still others seem to be made of tubes that connect together in intersections called a tube nexus. This may have been because of the climate further from the equator, for example tubes might have been used to travel through in colder areas.

The Martian Faces are mainly discussed in books 11 and 12, a reprint of published peer reviewed papers. These differ according to where they are. The Cydonia Face, Nefertiti, and King Face all fall on a great circle, this is hypothesized to have been an old equator that lines up with a known previous pole position west of Hellas Crater. The newly discovered Queen Face is in Cydonia but not near the old equator. If the faces were used to mark latitudes and longitudes then the overall system remains obscure. For example there is a large hyperbola shown close to the old equator. Another is far from this equator, but drawing a line from it to Nefertiti gives a right angle to this old equator. Joining these two hyperbolas and the King Face gives an Isosceles Triangle. The hypothesis of these mapping system is highly speculative at this stage.

Canals, lakes, and water channels also vary in different areas. West of Cydonia there is an extensive array of hypothetical canals, also east and west of Elysium Mons. Some of these connect to larger lakes which may be artificial. Some hypothetical dams have water channels to direct water into a dam, and to collect an overflow to another dam.

There are also darker areas often bounded by walls or geometric shapes. These may have been farms, why they appear in some areas like around Cydonia and in Isidis remains unanswered. Other areas contain hypothetical artefacts but no farm formations, so these creatures would have used a different way of collecting food.

The idea of these books then is not just to prove artificiality, but to try to prove a global hypothesis of how the whole civilization functioned. Once the evidence becomes plausible enough, and the shock wears off, this larger question is much more interesting. Each section is labelled with the title hypothesis to make clear these notions are being proposed along with the evidence there. The sections all have many keywords connecting to the index. If you see a connection to a kind of formation then it is easy to find similar formations. In seeing the global hypothesis the different pieces of the puzzle are more likely to come together, for example the hypothesis of dams sounds less plausible if it is not connected to the hypothesis of buildings and farms. Together they give the ideas of habitation, food, and water. The conclusions can be controversial. However there is so much evidence it was better to put it all together into a more comprehensive hypothesis. Otherwise people are looking at isolated formations like faces without seeing the overall context in which they appear.

Images, main section

Cymd460a

Hypothesis

This is probably a highly eroded dam, the walls having broken off.

Cymd460a2

A parabola is shown.

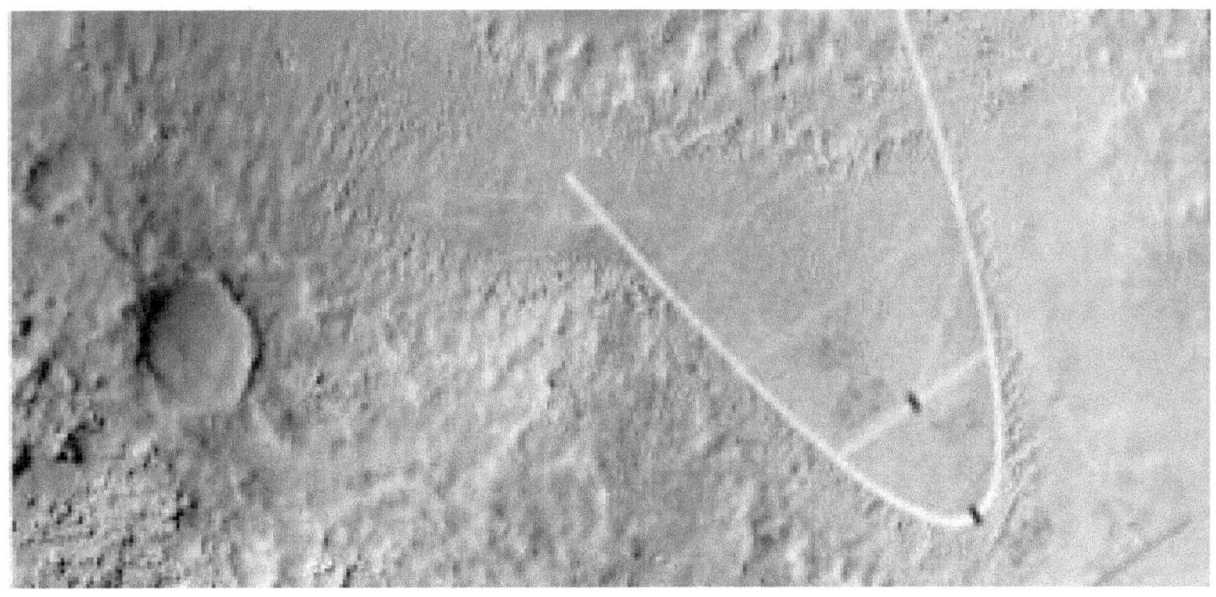

Cymmh464d

Hypothesis

A shows these walled structures, in them the pale areas may be a degrading floor. Some of these objects may also have been furniture made of stone or cement. B shows more of these that may be degraded or buried. C at 2 o'clock shows a long very straight wall extending to the right. C at 4 o'clock shows more structures between the walls like furniture. The rooms at D show many small objects in the walls, these may also be from a collapsed roof. The rooms around E, F, and G may be partially buried like at B.

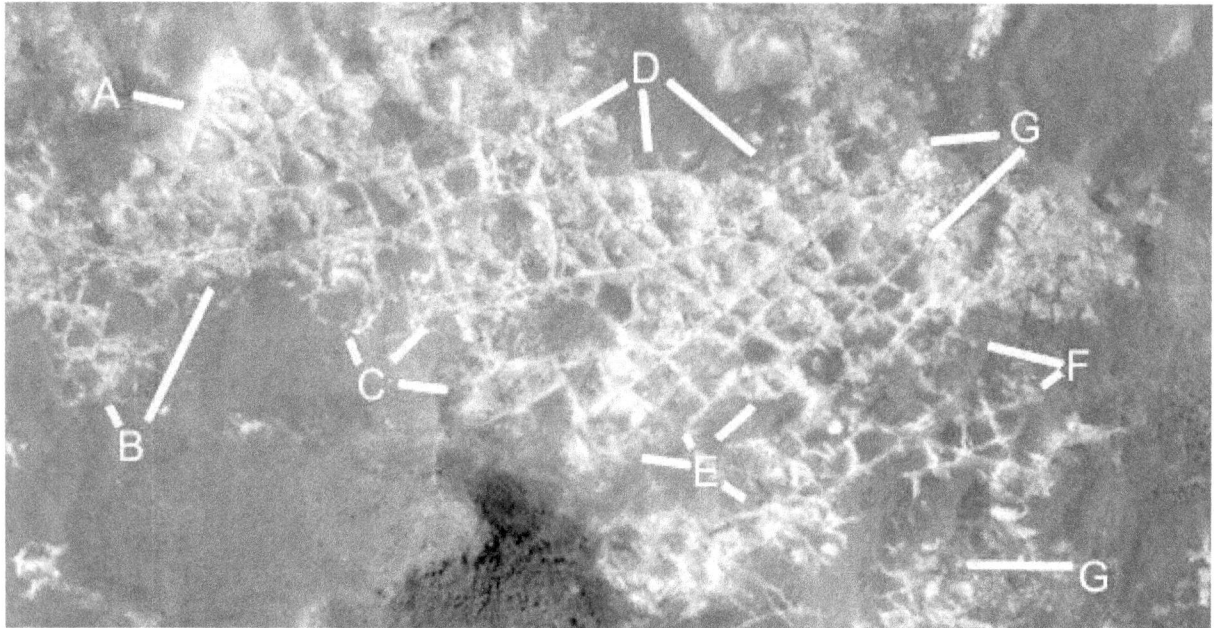

Cymhh464g

Hypothesis

A shows a ridge that may be natural or associated with these formations. The ridge at B is dark but continues on as a pale wall casting a shadow on its upper side, down to G. C shows some small rooms. D, E, and F show some fainter ridges, perhaps worn down walls. At G and H the walls are triangular.

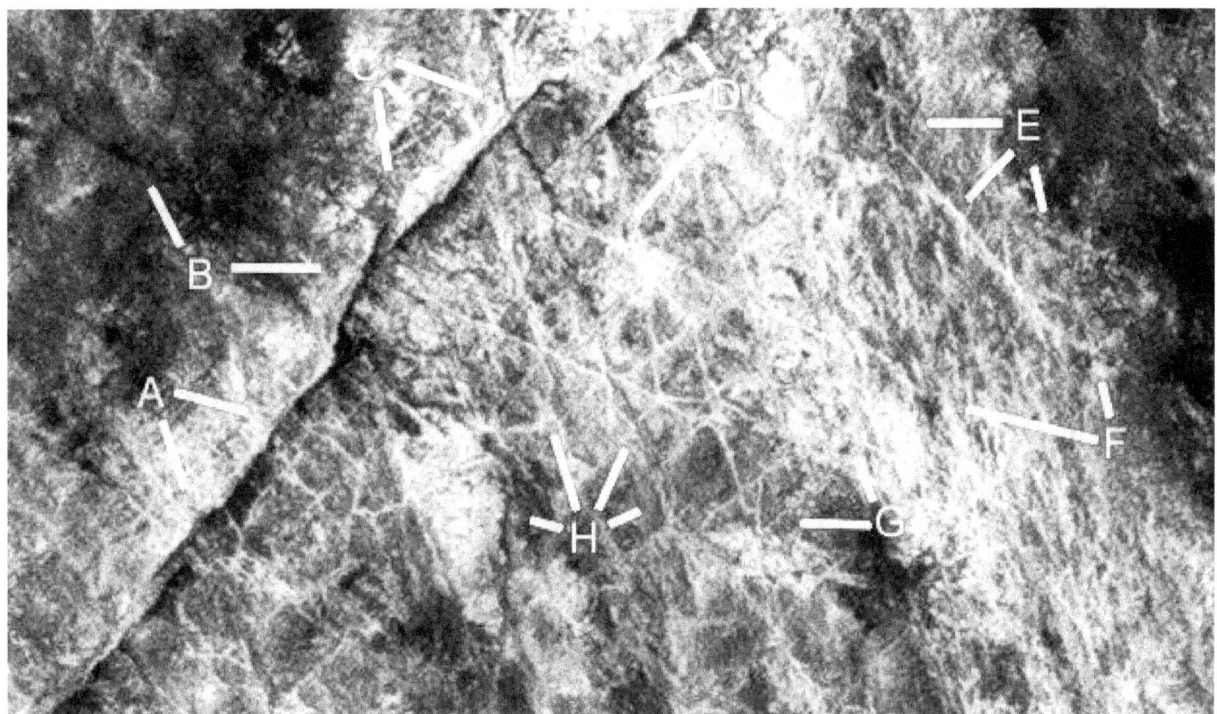

Cymhh464g2

Hypothesis

Here some of the walls have a line on them to show how straight they are. The triangles appear to be 50, 60, and 70 degrees. This is hard to explain geologically where these angles would appear over and over, one triangle is marked with these angles.

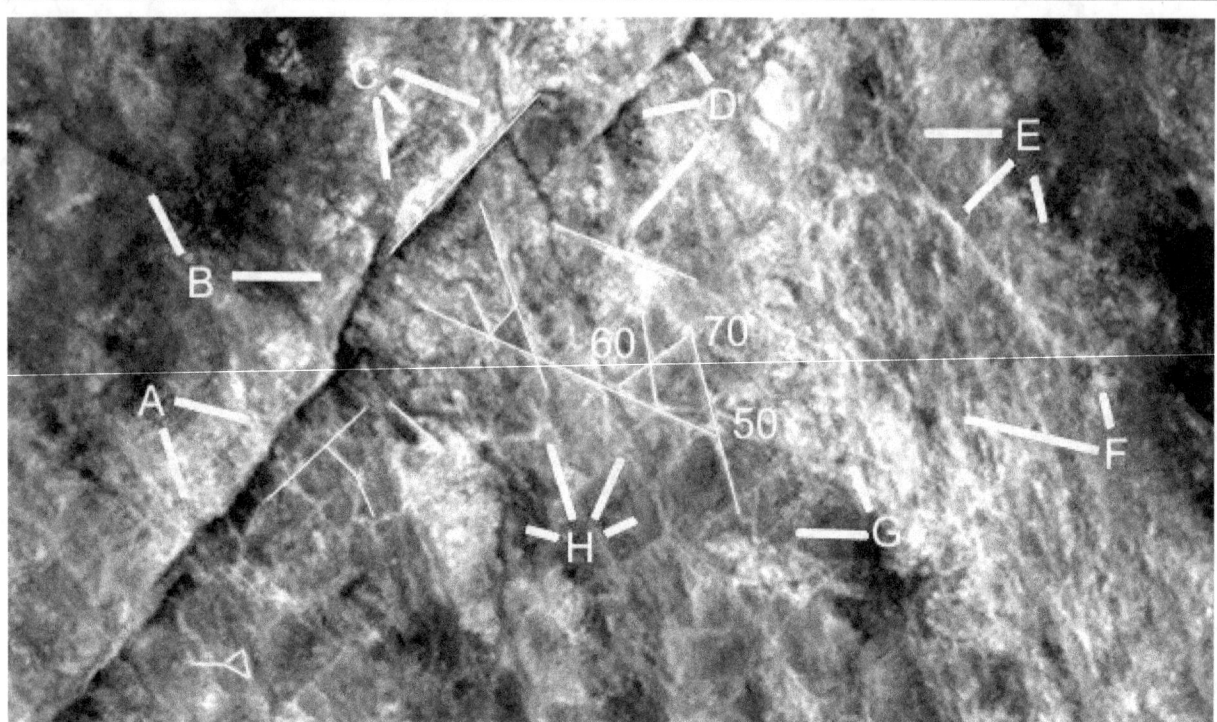

Cymhh464i

Hypothesis

A and B show many rooms and walls. C at 11 o'clock may show a nexus where the walls come together into a circle or crater. The walls seem to be directed towards this crater though a meteor would fall in a random position. D shows many small rooms extending up the image, these may be buried or under an intact darker roof. These areas could be explored to see how many intact and sealed rooms survive. E at 8 o'clock shows another round nexus where the walls appear to converge, perhaps a meeting place. At 10 o'clock there may be a bridge over a cavity. At 3 and 5 o'clock the rooms may be partially buried. F shows another nexus at 9 o'clock the other rooms may have eroded away. G shows a large array of rectilinear rooms at 10 o'clock extending over to A. At 4 and 8 o'clock the rooms are more irregular.

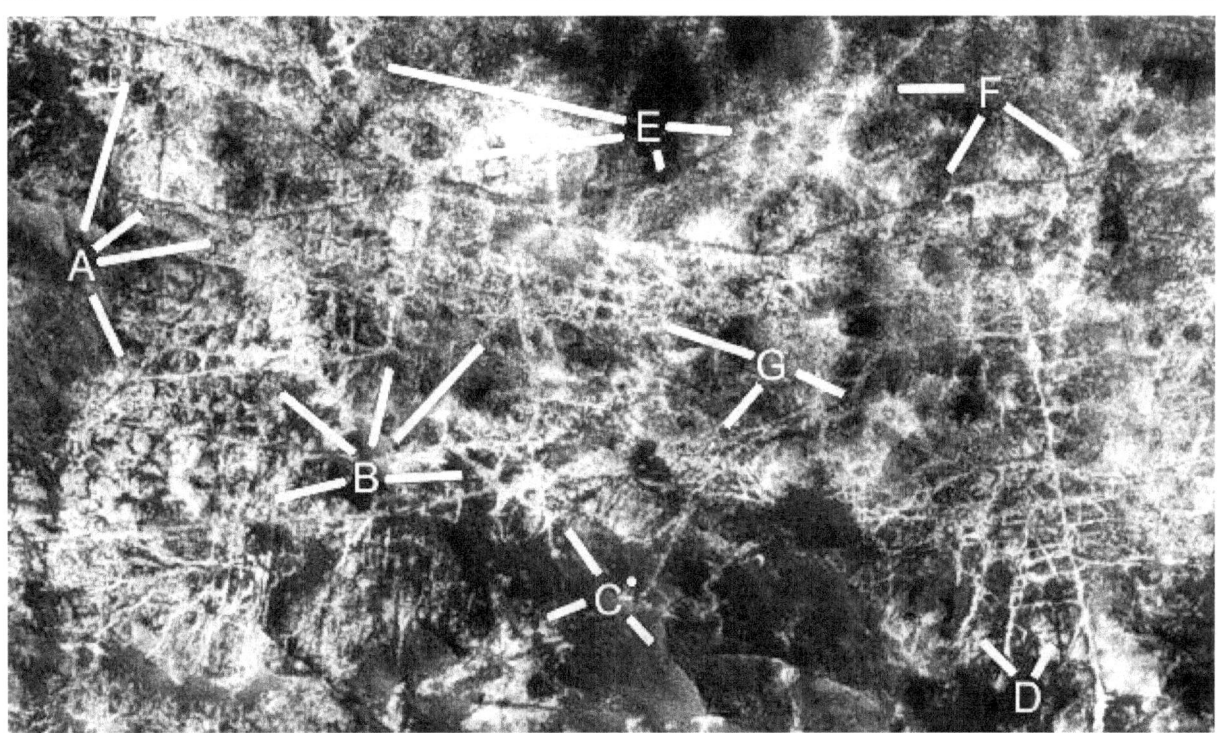

Cymhh464i2

Hypothesis

Straight lines are overlaid to show how straight these walls are.

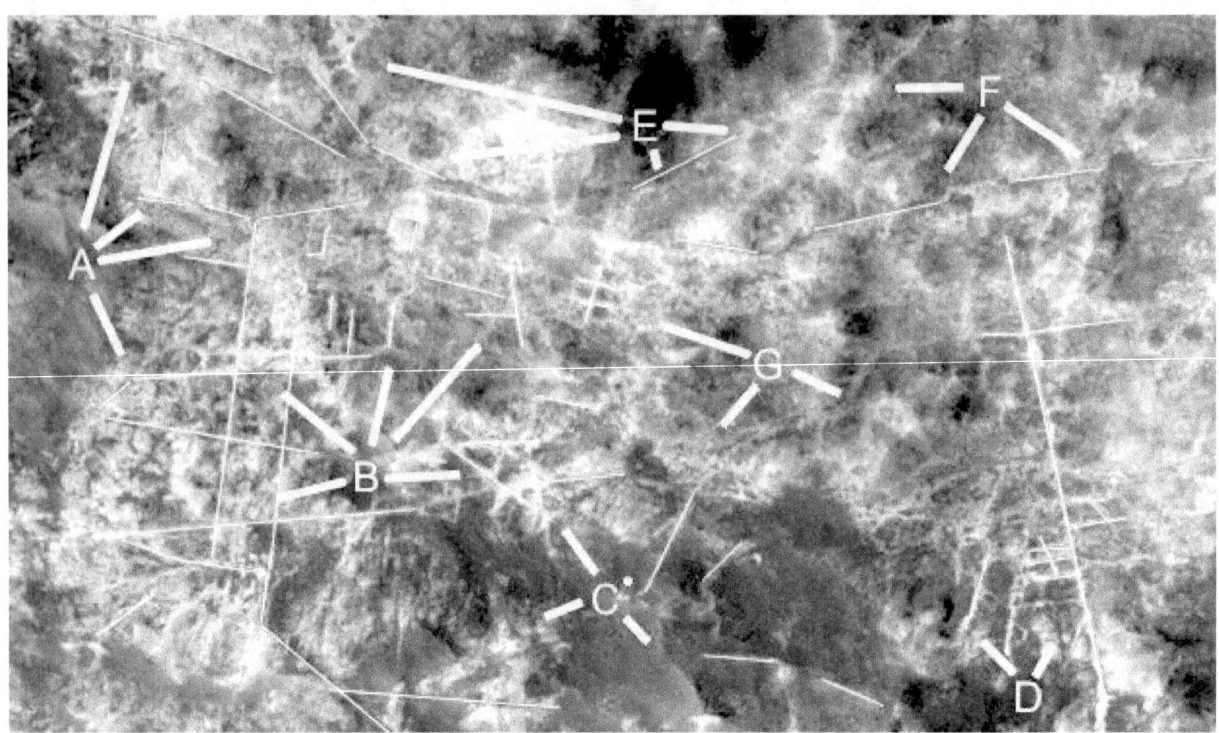

Cymhh465e

Hypothesis

A shows a long straight wall at 8 o'clock continuing down to C. Between 4 and 6 o'clock there are rooms with some objects in them, perhaps furniture. Around B the rooms may be partially buried or the ceilings are intact. C shows many more roads or walls. D may be a hollow where the walls of rooms have collapsed. There may also be rooms under E and F.

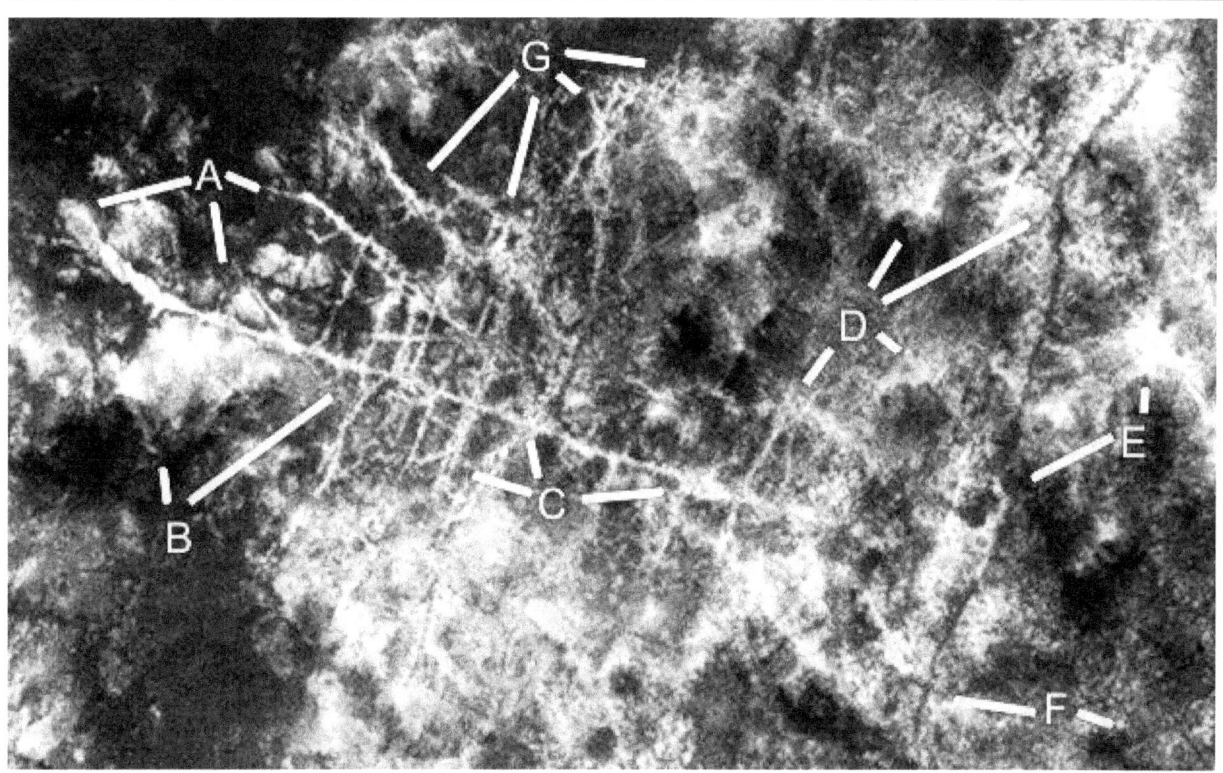

Cymhh465e2

Hypothesis

The lines show how straight the walls are.

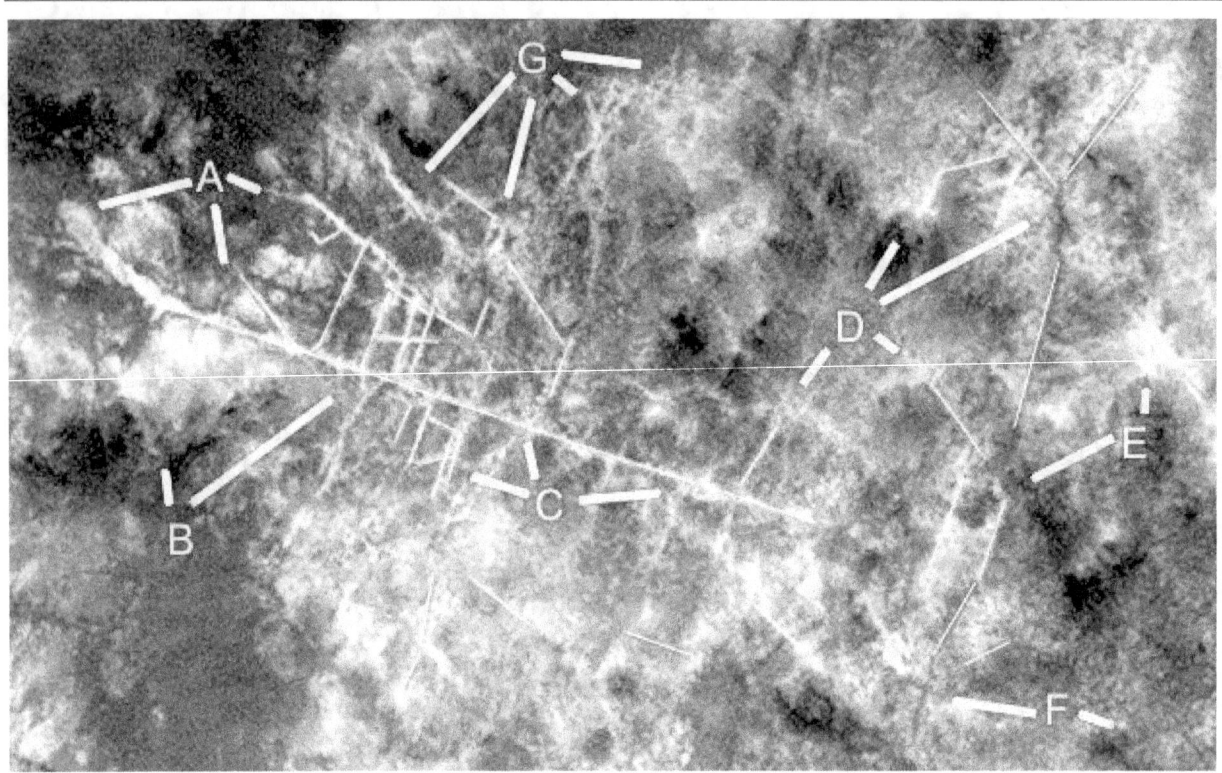

Cymhh465g

Hypothesis

A shows many small rooms around the same size. B may be intact roof material, this changes to open rooms between C and G. C at 11 and 1 o'clock shows larger rooms continuing on to F. D shows rooms with very straight walls, they seem to nest inside each other. E shows a hollow or the rooms may be buried by the darker soil. G shows more small rooms

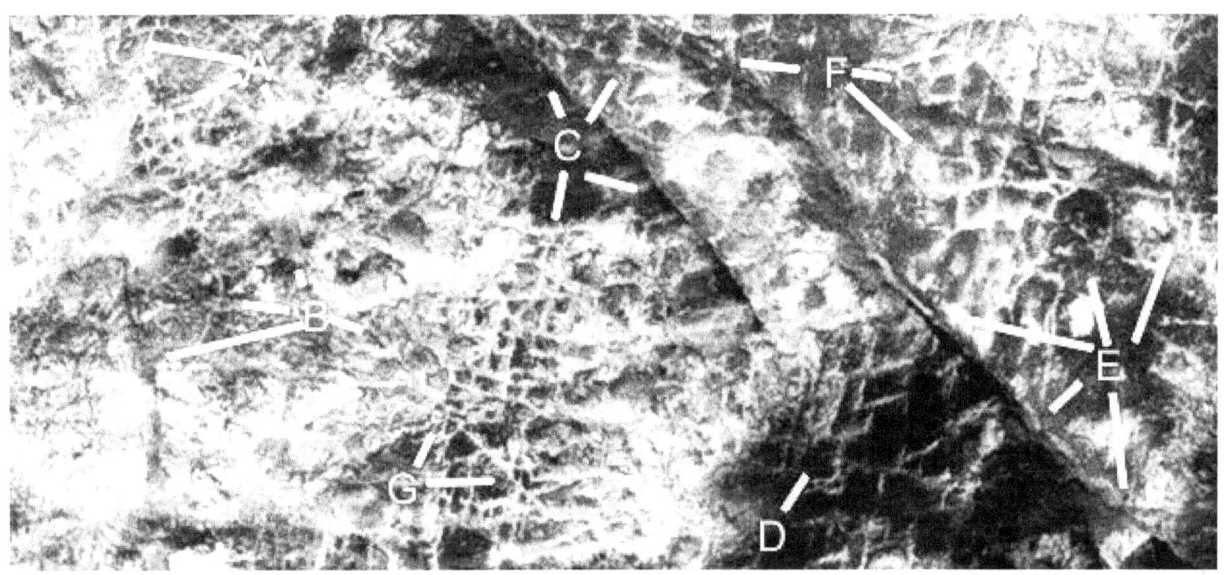

Cymhh465g2

Hypothesis

The lines show how straight the walls are.

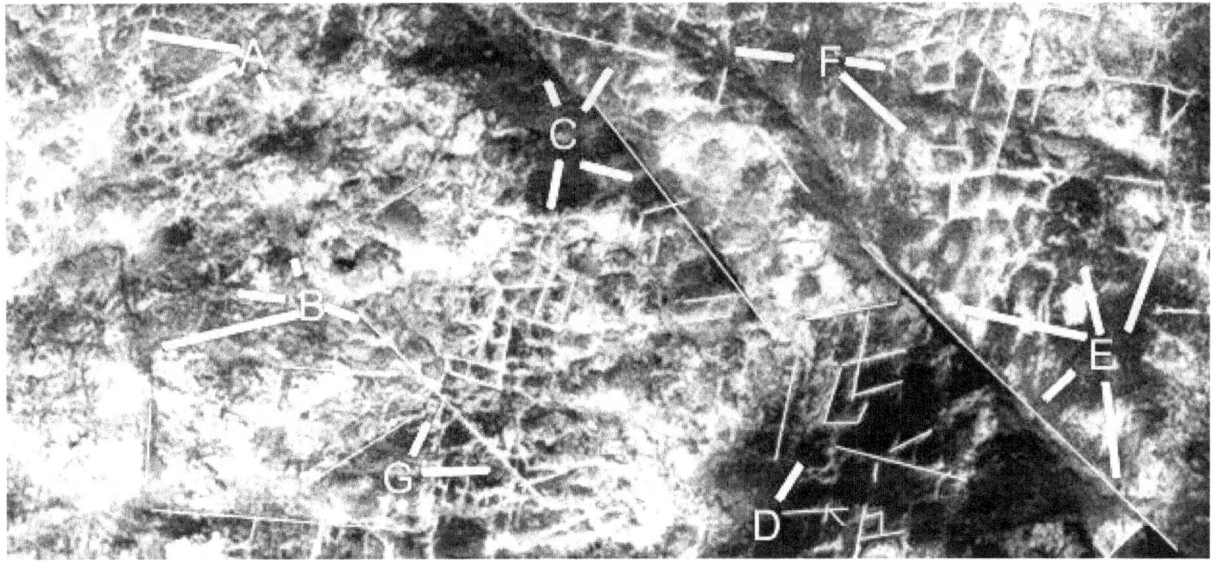

Cymhh465h

Hypothesis

This shows some more triangular walls, A follows a ridge to some faint rooms on the right. B shows some larger rooms on the left and a nexus at 4 o'clock. C shows a cavity, perhaps with some intact ceilings at 2, 4, and 7 o'clock. At 12 o'clock the rooms are more irregular in shape. Between D, E, and G the rooms are trapezoids like two joined triangles, as well as triangular. F and G show more walls.

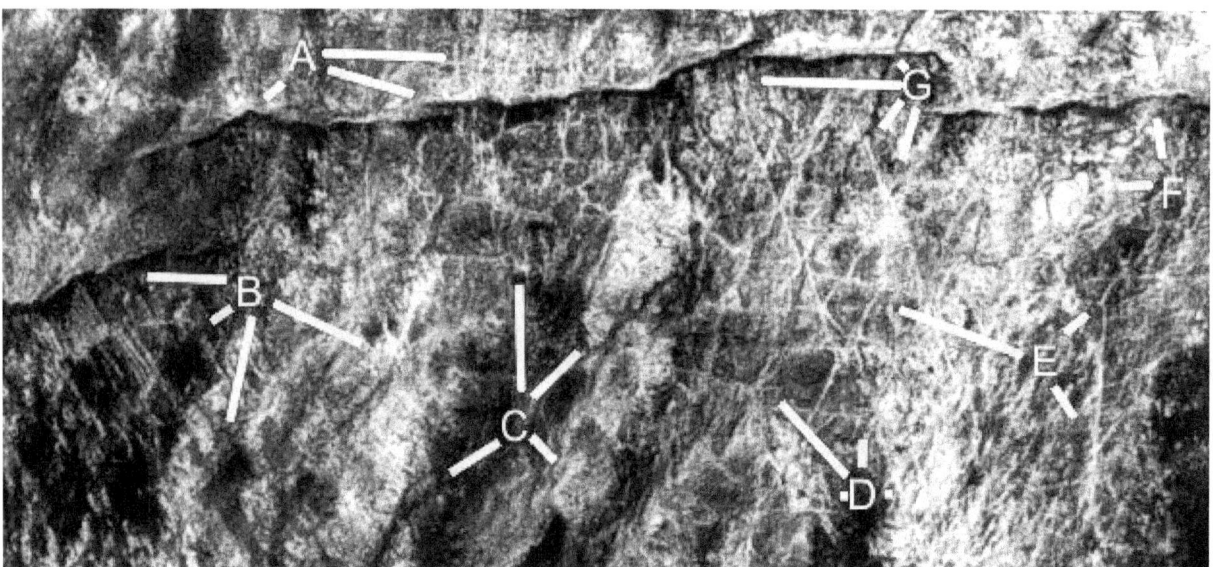

Cymhh465h2

Hypothesis

The lines show how straight the walls are, also two curves are parabolic in shape.

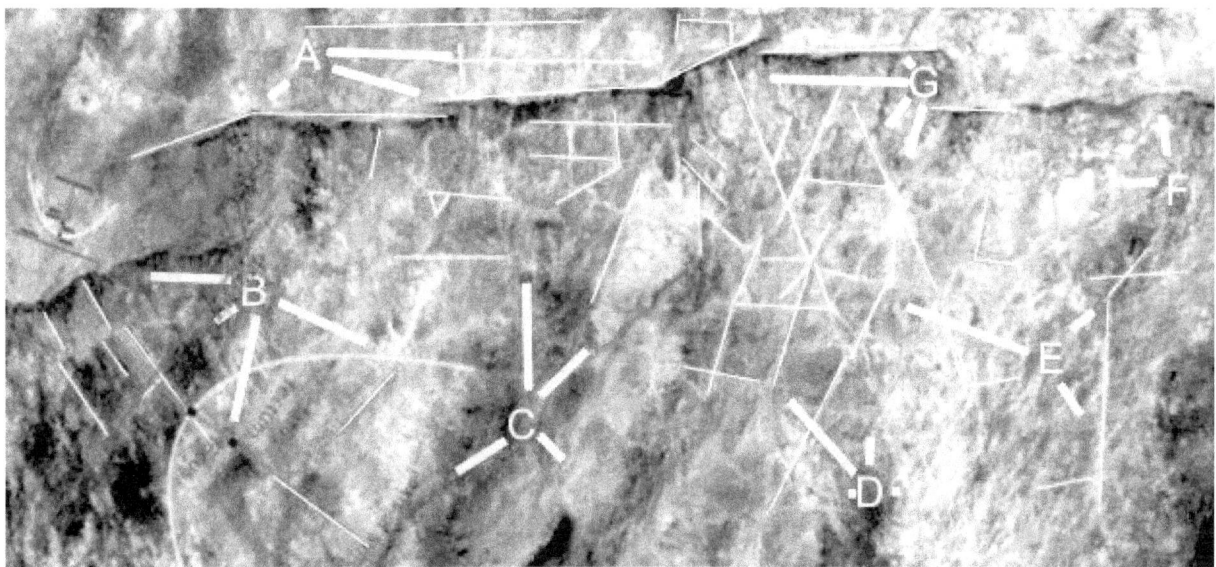

Cymhh466d

Hypothesis

Between A, B and E there is a large hill that may contain intact rooms. Some are shown at A, also at B at 7 o'clock. At 3 and 9 o'clock the edge of the pit is smooth in shape, C shows layers underground which may have been used to build the lighter walls. If this was volcanic ash it may have been used to make cement, some may have been more easily eroded into dust and blown away. D may show eroded rooms at 4 and 5 o'clock.

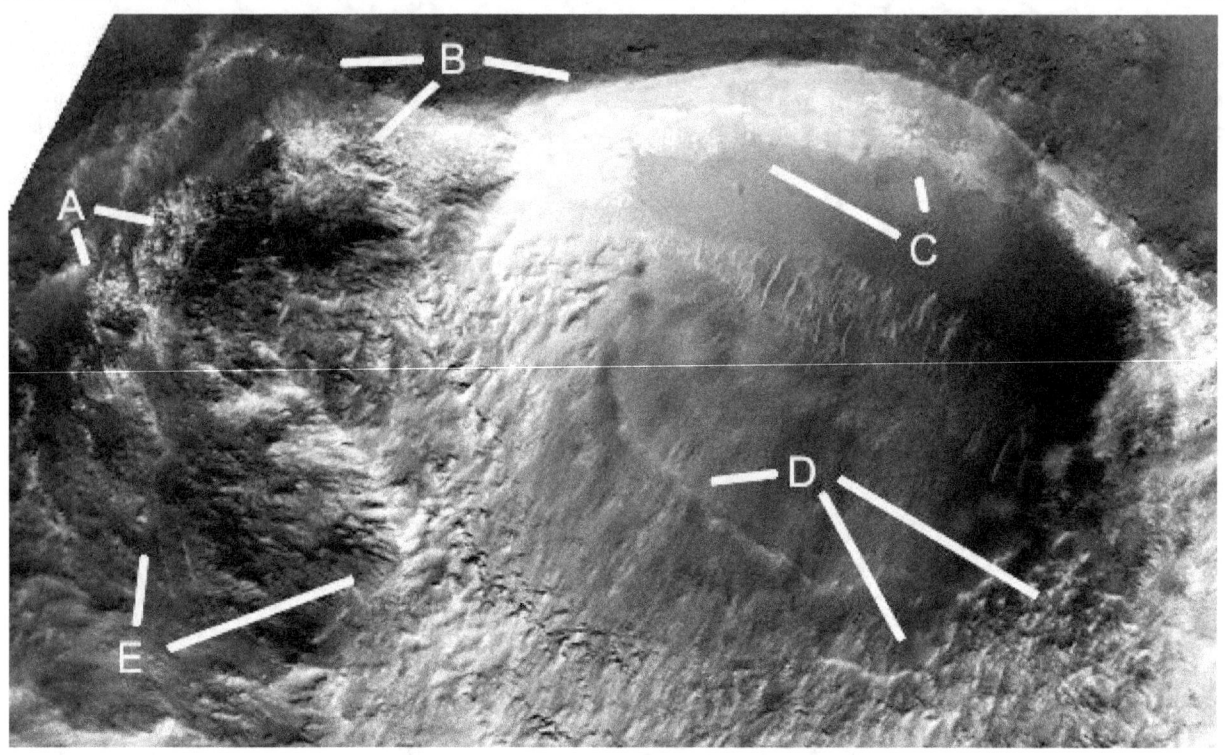

Cymhh466d2

Hypothesis

The edge of the pit is a parabola.

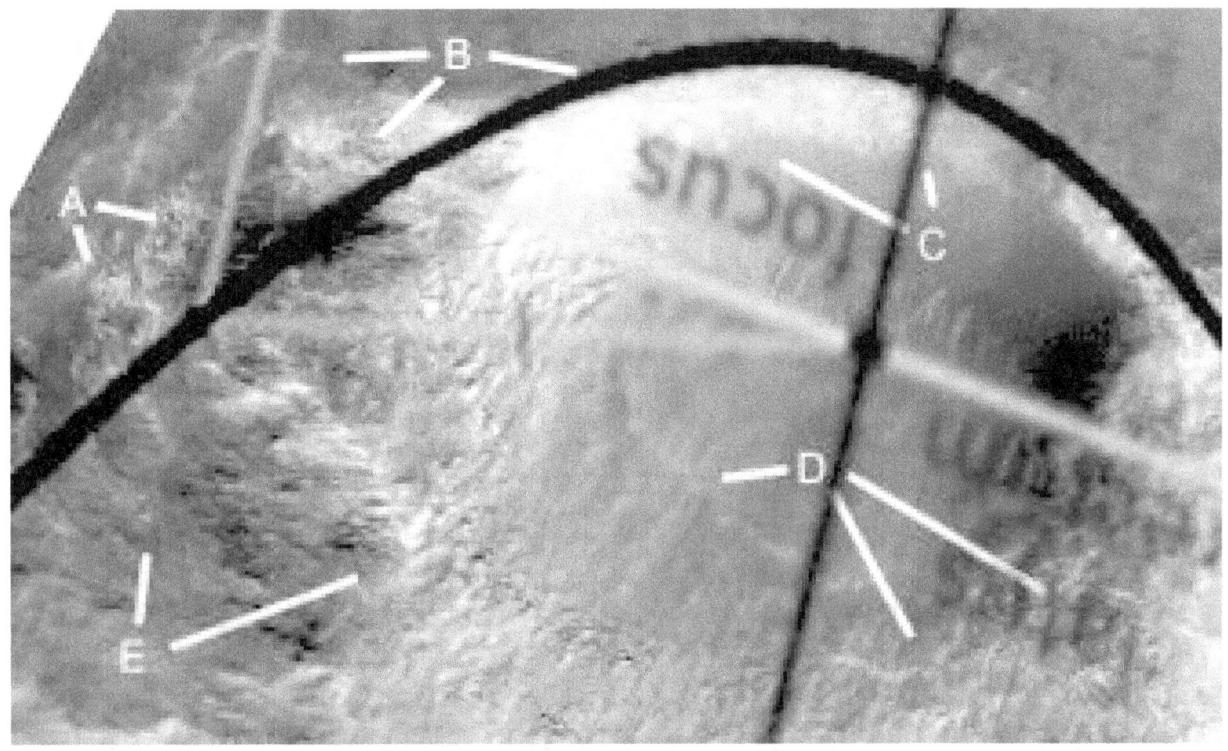

Cymhh466k

Hypothesis

A shows some walls at 11, 1, and 3 o'clock, partially buried by the dark soil such as at 8 o'clock. Between A and B there are more walls, at B there may be a nexus of walls or tubes.

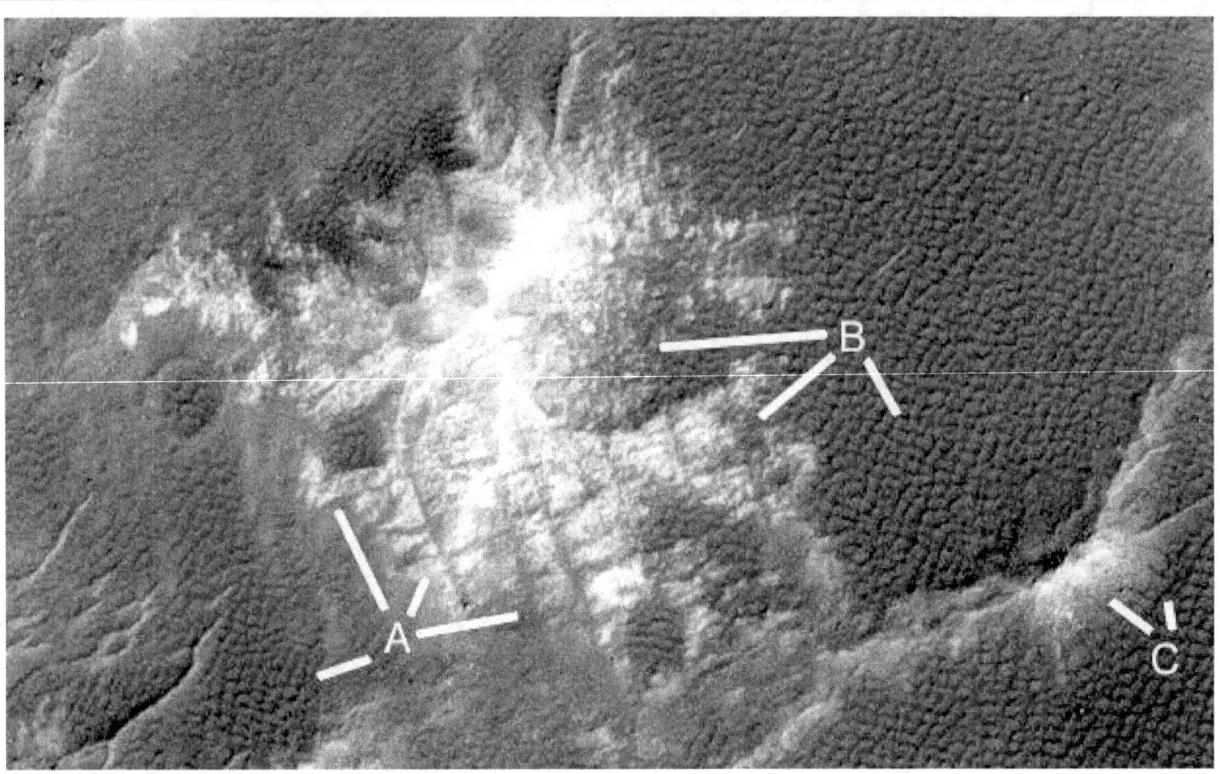

Cymhh466k2

Hypothesis

There are many straight grooves here at right angles to each other, perhaps the walls eroded away leaving the foundations. Some dams have been like this, eroded away with a parabolic groove left in the ground.

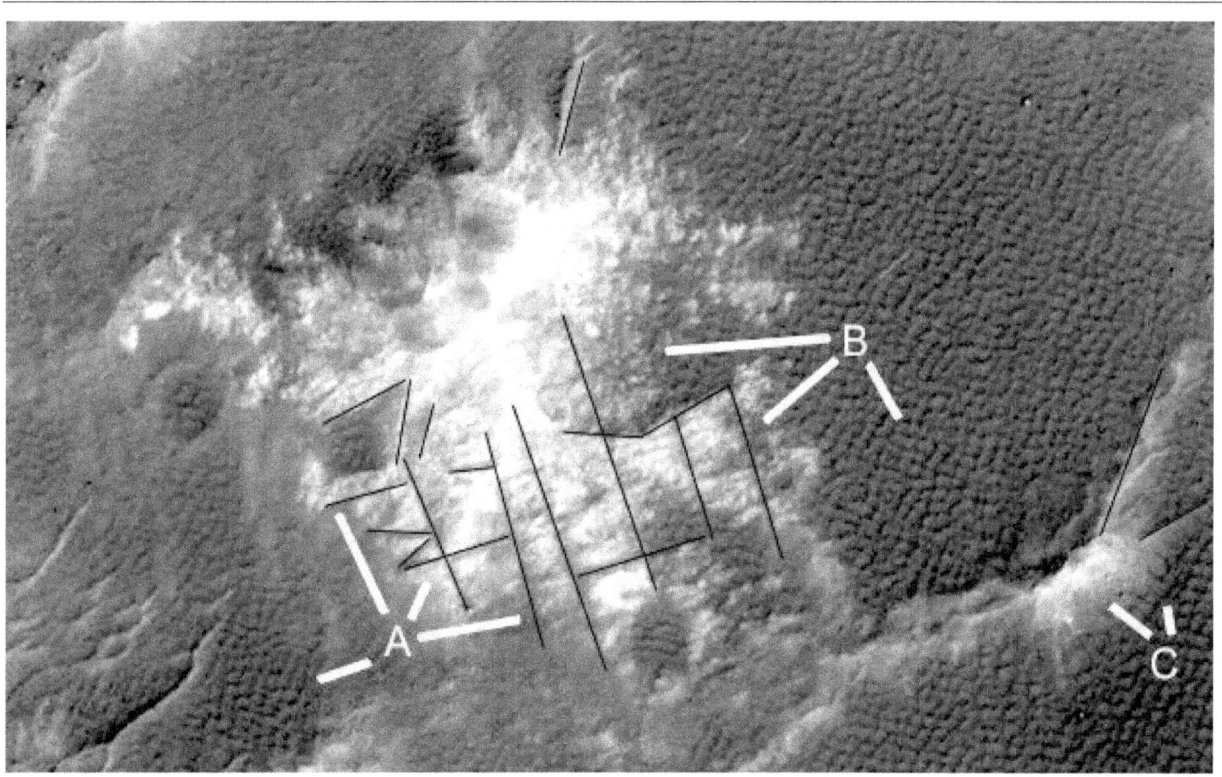

Cymhh467

Hypothesis

A may show some collapsed hollow hills. B shows some straight ridges, perhaps interior supports of this larger formation. From C to D is a curved interior support. E may be a collapsed section, F shows some tubes or walls.

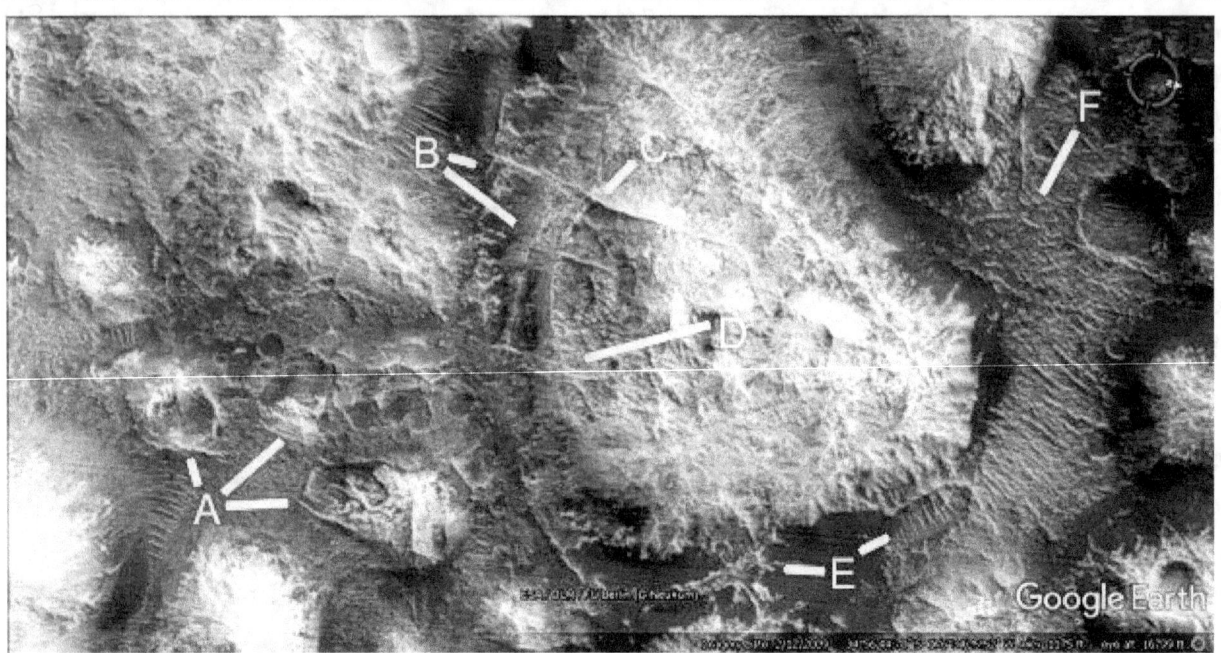

Cymhh467a

Hypothesis

There are two parabolas in this formation, as well as the straight walls.

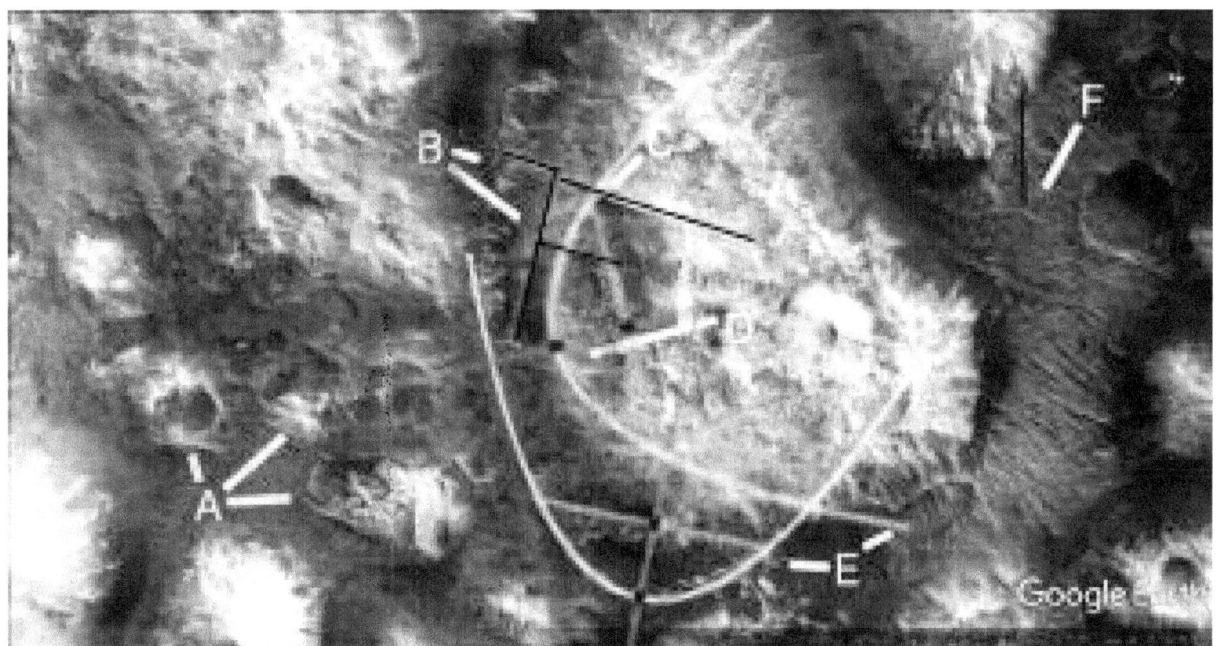

Cymhh469g

Hypothesis

A at 10 o'clock shows a hill with room like shapes on its lower side, at 3 and 5 o'clock are more rooms. B and C show many walled rooms. D shows rooms that may be partially buried by the dark soil, or they ended in this open area. E shows more degraded rooms, F at 10 o'clock shows a nexus where many walls converge to it. At 3 and 4 o'clock there are perhaps rooms under the dark soil. G at 10, 12, and 1 o'clock as well as H at 12 o'clock follow this edge of the rooms, this section may be an intact ceiling with rooms under it.

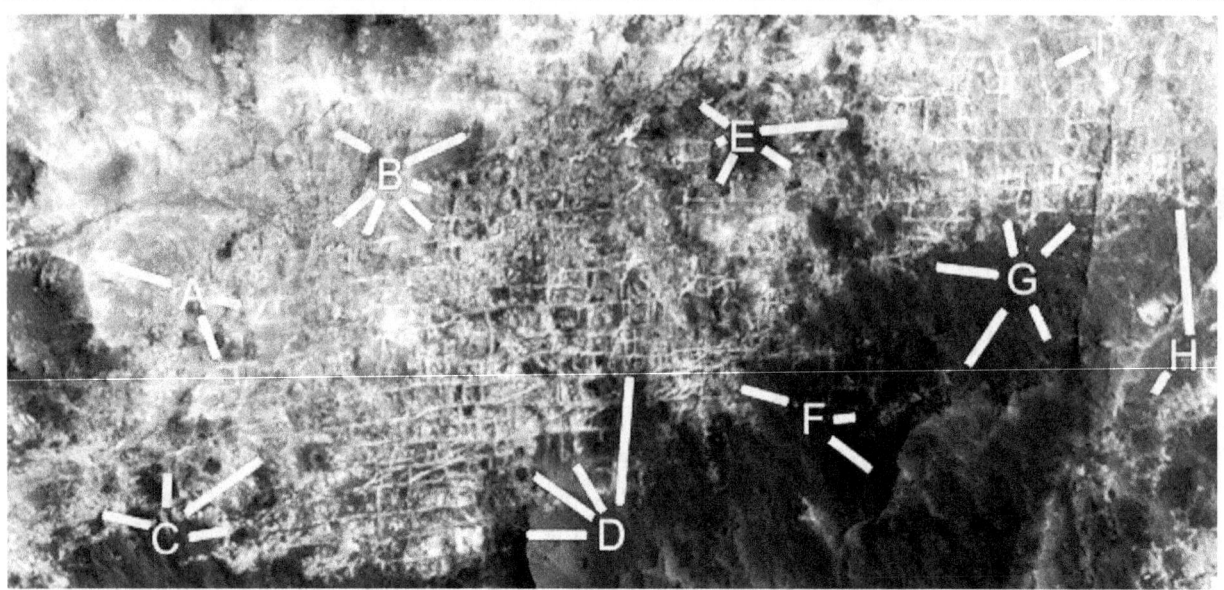

Cymhh469g2

Hypothesis

There are many lines here showing how straight the walls are, but many more could have been drawn as well.

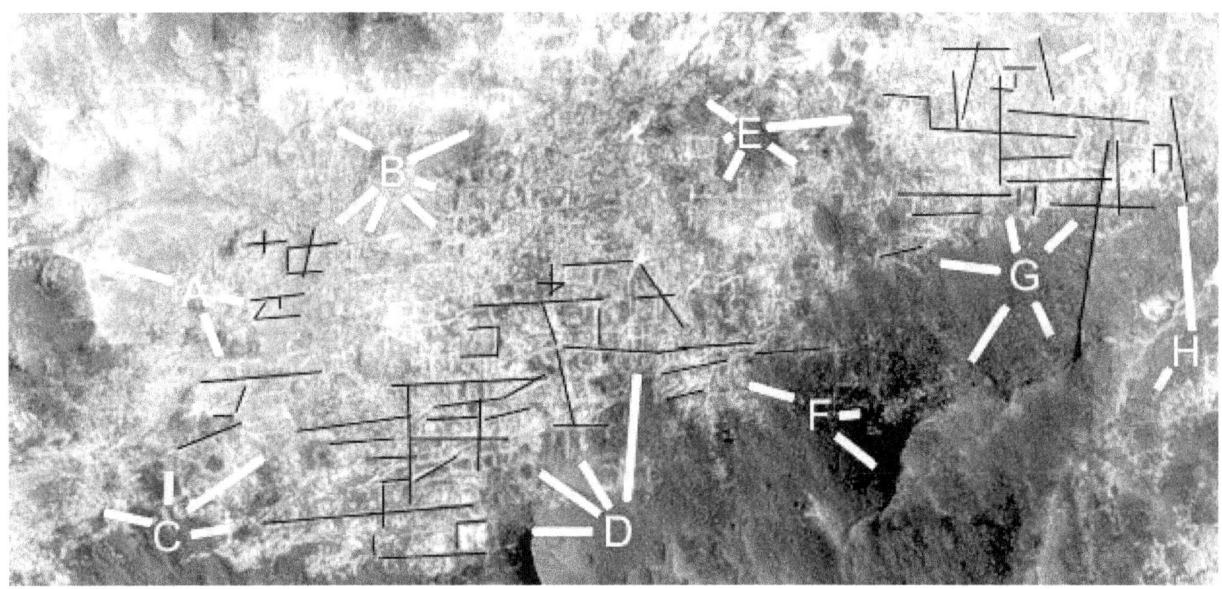

Cymhh469h

Hypothesis

The walls around A seem to be the same as to the right, but covered in dark soil. B may be more intact ceiling material with the tops of the walls showing through it. On the left of C may be more intact ceilings, the shading implies they are domed. To the right at 3 o'clock appears to be a nexus of walls converging. D shows many more walls, E at 11 o'clock shows with the shadows how high the walls are. At 4 o'clock is another nexus. F, G, and H may be more intact ceilings. From I across to D the walls seem to be catching the sun's light and reflecting it upwards, this indicates the albedo of the walls.

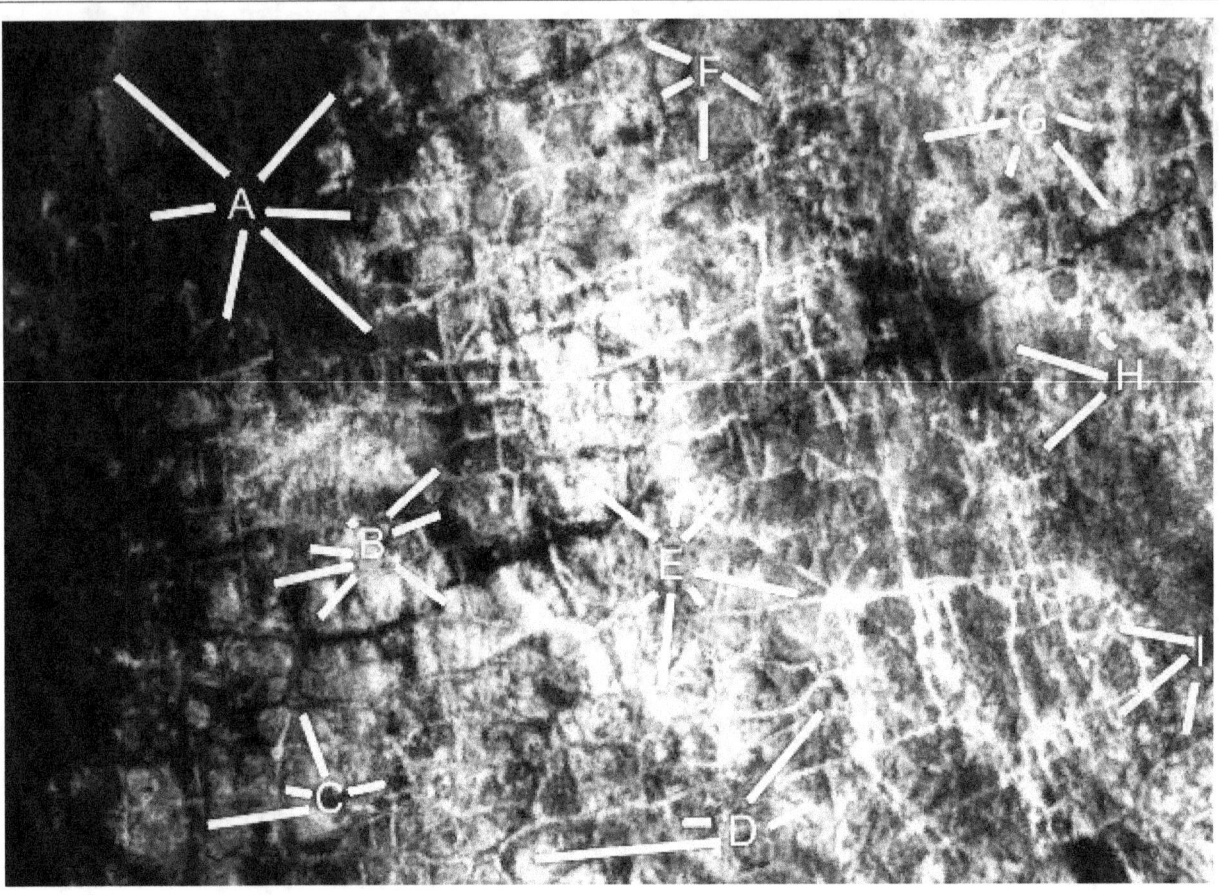

Cymhh469h2

Hypothesis

This shows how straight the walls are.

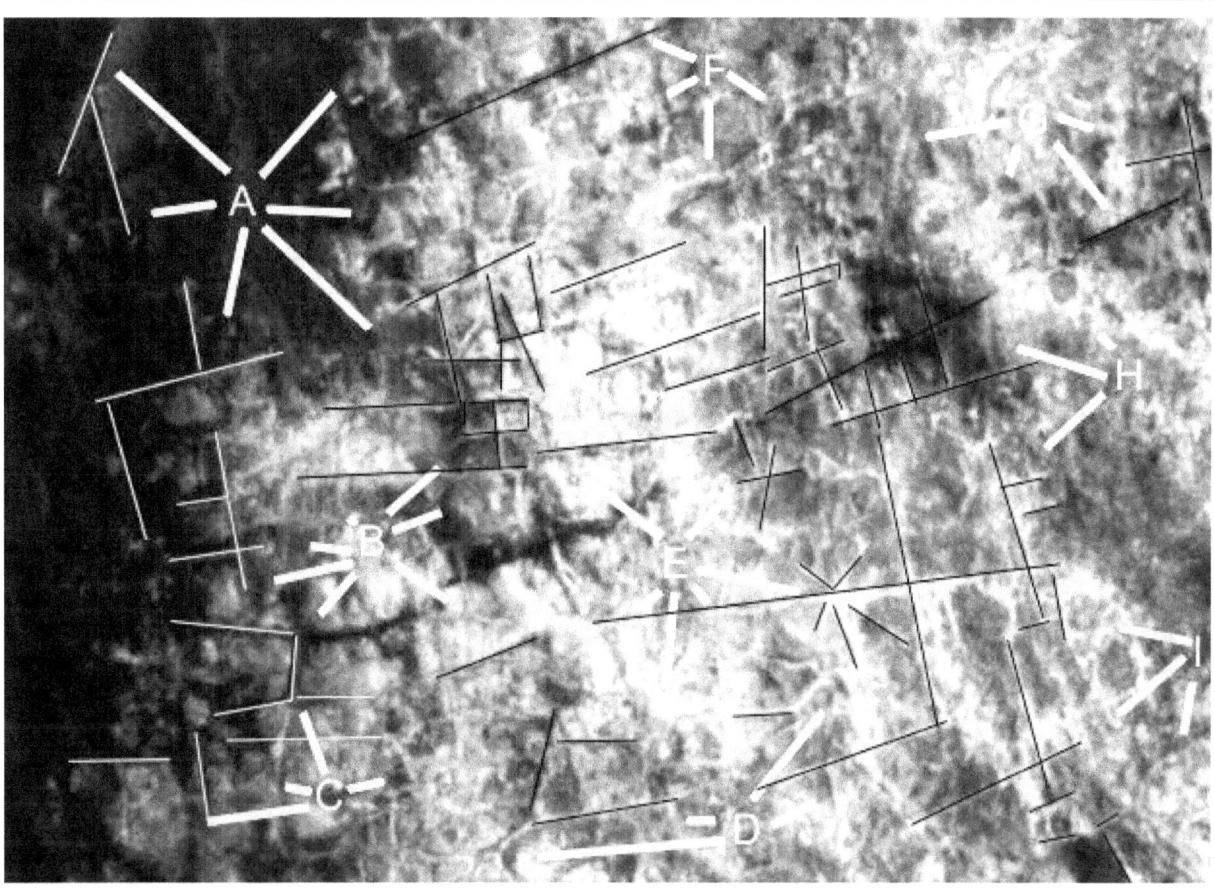

Cymhh469i

Hypothesis

A shows the walls in more of a shadow to the right, indicating their height. At 9 o'clock there may be an intact ceiling with rooms under it. At C the walls are more irregular, at D the sun might indicate the ceilings have collapsed into a cavity. E and F may also have intact ceilings, G may be where the ceilings have collapsed.

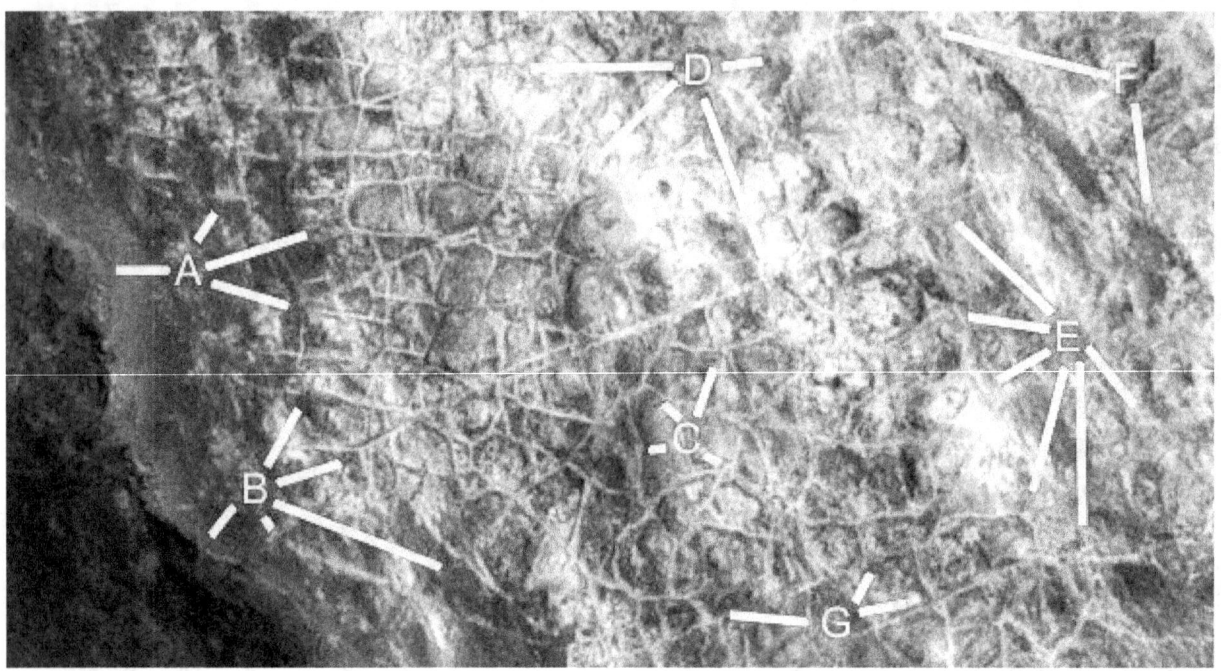

Cymhh469i2

Hypothesis

This shows how straight the walls are.

Cymhh469j

Hypothesis

A shows a distinct room with shadows at 12 o'clock, at 2 o'clock is a rounded dome. In this area the walls seemed to have intact ceilings, some like at 12 o'clock have lost their roofs exposing the interiors. At 4 o'clock is an unusual object. B at 7 and 8 o'clock have clearer walls, the section at 2 and 4 o'clock may be a large intact roof. Around C the ground is lower than this roof, 4, 5, 8, and 9 o'clock show protruding walls. D, E, and F show more wall variations, F at 10 o'clock shows finer wall structures. G shows distinct walls, at 11 o'clock one curved wall connects to a straight wall towards C. At 4 o'clock may be the remains of a ceiling.

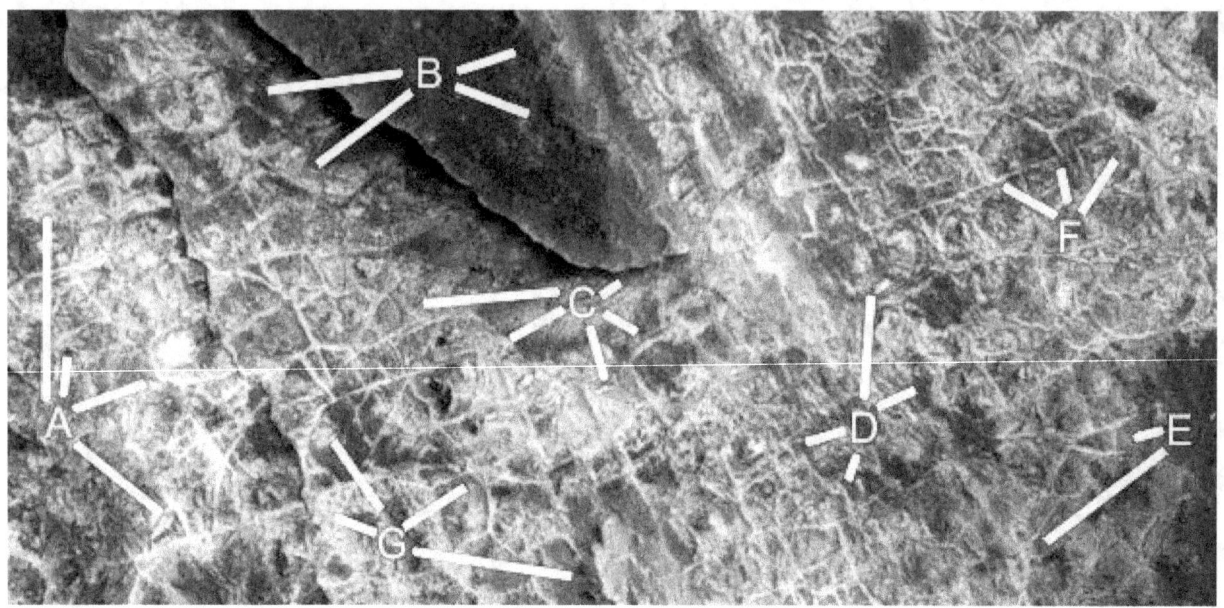

Cymhh469j2

Hypothesis

The lines show how straight the walls are, a semicircular shape is also shown. The walls appear to converge to the center of the circle.

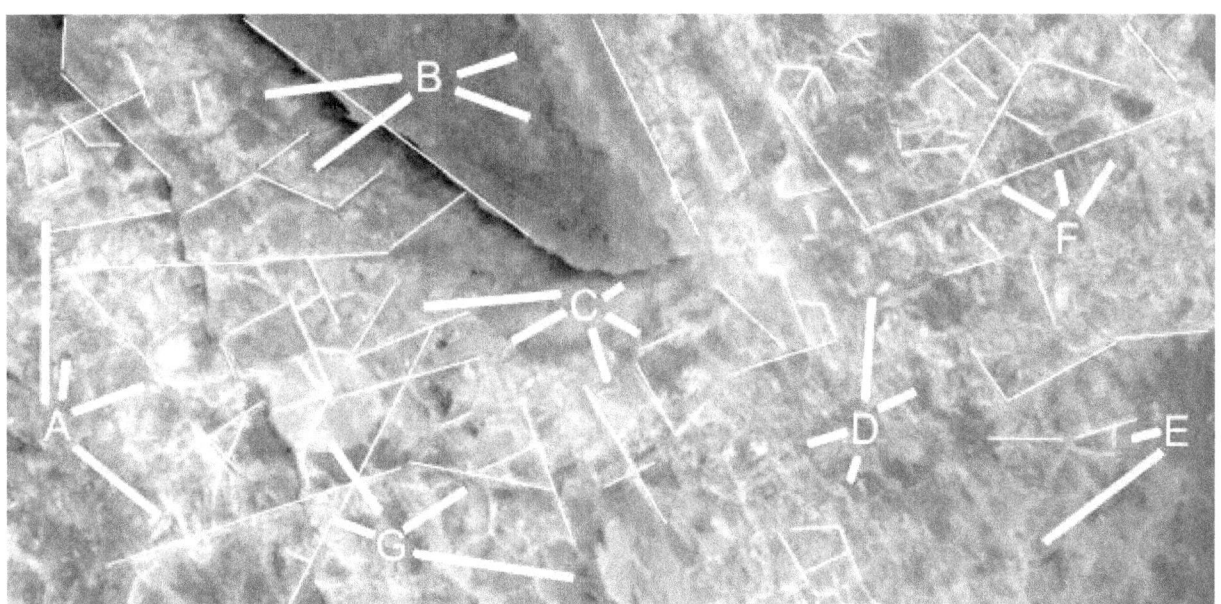

Cymhh469l

Hypothesis

There are many objects inside these rooms, perhaps furniture. A shows some walls partially buried in the dark soil, covering more walls above A. B appears to show blown dark soil across the walls at 9 o'clock, this is so prevalent it may be from a disintegrated roof. At 4, 6, and 7 o'clock the walls are more distinct though the dark soil is in the rooms. C, D and E show the edge of another dark soil area burying rooms. F is a higher area perhaps with intact pale ceilings.

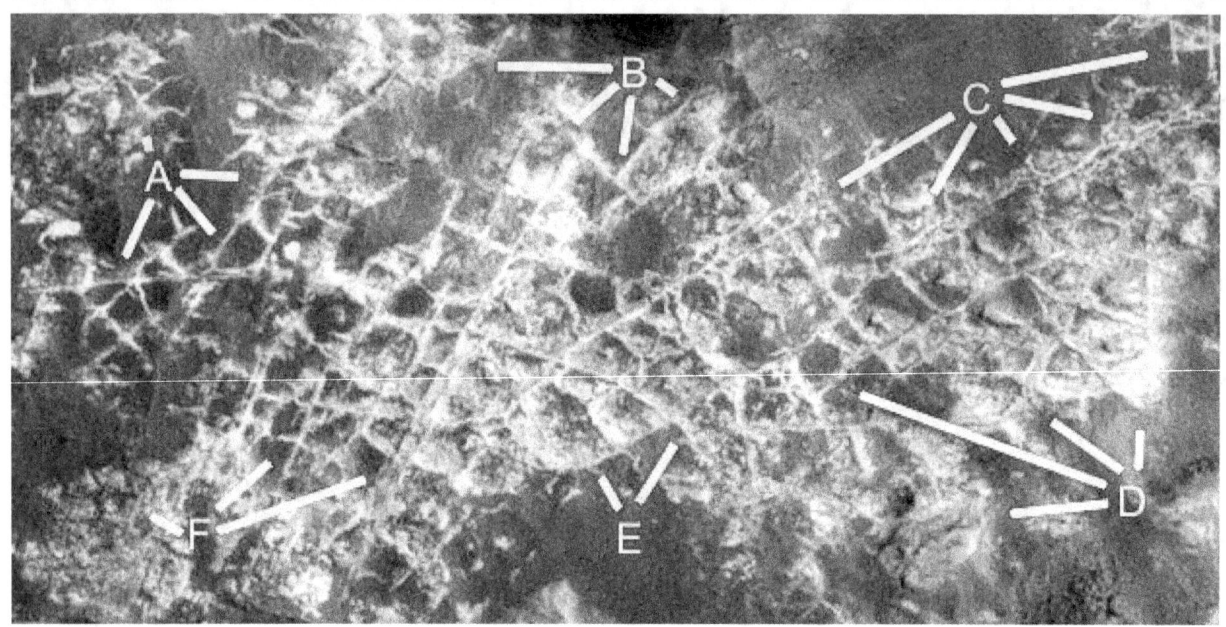

Cymhh469I2

Hypothesis

The lines show how straight the walls are.

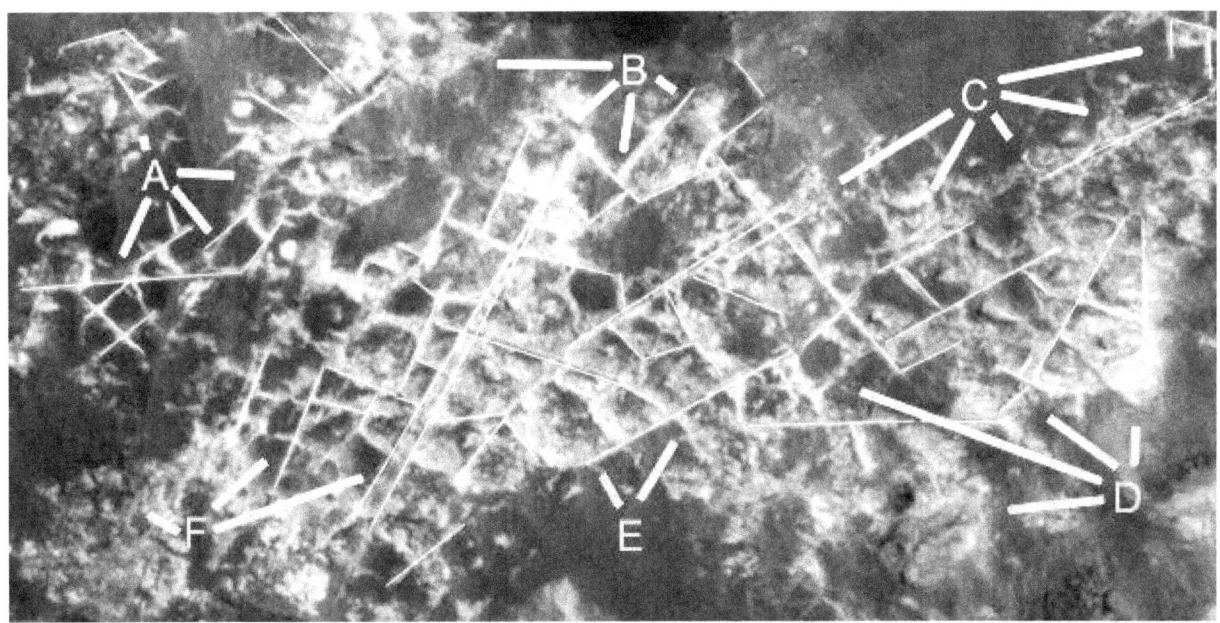

Cymhh469m

Hypothesis

A shows some faint walls, perhaps partially buried. B shows walls at 6, 7, and 8 o'clock without the dark soil. At 1 o'clock may be ceiling material, at 4 o'clock may be a dome. C shows more irregular rooms, D may be the edge of where some rooms are buried. E may be a hollow of eroded rooms, it points to different kinds of walls. F shows clear walls at 10 and 11 o'clock, at 4 o'clock the walls have dark soil on their floors. G between 2 and 4 o'clock shows small domes or objects like furniture on some rooms.

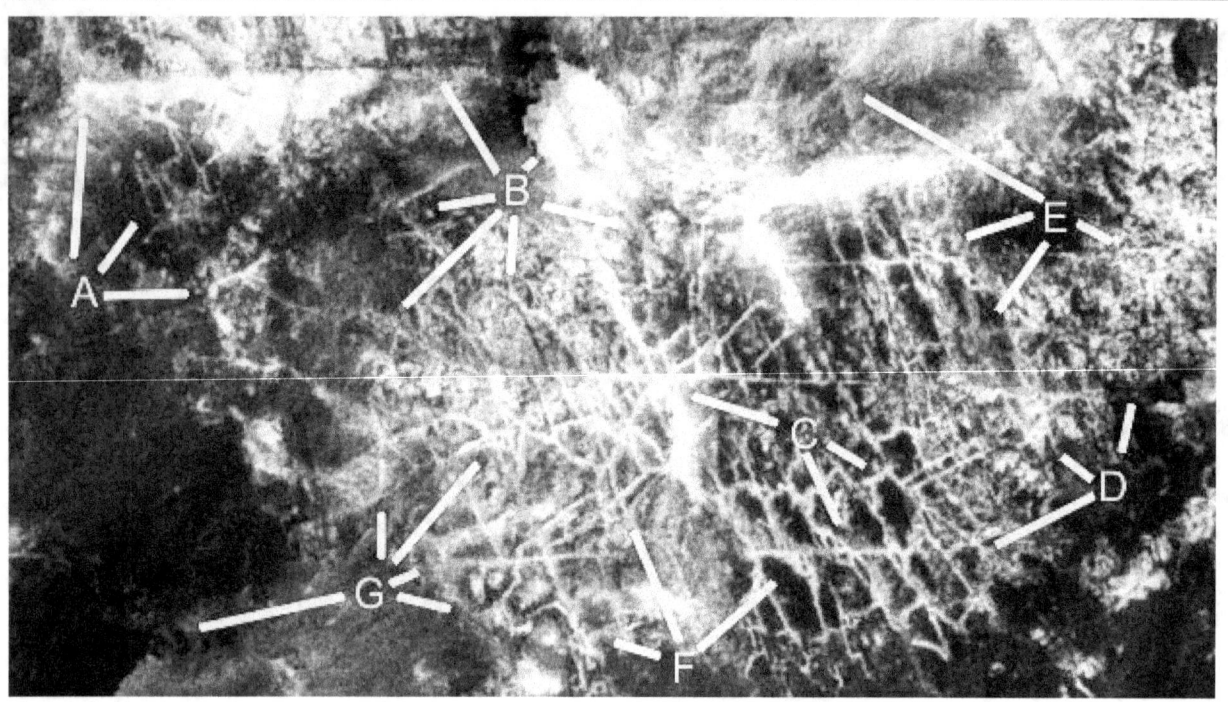

Cymhh469m2

Hypothesis

The lines show how straight the walls are.

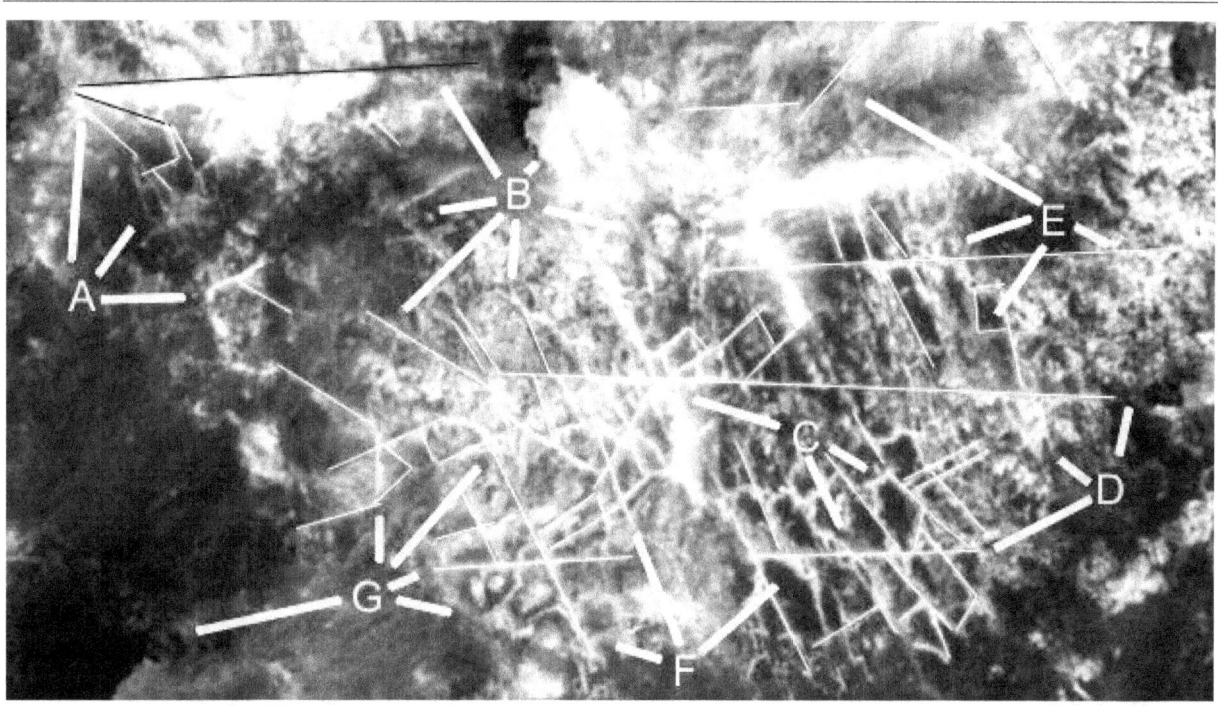

Cymdhh470b

Hypothesis

A may shows some walls that are highly eroded or buried. B appears to be higher so these may be intact ceilings. C and D may be eroded wall material that has scattered as soil. E may be buried walls, F appears to be the edge of a buried area as well. Between C, F, and G the walls are the clearest.

Cymdhh470b2

Hypothesis

The lines show how straight the walls are.

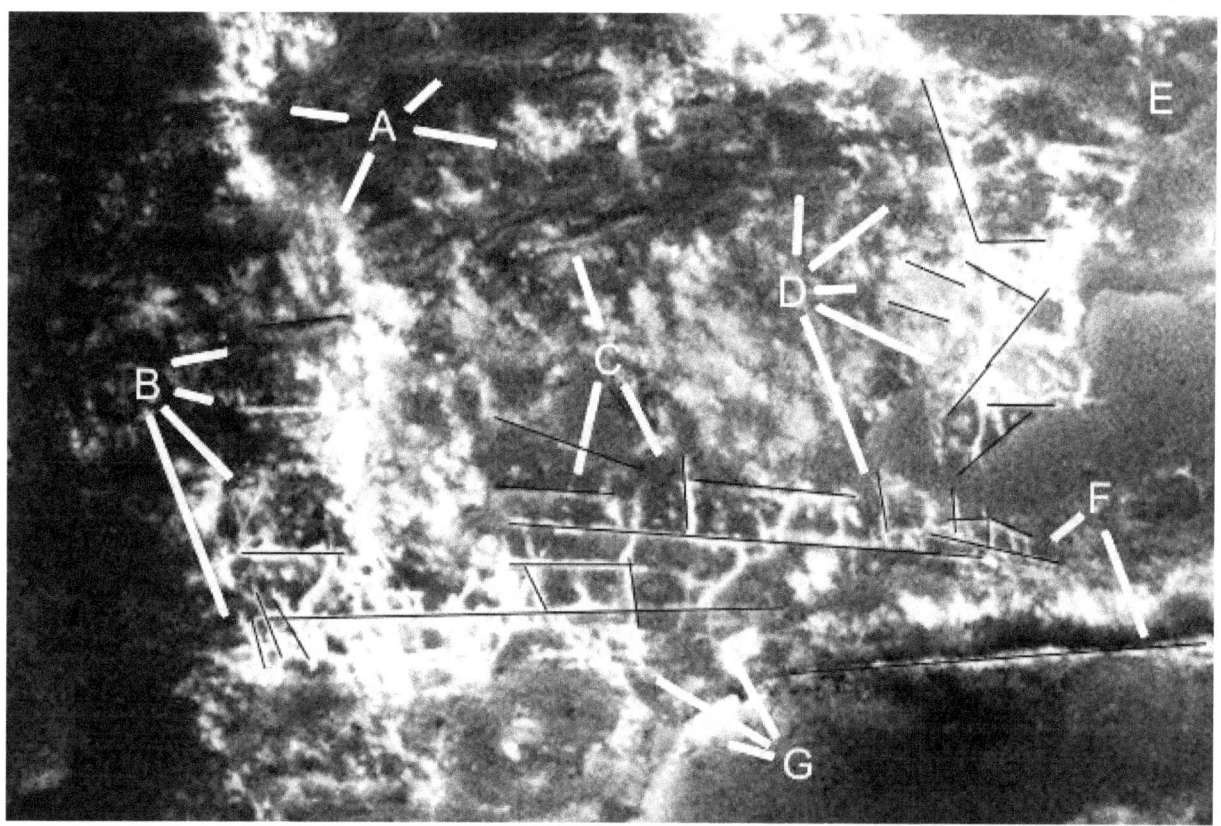

Cymhh471f

Hypothesis

A may be intact ceiling material with walls around its edges. B may be a dome at 2 o'clock, down to 3 o'clock the ground appears to be elevated and may have intact rooms under it. The material around C is very smooth like an intact ceiling, this may disintegrate over time and become dark soil. The rooms appear to be lower and under this ceiling particularly at 4 o'clock. At D the rooms also seem to appear form under a ceiling. E, F, and G show many rooms with different levels of ceiling degradation.

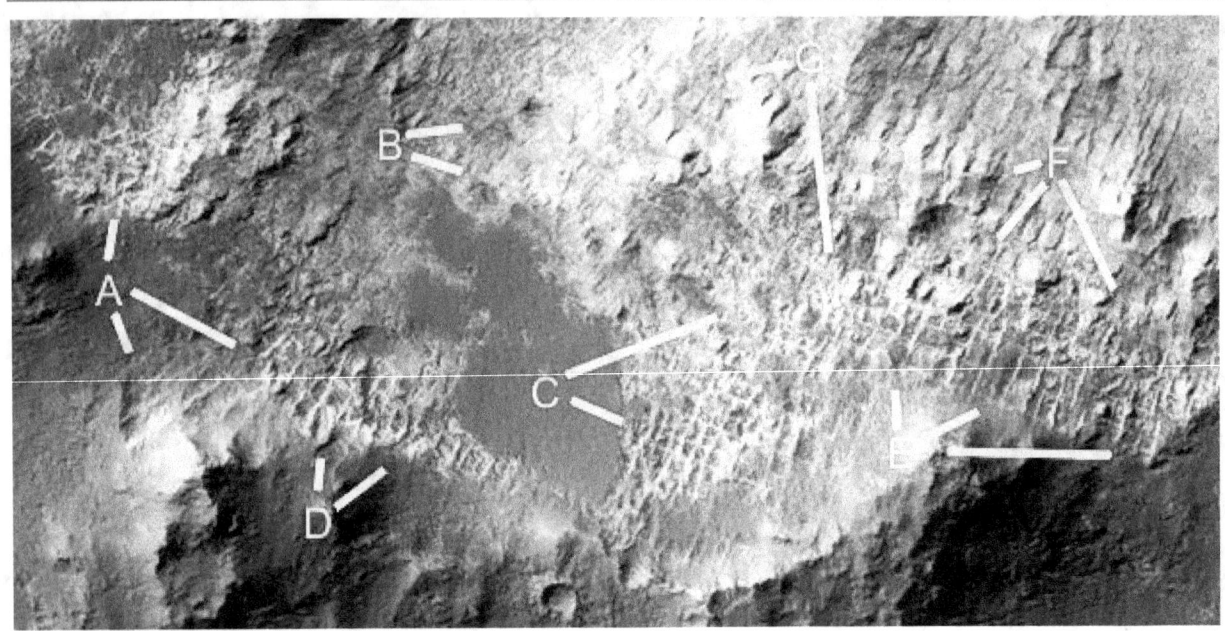

Cymhh471f2

Hypothesis

The lines show how straight the walls and grooves are.

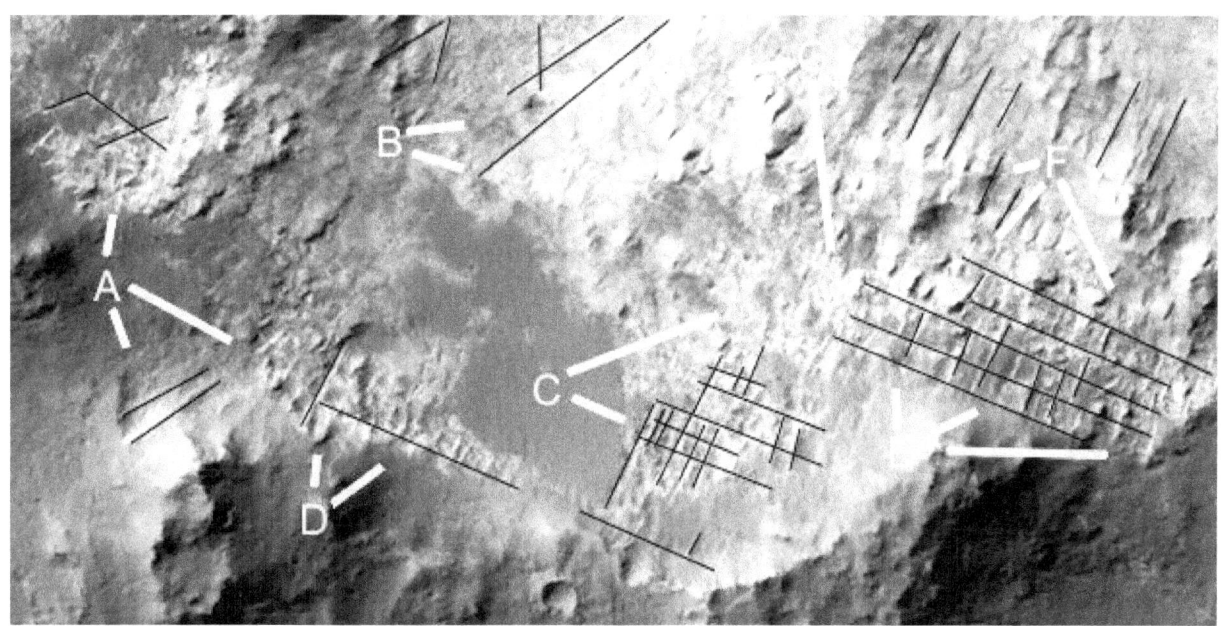

Cymhh471g

Hypothesis

The areas around A and B show many rooms, C and D may be a darker ceiling material. E looks like a hill composed of rooms, there may be some intact inside it. F shows some rooms appearing from under smoother ceiling material, G shows many more rooms.

Cymhh471g2

Hypothesis

The lines show how straight the walls and grooves are.

Prca480

Hypothesis

More of these tube shapes, A shows dark spots along it like it is breaking up. B at 9 o'clock is like a hollow hill as seen in many other areas, the dark patch on top may be the roof. B at 5 o'clock shows more collapsed areas. C at 7 o'clock shows the bank is well defined, at 4 and 8 o'clock the tube shape changes from dark to pale. At 10 and 4 o'clock the bank is also well defined.

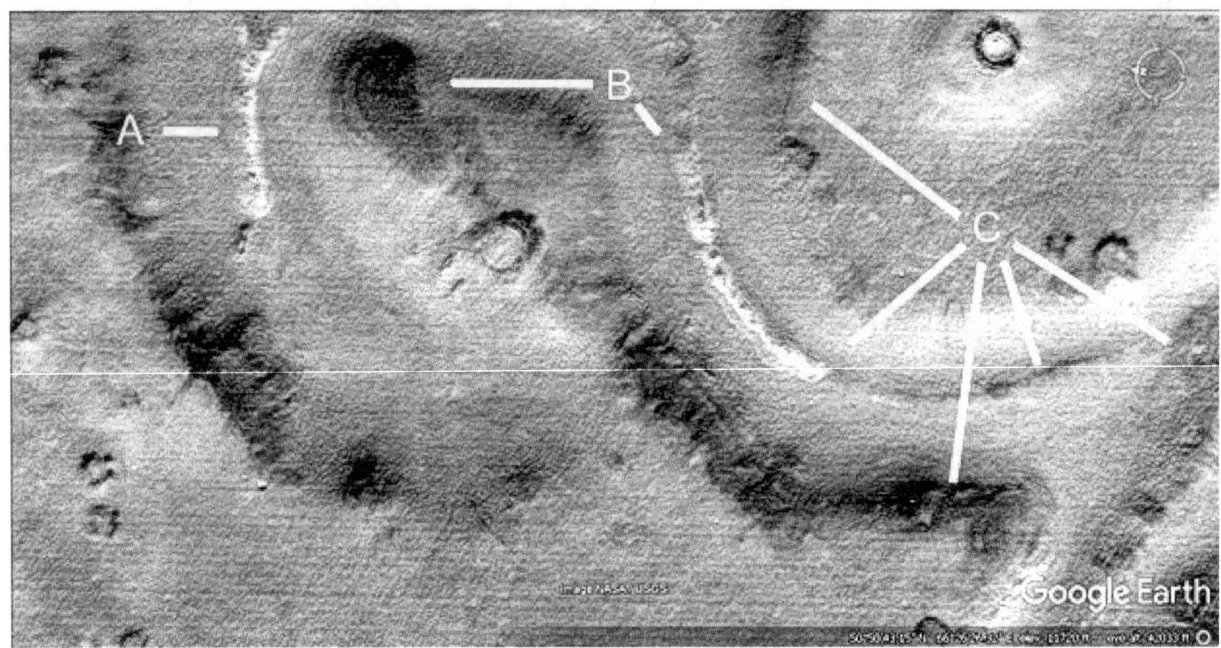

Prca480a

Hypothesis

This part of the tube shape is a near perfect parabola as shown, unlikely to occur by chance. The tube shape is also about the same height and width wherever seen, it does not vary much randomly like a natural formation from weather erosion.

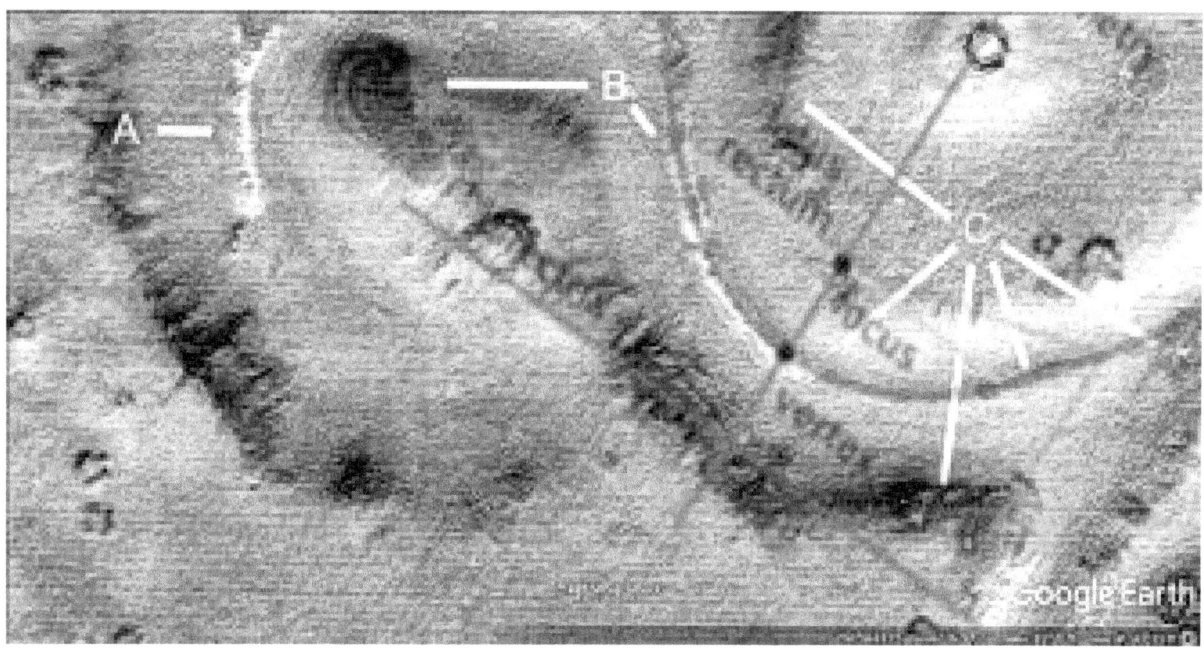

Prr493

Hypothesis

This is a road shape going from the hollow hill at A along B into the crater at C. This crater is then connected to another, shown by D perhaps by a hollow hill.

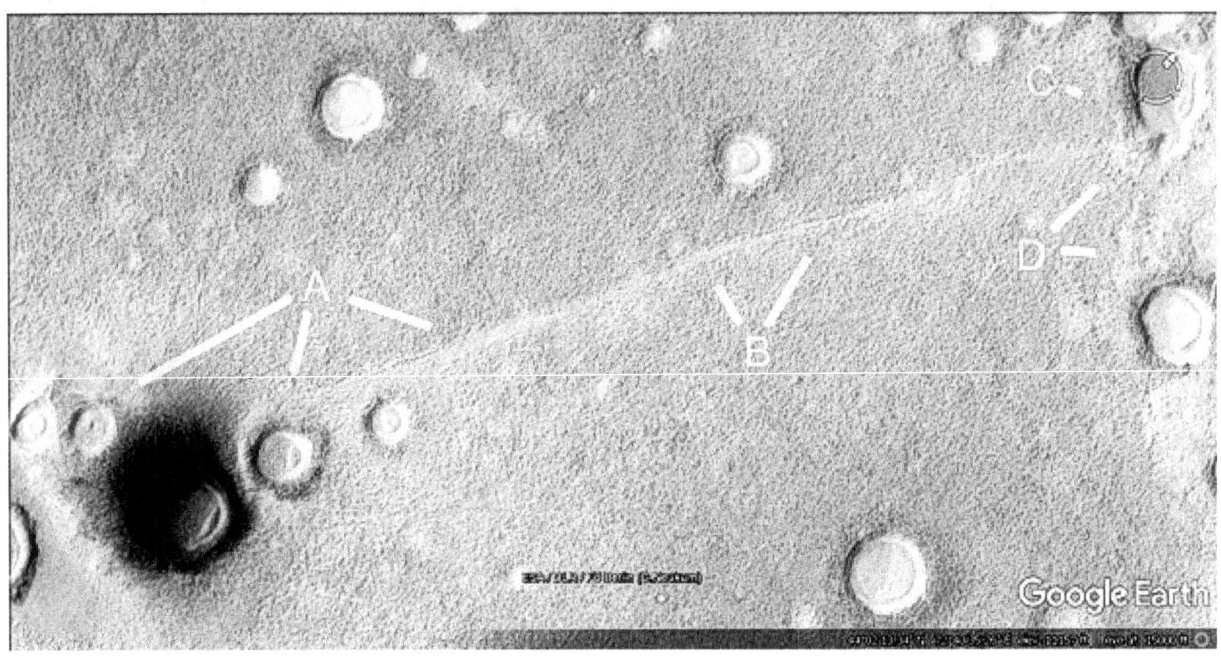

Prhh498

Hypothesis

The hollow hill has collapsed at A, B shows a straight wall still standing. C shows another road going into the hill perhaps with two lanes, this extends to D at 10 and 1 o'clock. There may be another road at 7 o'clock.

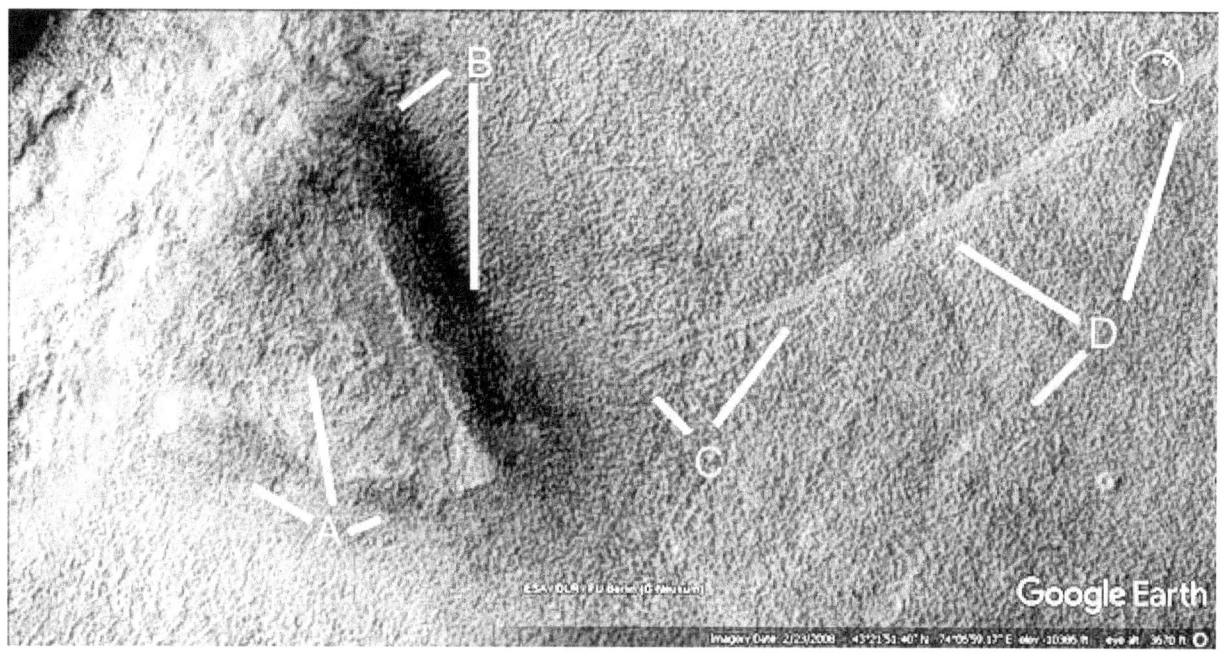

Prr499

Hypothesis

This is a closeup of a road, much smoother than the surrounding terrain like cement. It extends past A to B where a tube or raised road intersects it. C shows this tube going down from 10 o'clock, then possibly at 6 and 7 o'clock into the crater.

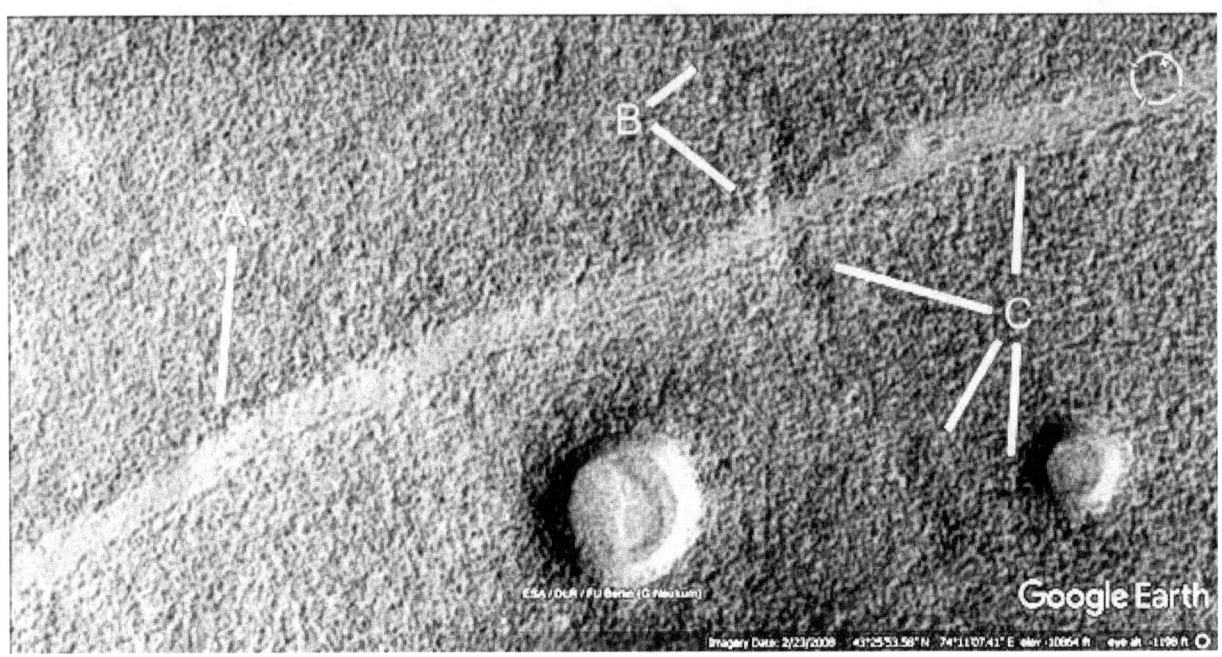

Prr508

Hypothesis

A shows the road continuing on over the pale material, B and C also show pits like altered craters perhaps with the same road material to act as dams.

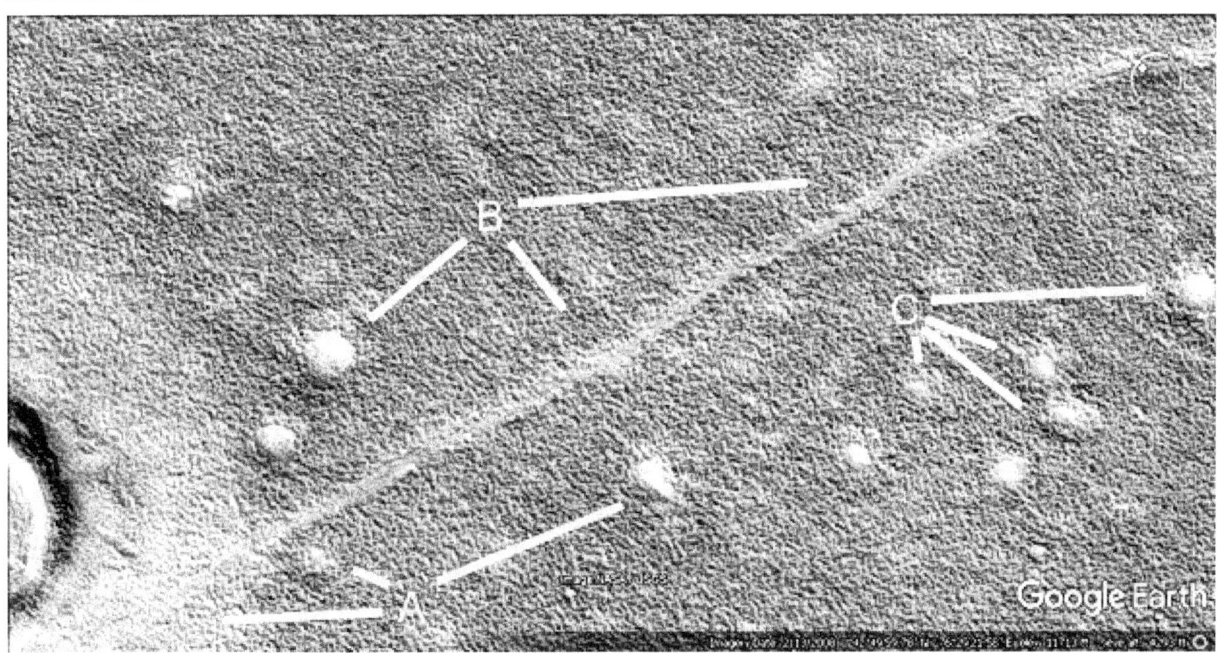

Prr509

Hypothesis

A shows a broad connection between the crater at 10 o'clock, there is a dark line like a road at 8 o'clock and 7 o'clock shows the edge of the smooth cement like section. B at 8 and 10 o'clock shows a hill with some patched areas, at 2 o'clock this other hill has a light stripe cross it like a patch. It may also connect directly to the road leading off through F at 7 and 8 o'clock. B at 2 and 4 o'clock shows more possible roads. C shows another road going to a small crater. F at 9 o'clock shows how a road goes up to the large crater. E shows a small dark area in the road at 7 o'clock and a sharp bend at 4 o'clock. The pale material at 10 o'clock may be another hollow hill surrounding the crater.

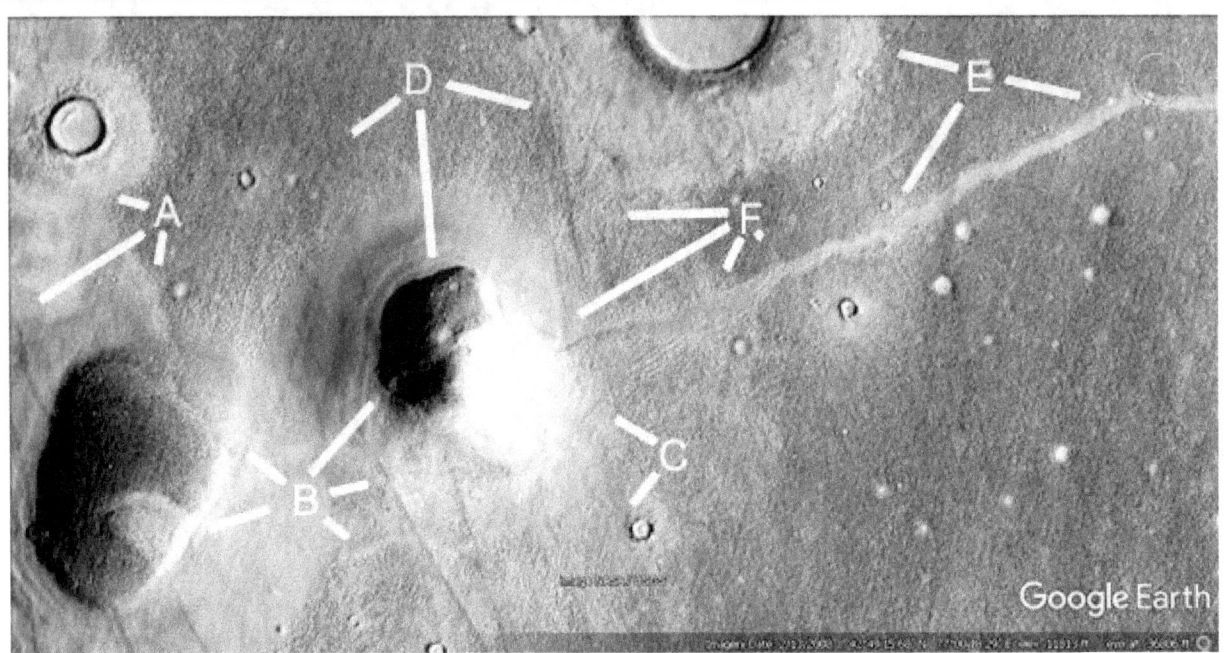

Prhh533

Hypothesis

A, B, C, and D show the road continuing and avoiding craters. This section is in better condition.

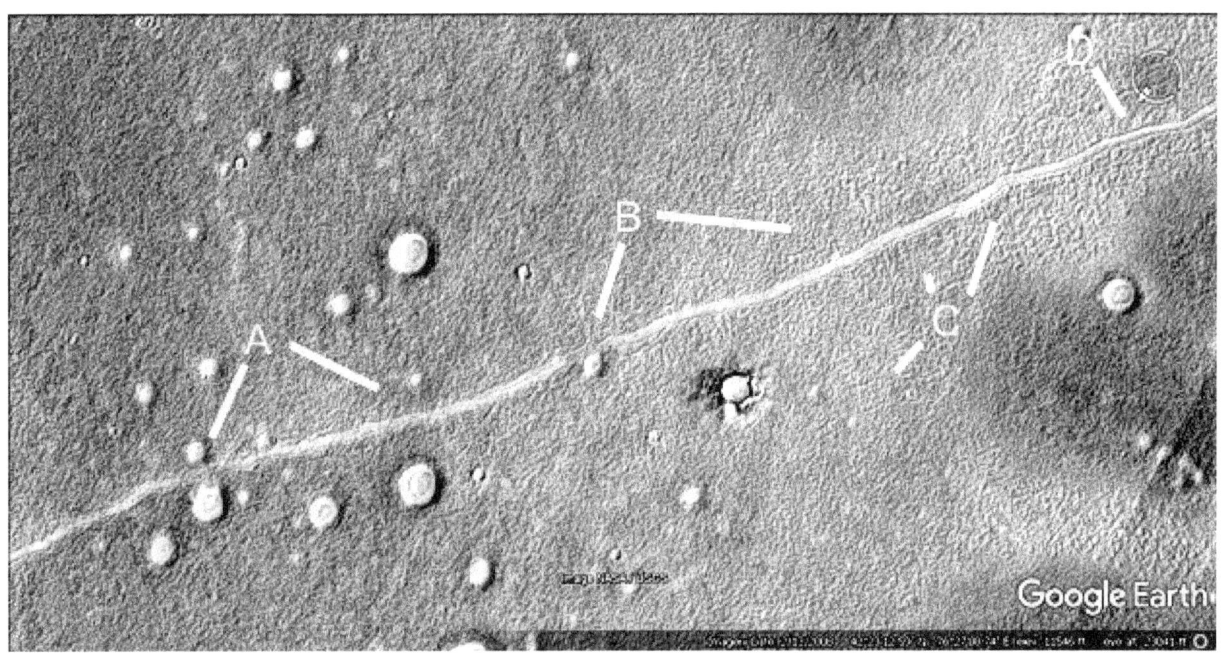

Prr533a

Hypothesis

This closeup of the road shows right angled shapes in it, perhaps like bricks or tiles. This impression continues along the road where it seems to vary in an angular rather than a smooth way. The center is very smooth compared to the surrounding terrain as shown by comparing A at 1 and 5 o'clock. B shows a shape like a gutter along the road's side. C shows a small pit at 10 o'clock that appears to be connected to the road, perhaps a former hollow hill, at 2 o'clock is an angular section on the side of the road.

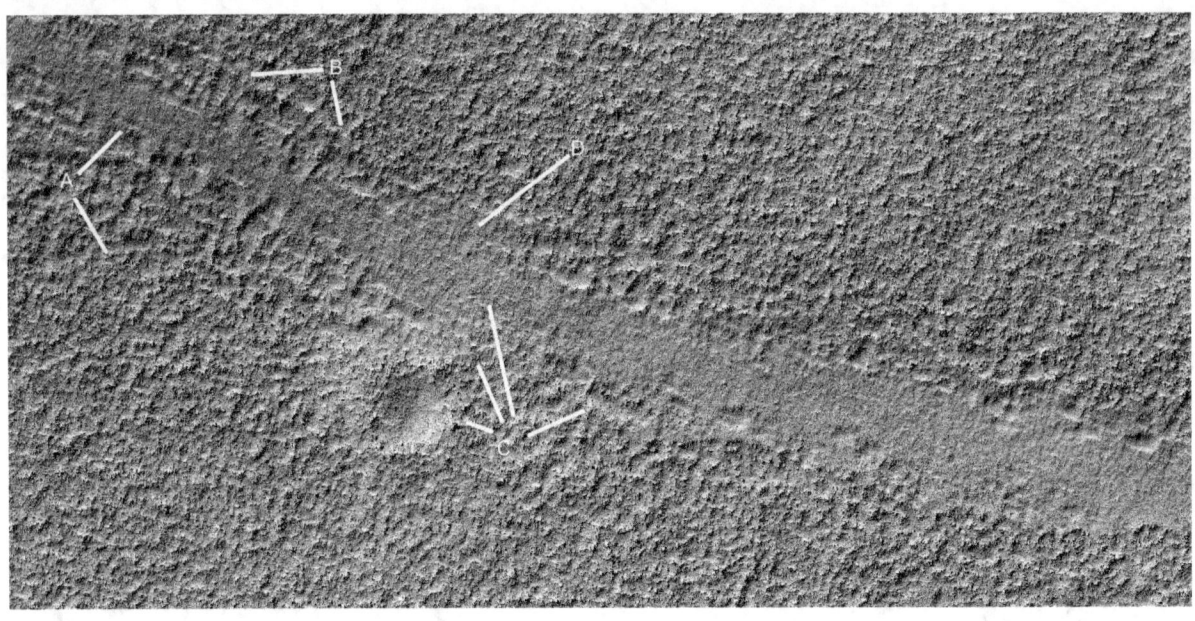

Prr533b

Hypothesis

Another closeup of the road shows a distinct difference between its smooth center at 7 o'clock and the random looking terrain at 10 o'clock. This might be cement but there is a persistent impression of angular structures in it.

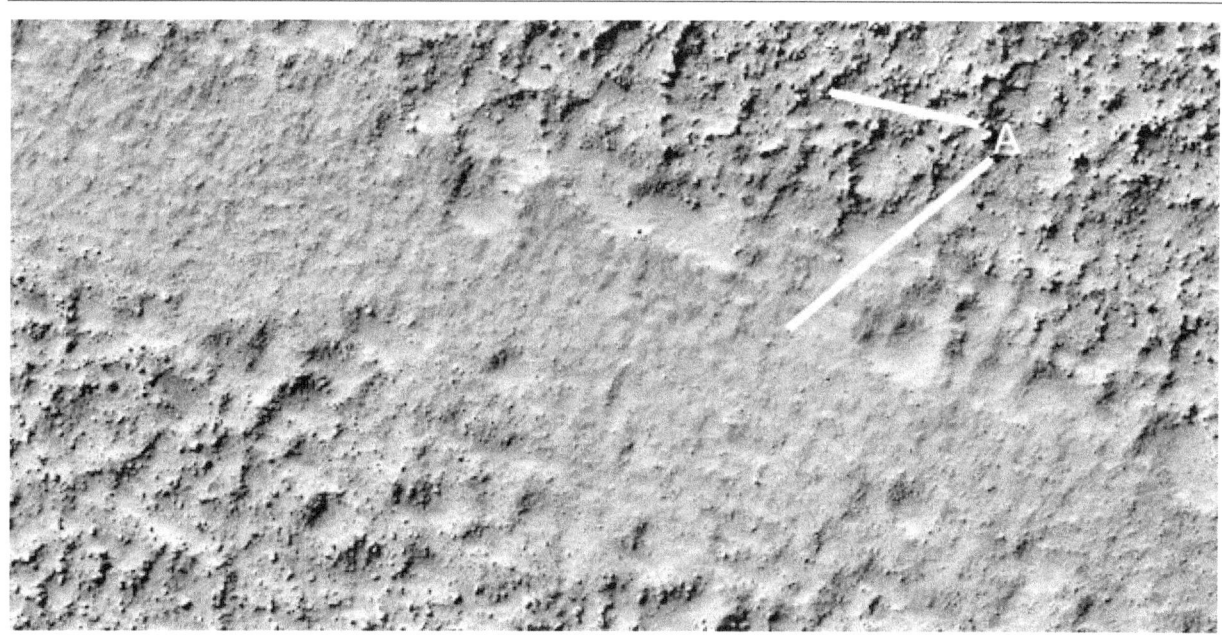

Prr533c

Hypothesis

There appear to be parallel lines going along the road like bricks lined up, seen at B at 5 and 8 o'clock. At 9 o'clock these pits are also angular as if a brick has sunk into the ground. A at 5 o'clock shows a smoother area like a road section, another at 4 o'clock that looks like the main road section at 2 o'clock.

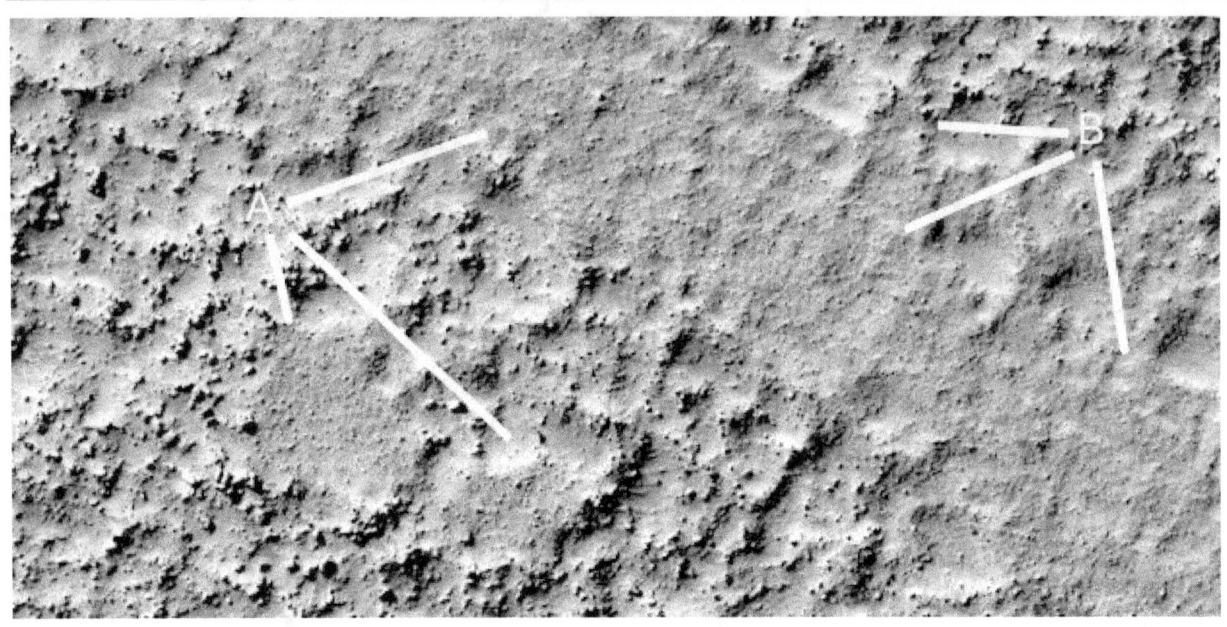

Prr533d

Hypothesis

A compares part of the road to the terrain alongside it, B shows regular shapes in it like bricks.

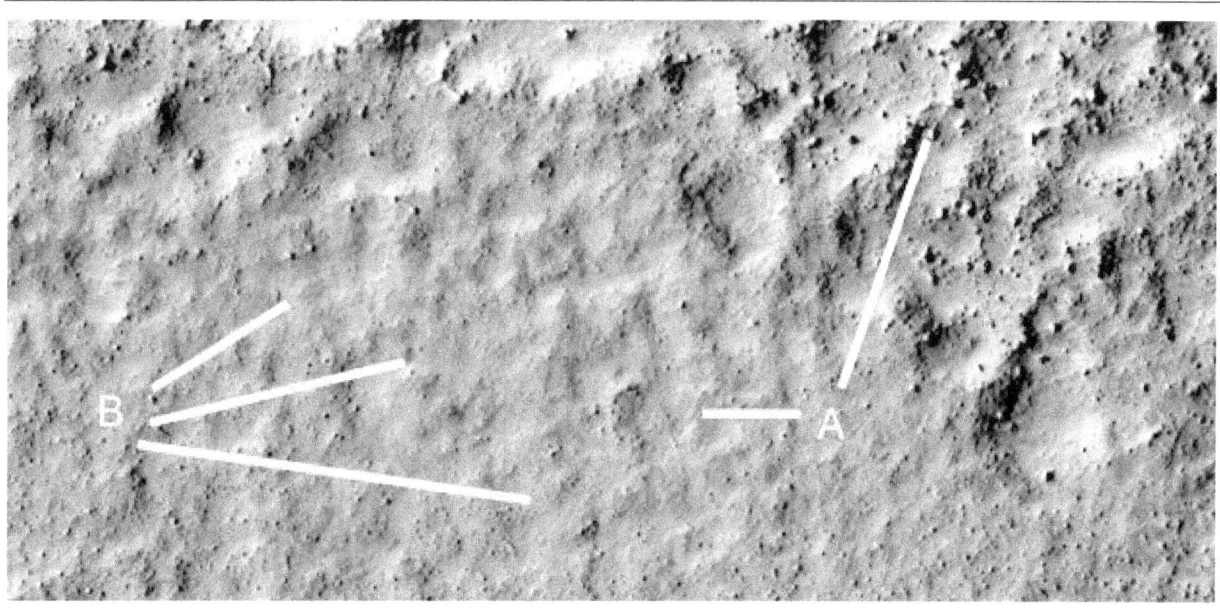

Prr546

Hypothesis

This is the same road, it comes out at A and seems to climb up the hill at 11 o'clock into a small cavity on its roof. It then goes to B where it meets a crater or pit, perhaps aimed at this to collect water. It then goes on through C.

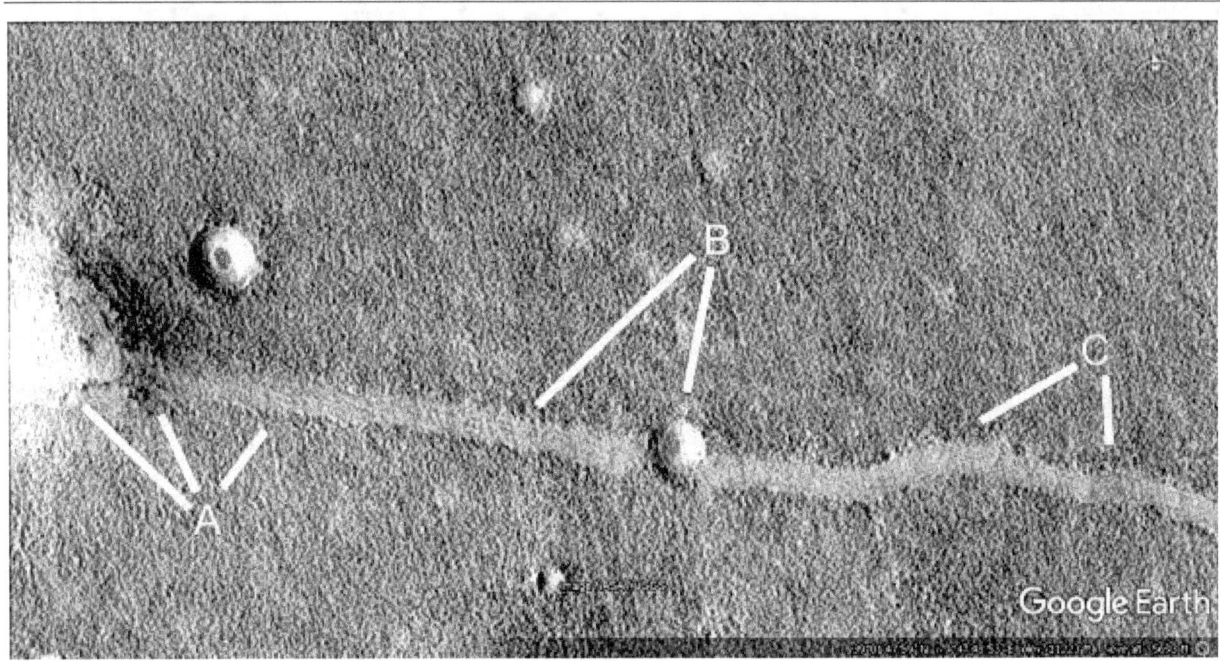

Prr561

Hypothesis

A shows two hollow hills where the roofs have collapsed. B shows some of these pits, also roads going from a hollow hill down to the crater. C follows one of these at 12 and 2 o'clock. D shows a main road, at 7 o'clock there may have been a hollow hill. E may be a tube or a road coming off the main road.

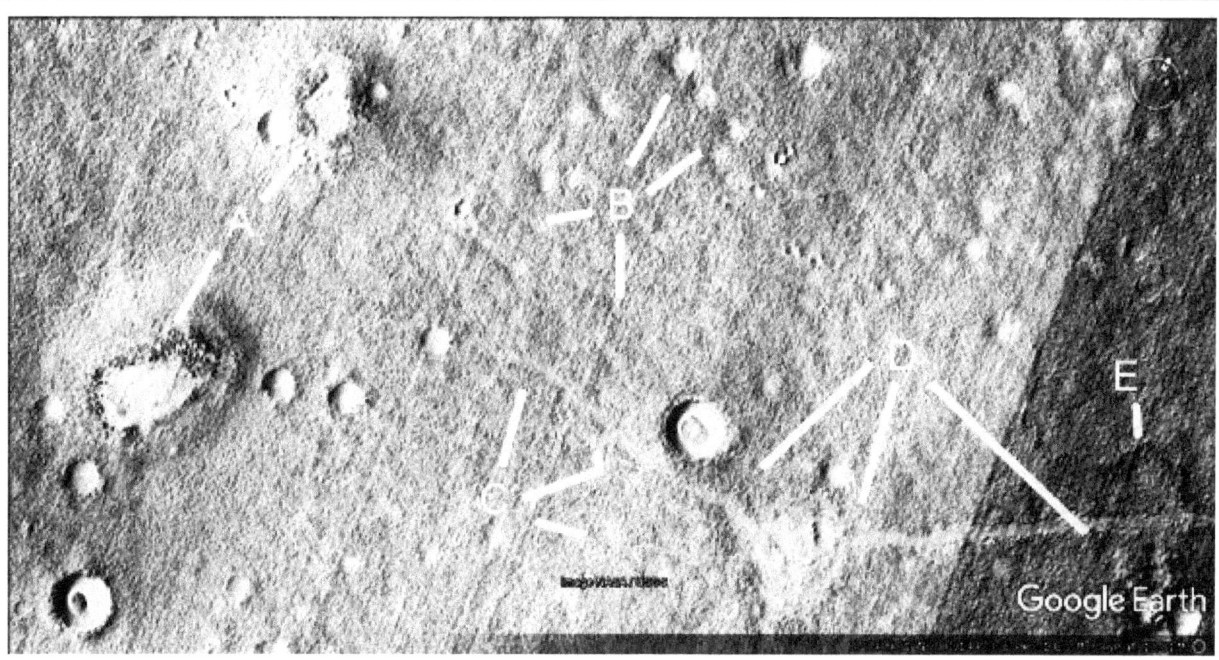

Prhh563

Hypothesis

A shows a road going into the hollow hill, B shows another road at 1 and 2 o'clock. At 4 o'clock is a collapsed hollow hill, another road is at 8 o'clock. C shows the skin of the hollow hill has peeled off at 10 o'clock, at 11 o'clock some settled areas are shown. At 2 o'clock there is another road coming out of the hill, D shows this and another road. E shows possible interior supports through the settled roof.

Prhh569

Hypothesis

A shows two craters that may have been repaired, at 5 o'clock the crater rim is an unusual shape and extends out from the hill. This should not form during an impact. At 2 o'clock the crater is surrounded by an unusually gentle gradient rather than a high rim, it also looks like the hill connects to this shape like it knew the impact would happen here, if natural. It implies then the hill section was built around the impact or repaired after this impact. B shows some walls or tubes with many right angles, C shows a rounded walled area at 9 o'clock, a dark wall at 10 o'clock, and a hollow hill at 2 o'clock. D shows more of these walls.

Prhh569a

Hypothesis

A parabola is shown.

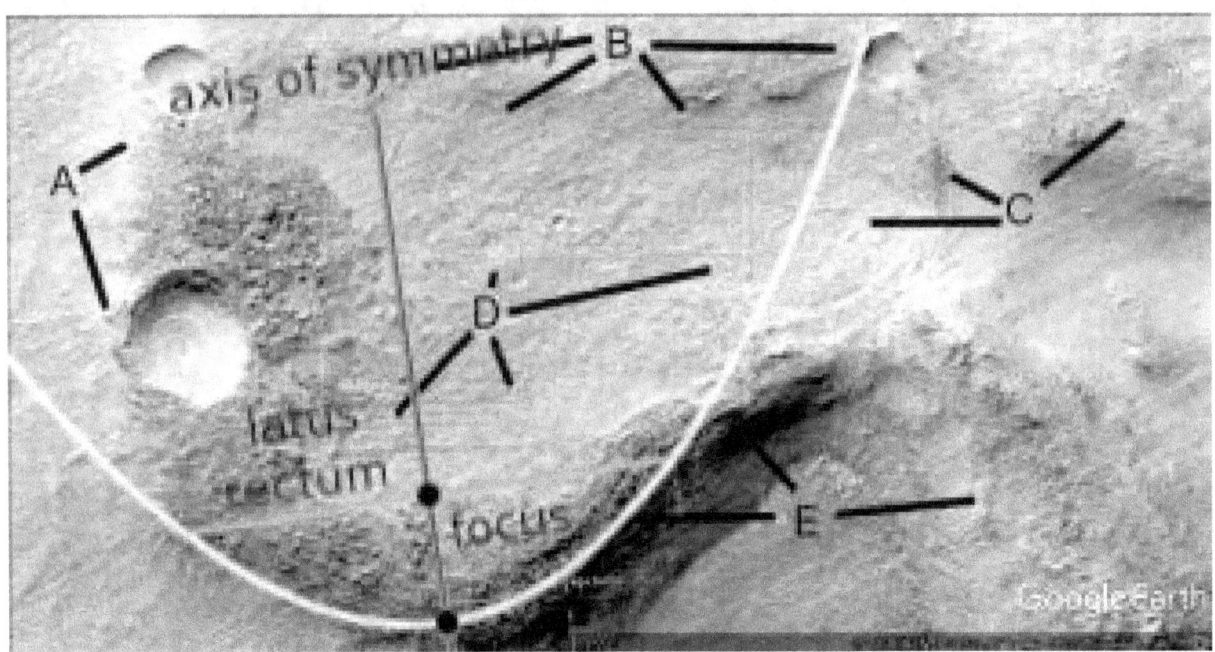

Prhh570

Hypothesis

A shows several tubes connecting to the crater going to the hollow hill at B at 3 o'clock. At 5 and 6 o'clock B shows more hollow hills. C shows the dome in the middle of the hill at 10 o'clock, at 11 there are tubes or roads going off along D to other hills. At 7 o'clock C shows another road or tube. E shows many tubes, from 4 to 6 o'clock the tube goes from the hills to the crater, at 8 o'clock it appears to go around it. F at 8, 10, and 12 o'clock shows another tube going into a crater then the hills, at 1 o'clock are more tubes.

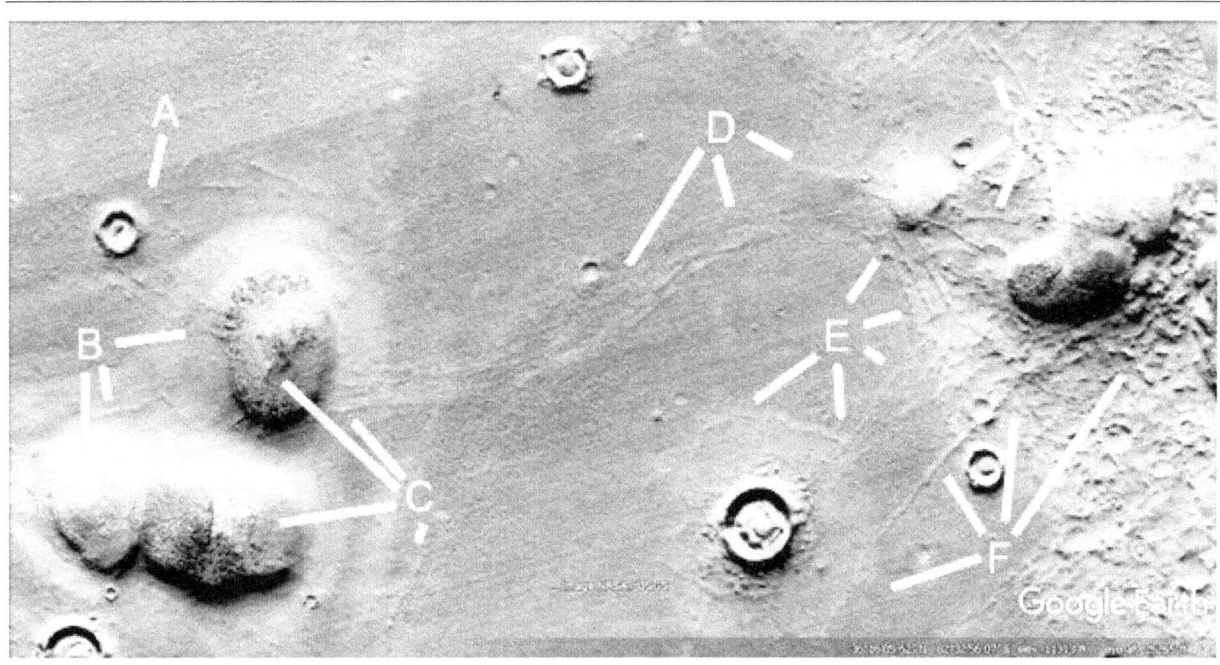

Prhh570a

Hypothesis

A parabola is shown.

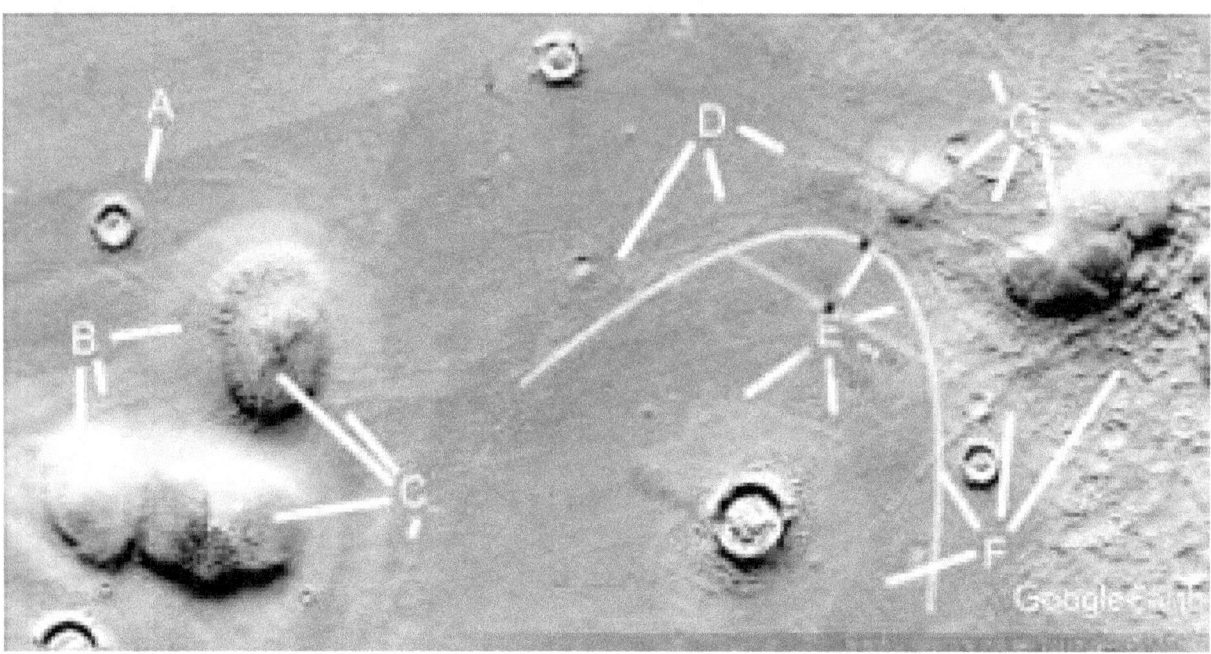

Prhh578

Hypothesis

A shows a tube between the hollow hills at 4:30, the edge of smoother ground surrounding them at 4 o'clock, a faint tube at 5 o'clock. B shows layers in the roof at 10 o'clock like a Cobler Dome or amphitheater. At 10 and 2 o'clock there are tubes connecting the two hills, another more degraded at 6 o'clock. C shows another tube, D at 9 o'clock shows a tube or patch on the roof, at 8 o'clock the skin may have broken off up higher on the roof, and more tubes at 5 o'clock. E also shows tubes.

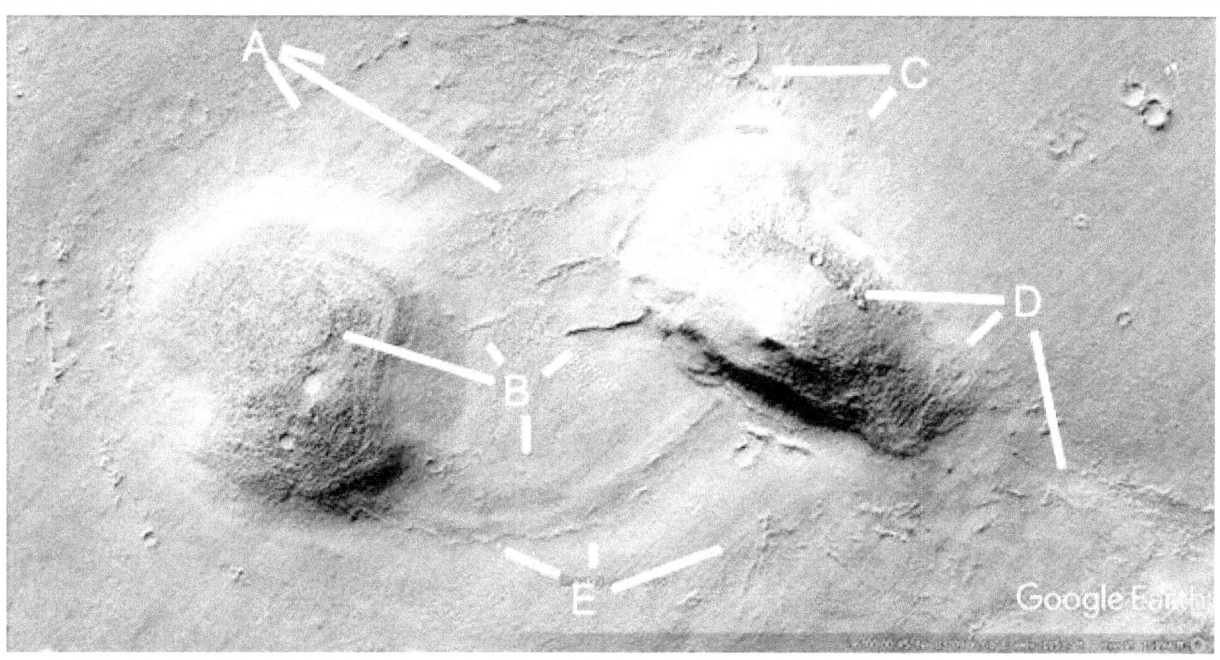

Prhh578a

Hypothesis

A parabola is shown.

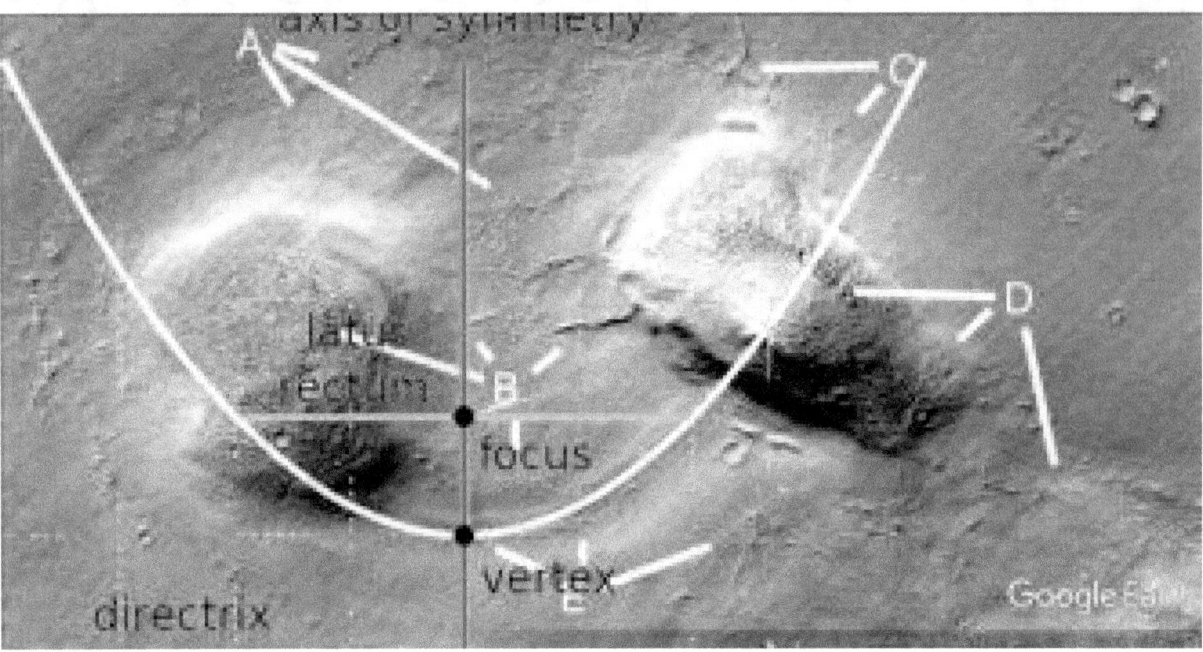

Prhh580

Hypothesis

A shows where the hill has settled at 10 o'clock, at 11 o'clock a tube comes out of the hill to the crater at C. At 11 o'clock a tube also branches off to go up to the crater, this crater rim is unusually shaped like it has been altered. B shows another small tube coming off the main one, it may connect to the crater faintly.

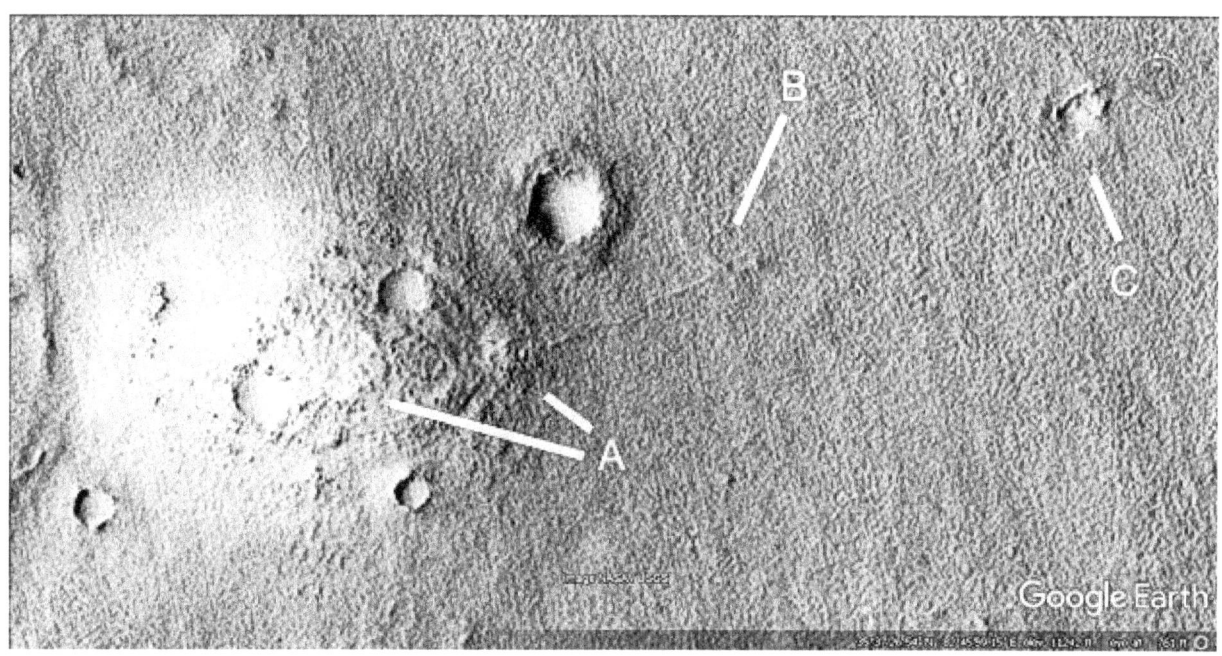

Prhh581a

Hypothesis

These also appear to be tubes, A shows a smooth section like cement where perhaps a hollow hill collapsed. B shows how the tube was hollow inside at 10 o'clock, at 3 o'clock it also seems to be hollow. C shows an enclosure that may have been a hill, D shows a flat area like cement at 8 o'clock, at 10 o'clock there is a right angled tube.

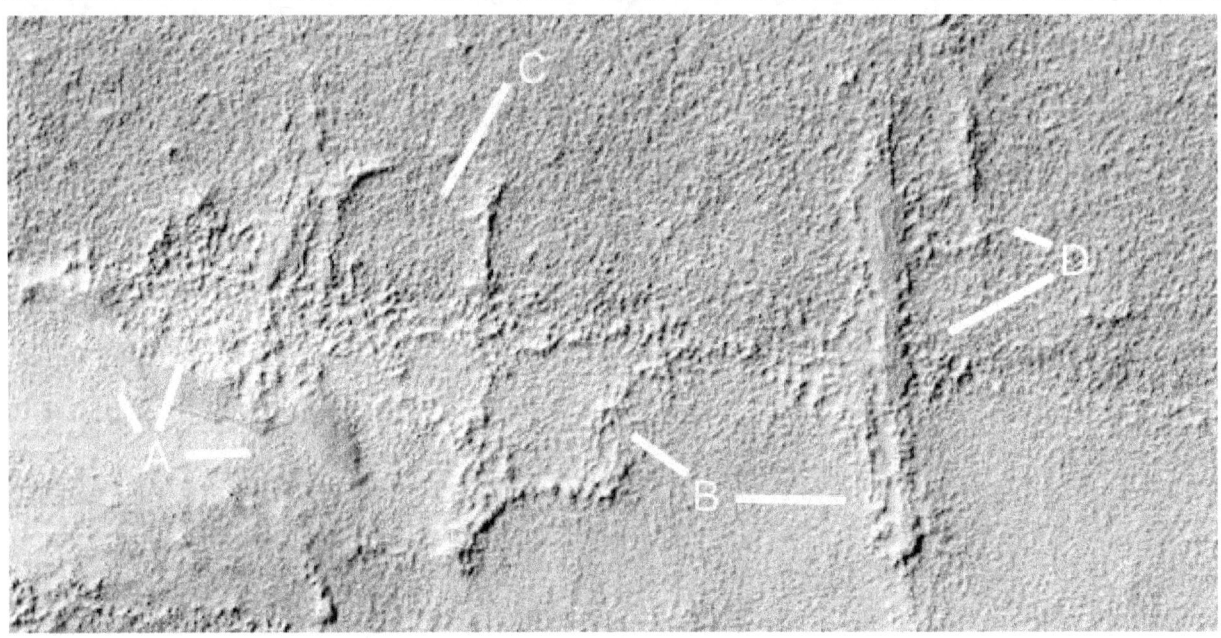

Prt586

Hypothesis

A shows some small hills connected to parallel tubes at 11 o'clock, another longer degraded tube at 3 o'clock. B shows several tubes going into the hollow hill. C shows a kind of tube nexus at 9 o'clock, at 4 and 7 o'clock there is a wall like the edge of a pit. D shows where the skin peeled off the roof at 10 o'clock, at 7 and 8 o'clock is a large degraded tube with regular shapes in it like arches. At 11 o'clock may be the edge of the pit. E shows tubes going into the hollow hill. F shows a tube nexus at 11 o'clock like a meeting place, there may have been other ones at 8 and 10 o'clock.

Prt586a

A parabola is shown.

Prt592

Hypothesis

A shows more tubes, B shows a tube at 6 and 8 o'clock going into the degraded hollow hill, this appears to have collapsed in the roof at 4 o'clock. C also shows the collapsed roof. D shows more tubes going into the hollow hill. E also shows more tubes. Between E at 10 o'clock and F at 1 o'clock is a hill with tubes around it. F at 8 o'clock shows a curved tube.

Prt593

Hypothesis

A shows a collapsed hollow hill at 7 o'clock with a tube coming out if it to 4 o'clock. At 10 o'clock may be another collapsed hollow hill, also a tube intersection at 2 o'clock. B shows two tube intersections. C may show a small tube nexus at 8 o'clock, perhaps another at 10 o'clock and some other tubes. D shows more tubes and a nexus at 10 o'clock. E shows some degraded tubes, F shows more tubes and a possible meeting place at 8 o'clock.

Prt594

Hypothesis

A shows tubes going into craters at 3 and 9 o'clock, also a possible former hill at 7 o'clock and a thicker tube at 5 o'clock. B shows more degraded tubes. C and F show some tube intersections around a small hill which is connected to a large flat tube up towards G and into the crater. This may be a large tube nexus or meeting place, also to the right of F there may be another tube nexus. D shows a collapsed hollow hill with some tube connections. E shows a tube nexus at 6 o'clock and some other degraded tubes, at 9 o'clock they show the regular grooves like between arches or interior supports.

Prt605

Hypothesis

A shows tubes going into the long ridge, each seems to terminate with a small hill on it. At 5 o'clock there is a small hill by itself. At 2 o'clock there is a tube intersection. B shows tube segments at 11 o'clock as if eroded, at 2 o'clock there is a tube intersection going into the crater. At 5 o'clock there are eroded tube segments that seem to have been in the groove parallel to the large ridge, there is also a tube intersection going to the main crater. C at 7 o'clock shows a wavy tube connected to a small crater, at 1 and 11 o'clock there are more eroded tubes. D shows a tube network at 8 o'clock, at 10 o'clock the tubes go through a long hill like a meeting place, at 12 o'clock there is a tube intersection.

Prt615

Hypothesis

A shows several tubes branching out from a tube nexus. B at 4 and 5 o'clock shows two branched tubes, also another at 6 o'clock and a hill at 7 o'clock. C shows a curved t at 6 o'clock with some tubes branching out of it, also a partially disconnected tube at 9 o'clock. D at 2 o'clock shows a hill with a tube coming out of it, at 4 and 10 o'clock the tubes are beginning to break into tube segments. E shows a tube nexus at 2 and 7 o'clock, an eroded hollow hill at 8 o'clock and a tube going into a hill at 10 o'clock. F shows more tubes, G shows an unusually shaped hill like a golf club with an extended tube like the club shaft.

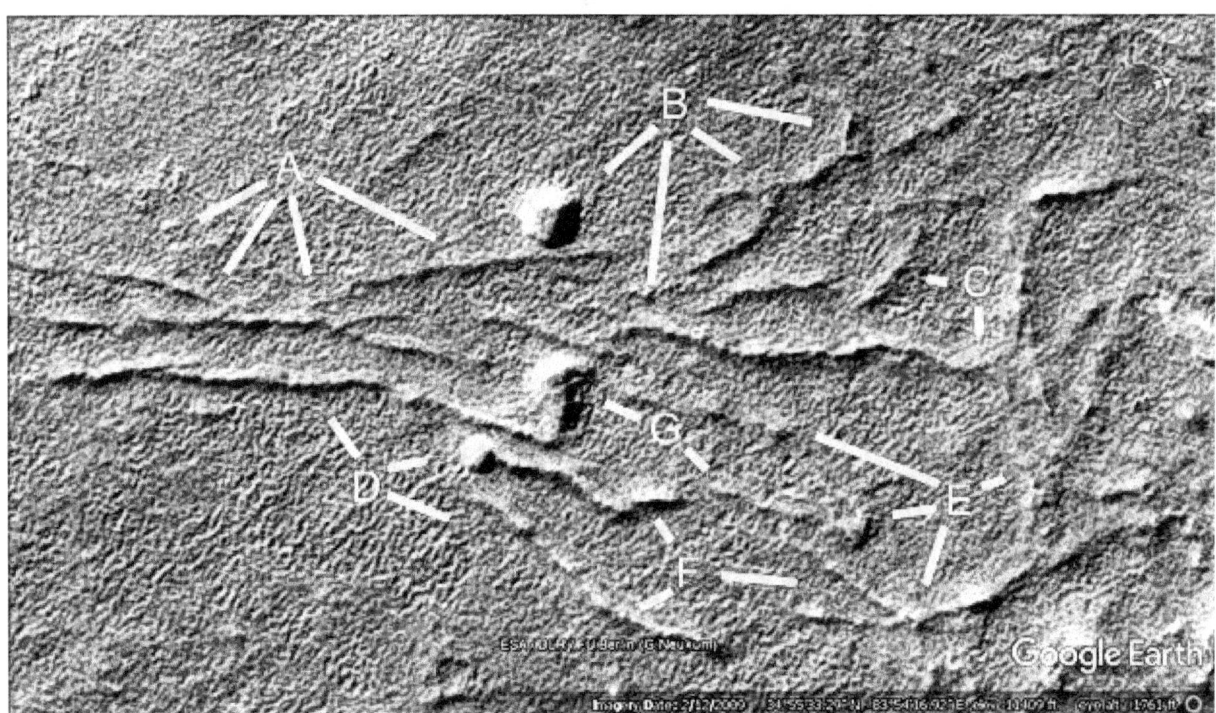

Prt641

Hypothesis

A shows a curved tube going from the walled hill at 4 and 5 o'clock to the small crater at 1 o'clock. B at 8'clock shows the walls of the hill, at 7 o'clock a tube comes out of the hill, at 1 and 4 o'clock are two more hollow hills. D shows the curved tube, it connects to another tube shown by B at 8 o'clock. At 9 o'clock is a small tube from the larger one, at 10 o'clock the smaller hill appears to have collapsed. This main tube continues up through E to the right.

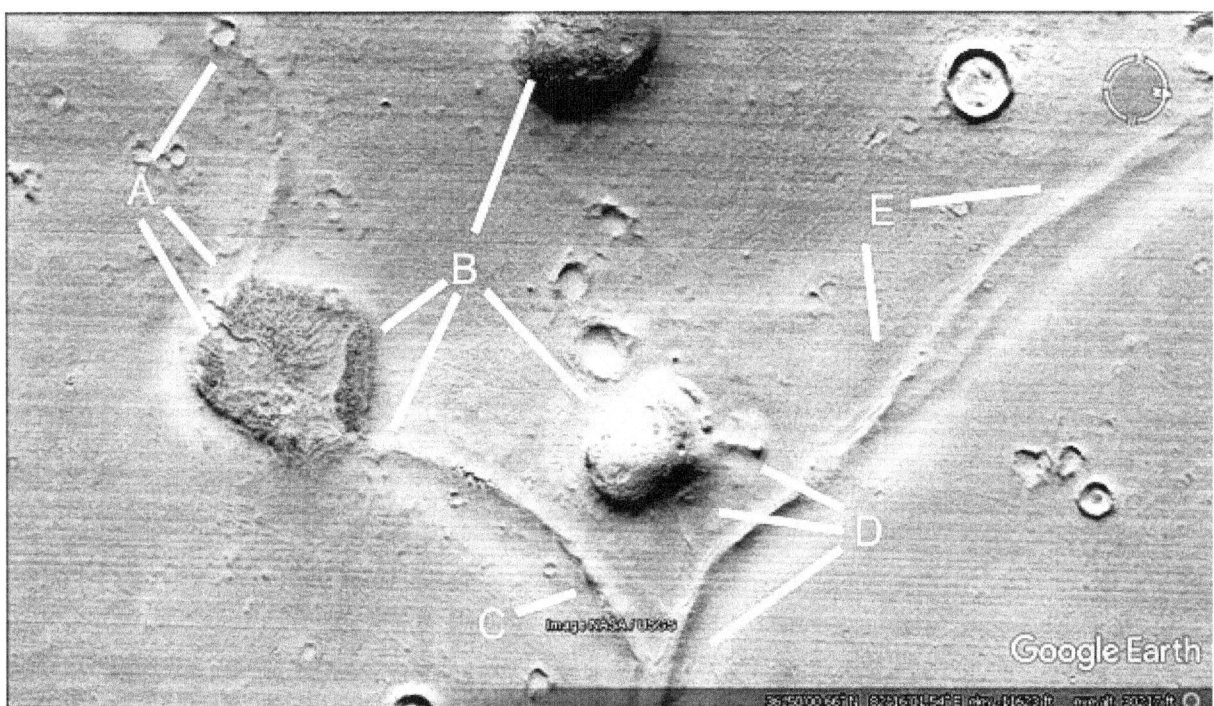

Prt641a

Hypothesis

Two parabolas are shown.

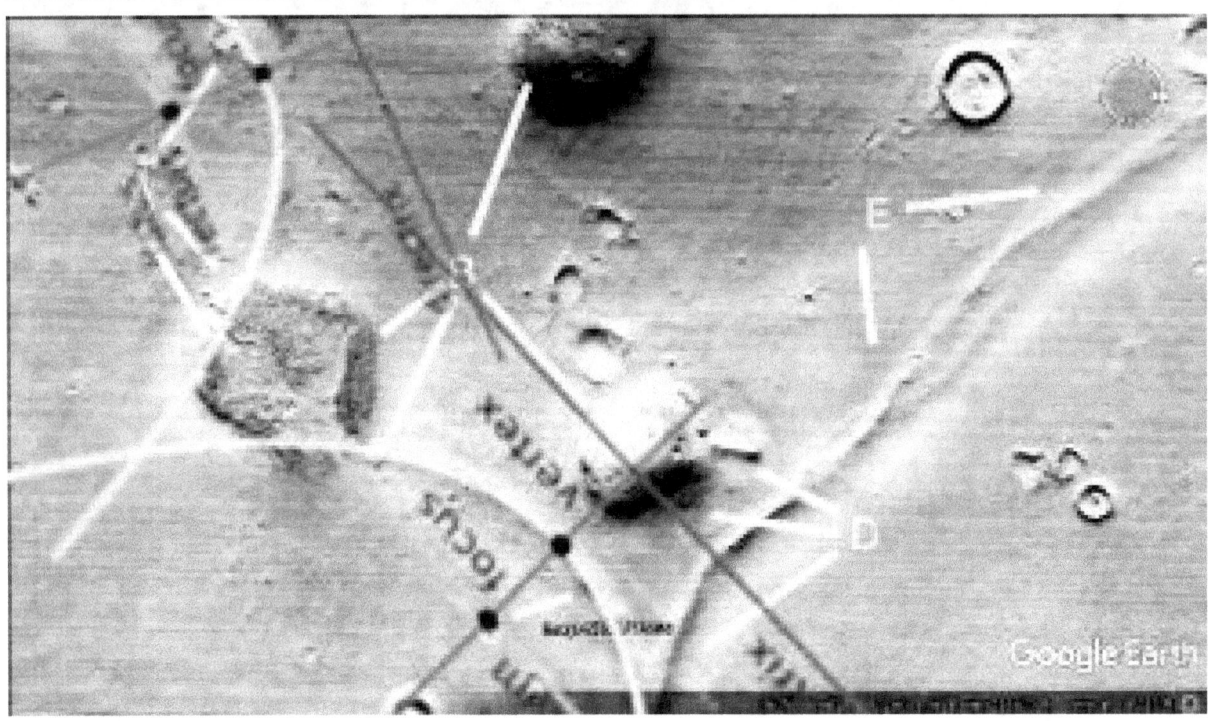

Prt657

Hypothesis

A shows approximately four parallel tubes from 12 to 2 o'clock going over to C from 10 to 2 o'clock, then through the crater at G at 10 o'clock to F. A at 4 o'clock shows a small tube coming out of a collapsed hill. B to H show up to nine parallel tubes. C shows an area free of tubes, at 9 o'clock there is a degraded hill with a tube going through it, another two are 6 and 2 o'clock. D may show an eroded nexus or collapsed hill at 7 o'clock with the tubes from the crater at 8 o'clock going through it. Four of these turn upwards at 4 o'clock. E shows a tube going into the crater from 6 to 7 o'clock.

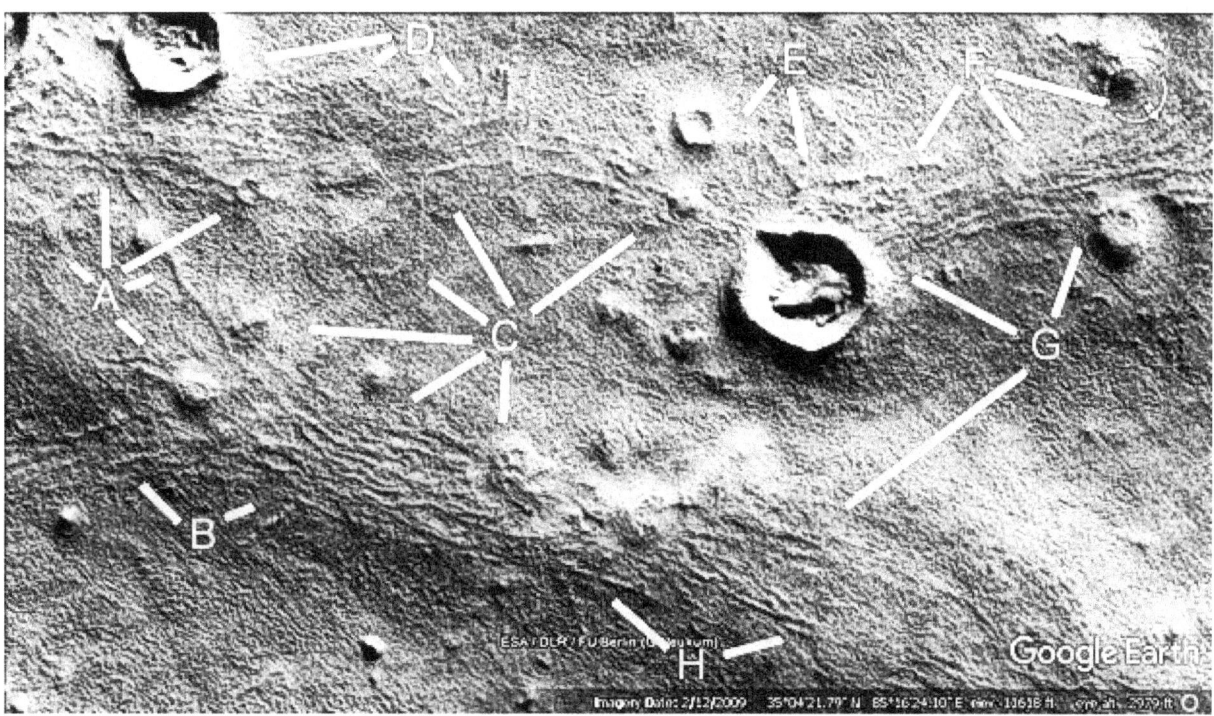

Prt662

Hypothesis

A shows a wavy tube, B shows a clear area surrounded by tubes like a field. C shows tubes going into a crater at 6 and 8 o'clock, at 1 o'clock they go into a rounded area, also shown by F at 10 o'clock, under a nexus. D shows more tubes going into this nexus. E at 6 o'clock shows an intersection of tubes then this goes down, making a right angled turn into a hollow hill at F at 1 o'clock. E at 12 o'clock shows a T intersection, at 4 o'clock there are about four faint parallel tubes going up the image. F at 7 and 8 o'clock shows tubes going into three collapsed hills, also shown by G. H may be a large habitat, at 9 o'clock a tube crosses other tubes at 10 o'clock going up to I at 2,4, and 6 o'clock and a collapsed hill. At 10 and 11 o'clock faint tubes go into the crater. J shows more tubes going into the collapsed hill.

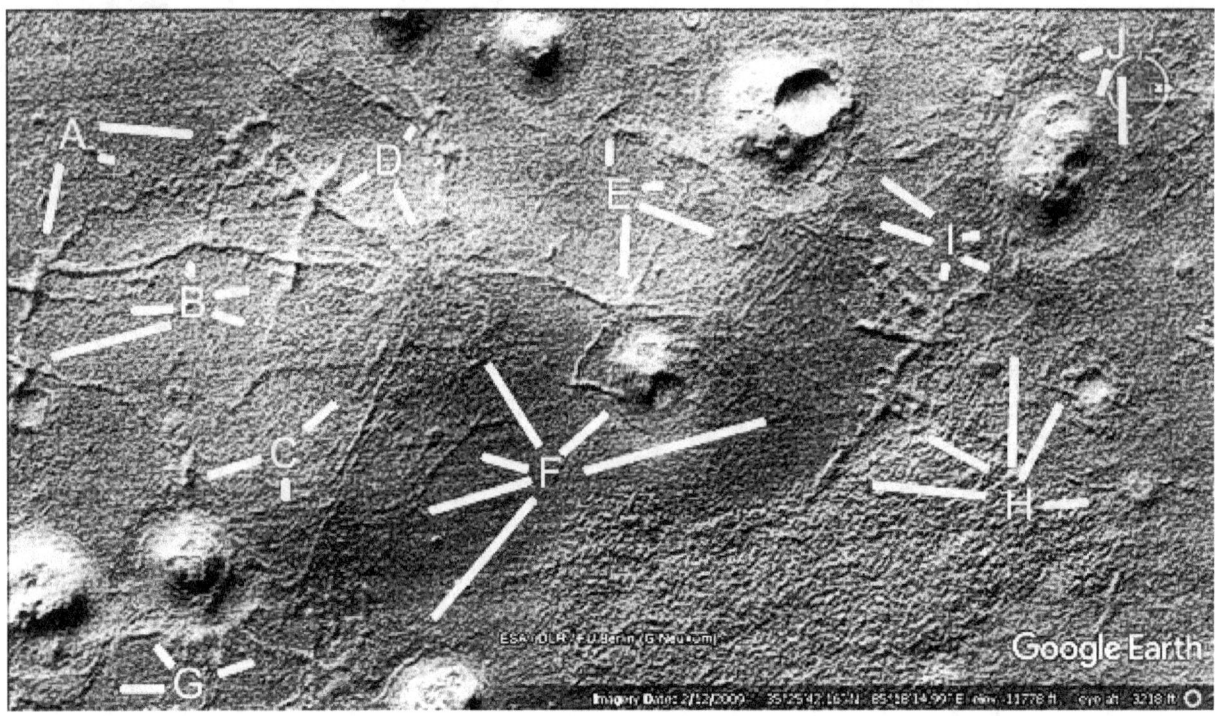

Prt671

Hypothesis

A shows tubes going to craters at 9 and 12 o'clock, a tube nexus at 3 o'clock and parallel tubes at 5 o'clock. B, C, and E show more tubes going into the nexus. E from 2 to 4 o'clock shows parallel tubes going into the crater coming out at D at 8 o'clock. From 5 to 7 o'clock tubes go into the crater. At 10 o'clock there is a small nexus, at 11 and 12 o'clock tubes go into a line of small hills. F shows two hills linked by tubes at 10 o'clock, parallel tubes from 11 to 7 o'clock going down to the crater. G shows a collapsed hill at 9 o'clock and a forked tube at 4 o'clock.

Prt681

Hypothesis

A shows tubes making a walled area at 3 o'clock, up to eleven parallel tubes at 5 o'clock going to the right and the small crater. B may show a collapsed nexus at 1 o'clock, tube intersections from 4 to 7 o'clock. There are many intersections here rather than a larger nexus, this would make it easier to change directions in the tubes. C at 9 o'clock also shows the nexus, at 6 o'clock is another walled or tubed field. At 3 and 4 o'clock may be collapsed hills. D shows more tubes, E shows A tube going into a small hill at 1 o'clock. From 7 to 10 o'clock is another intersecting tube. F and G show more tubes.

Prt682

Hypothesis

The tubes come together in a large nexus here, there also seems to be flat areas like cement over the tubes. These might act as a roof with rooms under them. A shows a tube crossing another at 2 o'clock, this connects to another tube at 10 o'clock. At 6 o'clock is the edge of the outer circular shape of the nexus. This may have allowed movement around the nexus without going into the centre, like an Earth ring road in many cities. B shows a continuation of the ring road at 3 o'clock, a forked tube at 10 o'clock and at 9 o'clock, and a narrow fork at 8 o'clock. C shows a larger tube at 10 o'clock where it appears to end on top of a small platform. At 1 o'clock the tube is hollow like the roof collapsed. D shows a tube ending at 11 o'clock, some tubes crossing at right angles in a mesh at 2 o'clock. E shows two tubes parallel to each other, further along one tube crosses over the other like a knot. F shows a small hill connecting to the tube at 3 o'clock, a loop of a tube at 5 o'clock with a central tube. From 8 to 10 o'clock is the flattened part of the nexus, whether from erosion or a roof. G shows a small nexus.

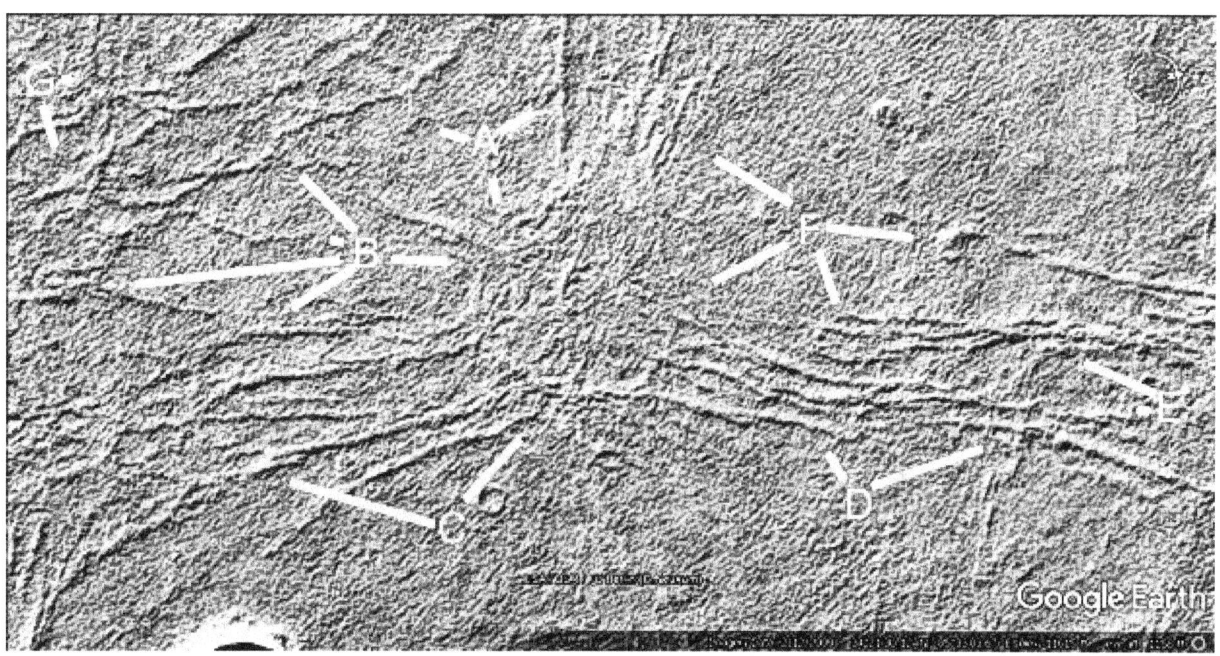

Prt684

Hypothesis

Another large nexus is shown at A at 3 o'clock, a small nexus at 7 o'clock. Several tubes cross each other at 2 o'clock. C at 2 o'clock shows two larger tubes coming out of the nexus, a forked tube at 3 o'clock, and a wider tube perhaps like a room at 5 o'clock. D shows tubes going into a crater at 4 and 7 o'clock, also a mesh of tubes at 1 o'clock. E shows the highest point of the nexus hill at 11 o'clock, from here many tubes come out of the hill. Some tubes come down the hill parallel at A at 9 o'clock, more eroded tubes connect to the nexus at 8 o'clock. F and H show many tubes at right angles like walled fields, at G the tubes fork many times. These tubes appear to be going into the nexus at ground level while the lower side has the tubes climbing the hill before entering.

Prt686

Hypothesis

A shows many parallel tubes from 7 to 12 o'clock, a parallel tube at 2 o'clock has a right angled intersection under it, then a second one at 5 o'clock. B has an intersection at 1 o'clock, also many parallel tubes at 9 o'clock with a cement roof covering them. At 8 o'clock is a fork, at 6 o'clock the tube looks to have collapsed. C from 2 to 5 o'clock shows a tube going into the crater, underneath it an eroded tube comes out the other side. Between C at 1 o'clock and D at 10 o'clock is a triangular array of tubes. At 7 o'clock a forked tube ends, the right fork is eroded but also goes into the crater. C at 11 o'clock turns sharply D at 3 o'clock shows a T intersection, E at 10 o'clock and F at 8 o'clock go into the crater. E at 1 o'clock points at parallel tubes going to this crater, from 5 to 7 o'clock is a rectangular mesh of tubes. F at 9 o'clock also goes into the crater, G from 11 to 2 o'clock connects to the tube going into the crater. F from 10 to 3 o'clock shows regular bulges like the tube is collapsing and exposing arches or pillars.

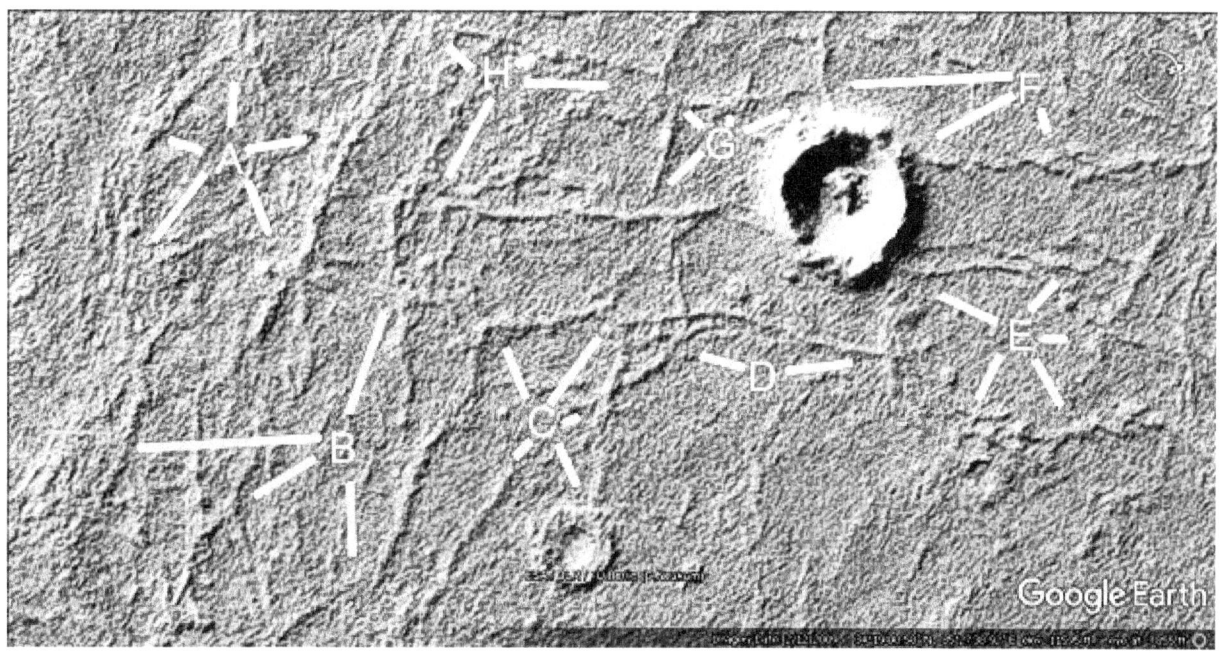

Prt687

Hypothesis

A shows some approximately rectangular meshes of tubes, also below A. B shows multiple parallel tubes like ring roads, at 11 o'clock a tube crosses each one. At 2 o'clock there is a circle which may have been a crater, but the tubes go across it now. This means the tubes came after the crater. At 8 o'clock is another tube connecting the parallel tubes. At 7 o'clock the tubes are much thicker, also at F at 2 o'clock. C shows a large walled field with parallel tubes or walls at 4 o'clock. D shows a squarish array of parallel tubes. E at 9 and 10 o'clock may show eroded parallel tubes, in better condition at 4 o'clock.

Prt697

Hypothesis

A shows a sharp turn in the tube at 1 o'clock, this goes into the crater at 10 o'clock. At 2 o'clock the tube has collapsed showing a cavity under it. B shows a tube going into the crater at 4 o'clock, at 5 o'clock a small tube goes into the hill. C shows an eroded road or tube at 8 o'clock, at 6 o'clock the tube goes into the hill. At 5 o'clock there is a dark line like a collapsed tunnel connecting to this tube. At 4 o'clock two other faint tubes go into the hill.

Prt704

Hypothesis

A shows tubes going into a crater at 5 o'clock, a small hill is connected to a tube at 6 o'clock. At 4 o'clock there is a flat smooth area like cement. B shows another tube going into the crater, many others are connecting to it. C shows flat cement areas at 7 and 8 o'clock, perhaps a small habitat connected by tubes. At 1 and 4 o'clock are small partially collapsed hills. D shows a cavity in the hill at 11 o'clock, at 10 o'clock a tube connects to it. Another tube at 8 o'clock goes into a small collapsed hill at 6 o'clock. E shows a collapsed hill connecting with a tube to the crater.

Prt704a

Hypothesis

Three parabolas are shown, the bottom one on the right is a standard parabola.

Prt708

Hypothesis

A shows more tubes connecting to the hill at 11 o'clock and the crater through 3 o'clock. A squarish array of walls or tubes at 7 o'clock. B shows this line of hills from 10 to 2 o'clock connected by tubes. At 7 o'clock is a small tube going into a hill, continuing on through 4 o'clock to 2 o'clock. C shows a tube going into a hill at 11 o'clock, this goes down to 6 o'clock at H into another hill. D shows a hill connected to a crater at 7 o'clock, some tubes at 5 o'clock going into the hill. At 1 o'clock the hill is collapsing, at 10 o'clock there is a tube. E shows a tube going into a hill at 6 o'clock, three parallel tubes come from hills at E at 8 o'clock over to F from 2 to 5 o'clock. At 8 o'clock a tube turns sharply to the left going into the hill below A. G shows more tubes going into the crater.

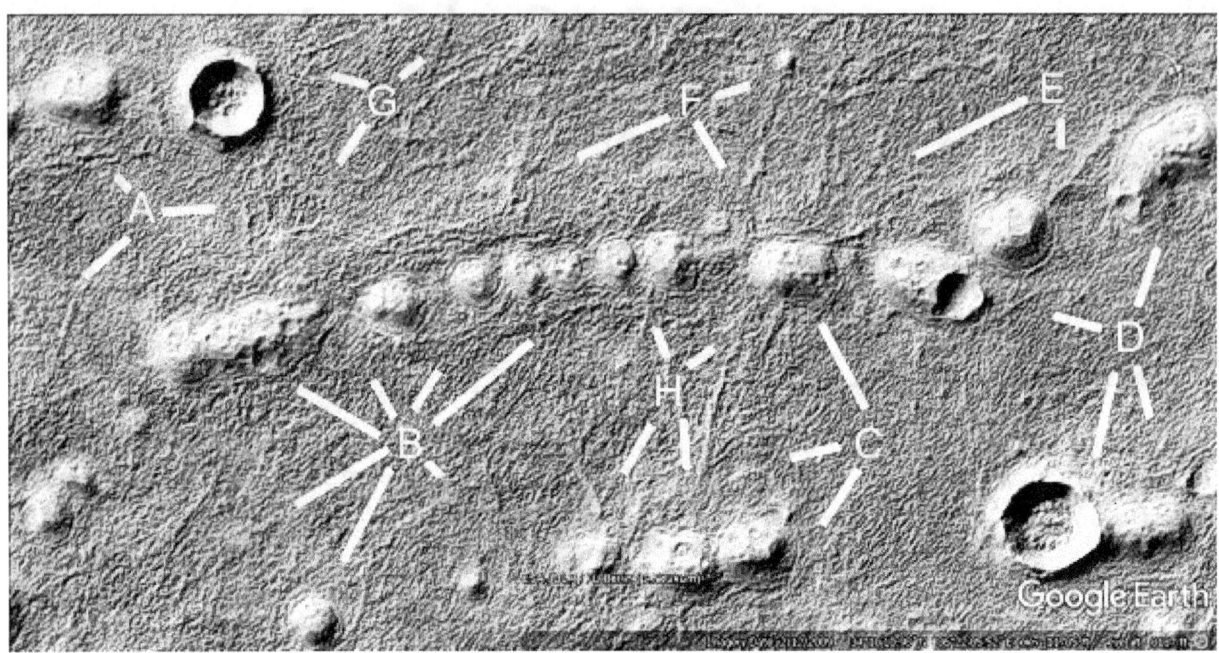

Prt712

Hypothesis

A shows a tube going into the hill, at 4 o'clock the hill comes to a point to receive the tube. B shows another tube, it connects at 10 o'clock, on the roof there is a tube at 12 o'clock. C shows a collapsed segment of the hill at a tube like formation perhaps an interior support exposed. D shows an intact segment of the hill.

Prt713

Hypothesis

A shows a collapsing hill at 3 o'clock, at 2 to 4 o'clock there are tubes connecting into a longer tube going up the page. More tubes are shown from 7 to 12 o'clock. B shows a tube. C shows a tube on the roof of the hill, another from 12 to 2 o'clock has regular breaks in it like the arches or pillars in it are exposed. D and E show more tubes. F shows a small tube going into the crater at 4 o'clock, perhaps a small hill was at 10 o'clock. G shows a curved tube going into the crater perhaps from a small collapsed hill. H shows more tubes.

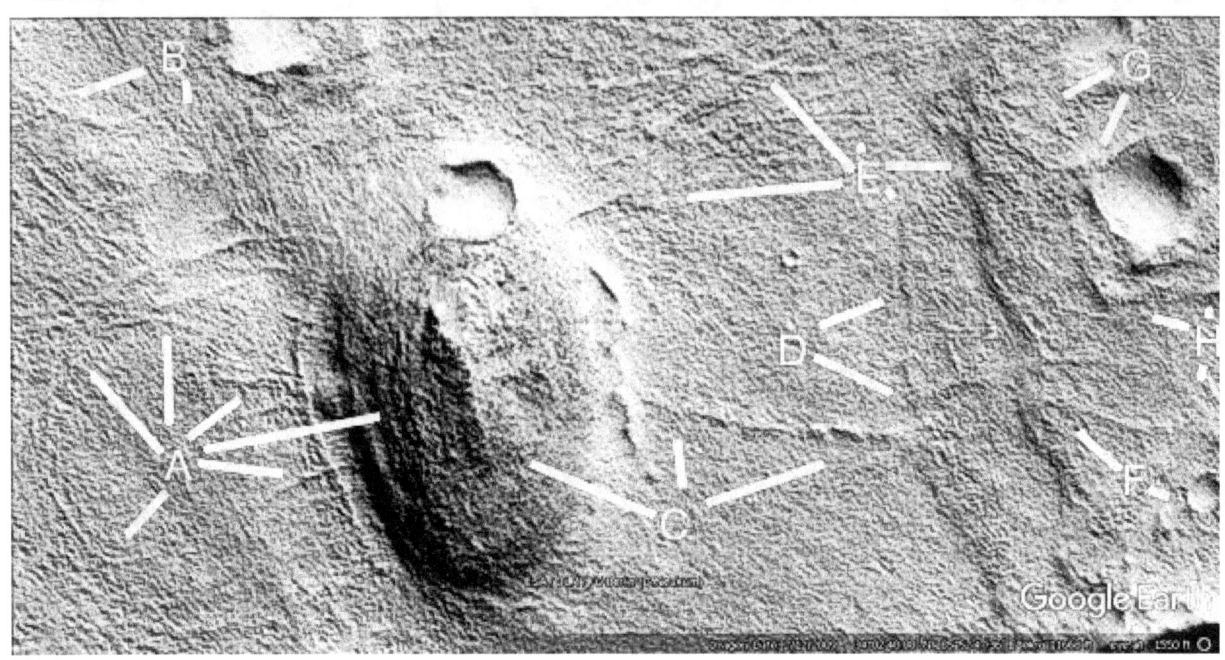

Prt713a

Hypothesis

The hill is shaped like a parabola on its lower side.

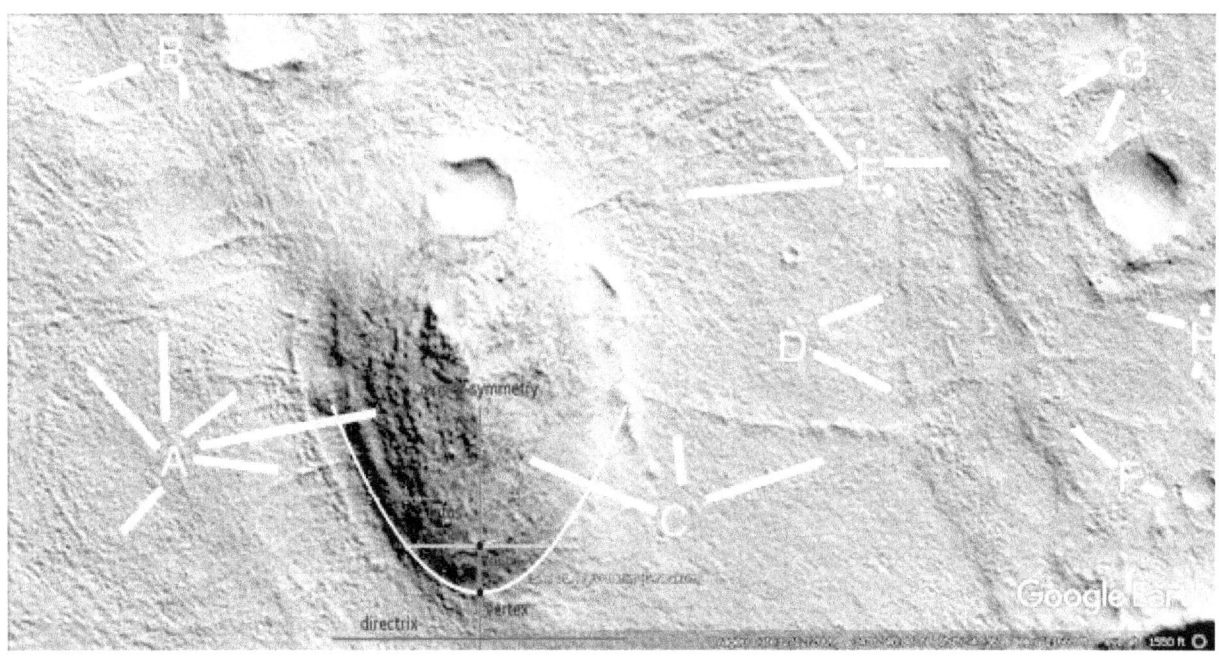

Prt714

Hypothesis

A shows a large nexus at 4 o'clock, it appears to have flat sheets of cement over it so some segments might be rooms. At 1 and 2 o'clock parallel tubes go to the nexus. B shows a squarish area surrounded by tubes, at 7 o'clock there are more like squarish walled segments. At 1 o'clock the crater appears to have been overed over on the right side or this can be an exposed room in the nexus. A wider tube is at 5 o'clock. C shows a T intersection of tubes at 1 o'clock, the tube goes down crossing a long hill at 5 o'clock going into a crater. Another tube crosses the hill from 6 to 7 o'clock. D shows another nexus at 2 o'clock again with flattened segments of a roof. At 4 o'clock this connects to a hill collapsing in many areas. Parallel tubes are shown at 1 o'clock. E shows more tubes, some going into a crater at 4 o'clock. F shows an arc of parallel tubes. G shows tubes exiting under the collapsing hill.

Prt719

Hypothesis

A shows a hollow hill with a cavity on its roof at 7 o'clock, this connects to a wavy tube going though D and E over to G from 7 to o'clock into another collapsing hill. B at 11 o'clock shows a tube going into a collapsing hill at 1 o'clock. Many tubes are show here, to the point where they become like a mesh of tubes at C and E at 7 o'clock. There is a hill connecting to this mesh at 4 o'clock. At 2 o'clock the tube terminates, this may be a collapsed segment of the hill to its right. F shows over ten parallel tubes going into the collapsing hill

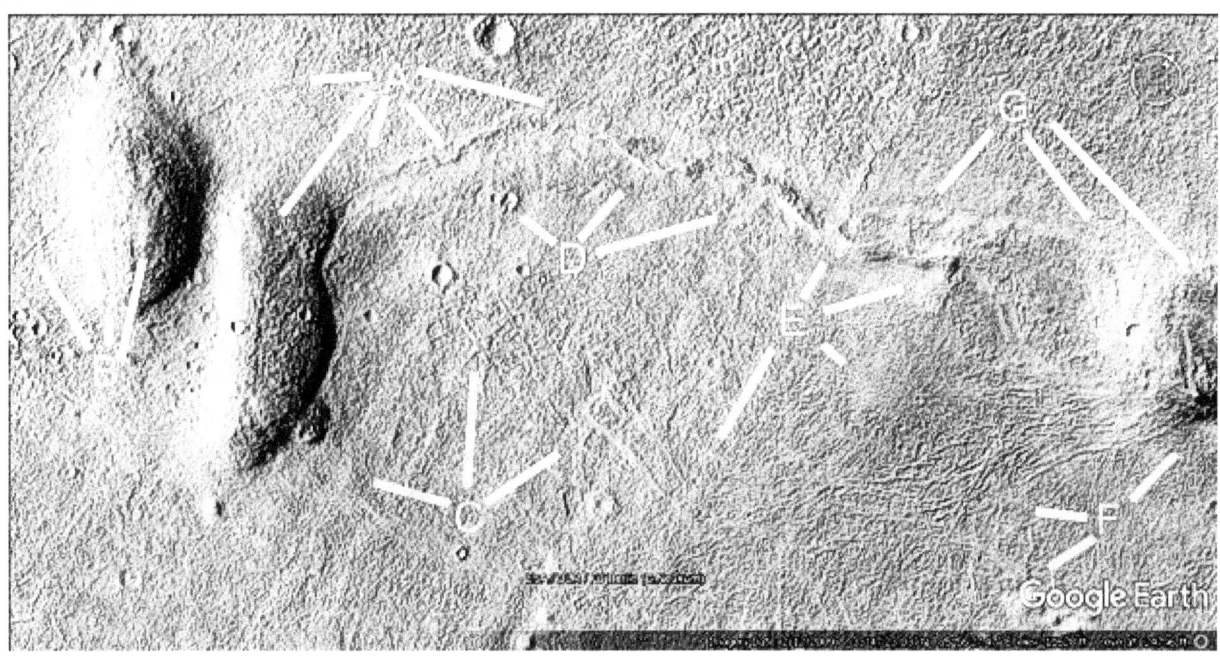

Prt738

Hypothesis

A shows a straight tube going into the collapsing hill from 11 to 2 o'clock. More tubes are shown from a tube mesh at 7 to a tube going up to the hill 4 o'clock. B shows more tubes connecting to a small collapsed hill, they continue on to E where there is a nexus at 1 o'clock. C and D show more tubes connecting to hills. F shows more tubes.

Prt742

Hypothesis

A, B, and C show many tubes connecting to each other and to the crater. There is a nexus at C at 4 o'clock, also D at 1 o'clock. Many tubes go into the collapsing hills from E at 8 to 4 o'clock and F. There are so many unusual connections here that virtually all the image could be studied.

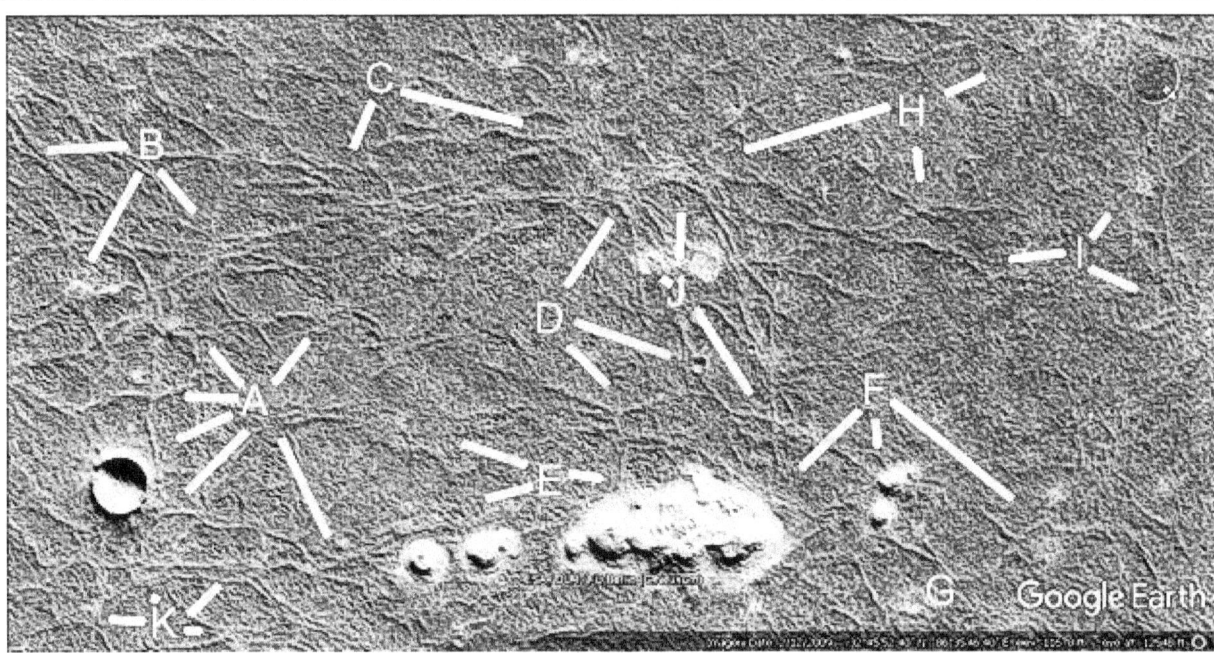

Prt746

Hypothesis

A shows three tubes going into a crater at 8 o'clock, one from the collapsing hill from 5 to 6 o'clock. B shows a wavy tube, as it collapsed it widened at 2 o'clock. C shows the roof at 12 o'clock is sitting higher than the rest of the hill, there is a band around this perhaps collapsing. At 1 o'clock there is a tube exiting the hill. At E at 10 o'clock this connects to another tube going to D from 2 to 5 also connecting to a crater. E shows a walled field at 8 o'clock, a tube from 2 to 1 o'clock goes to a crater and then over to D at 4 o'clock.

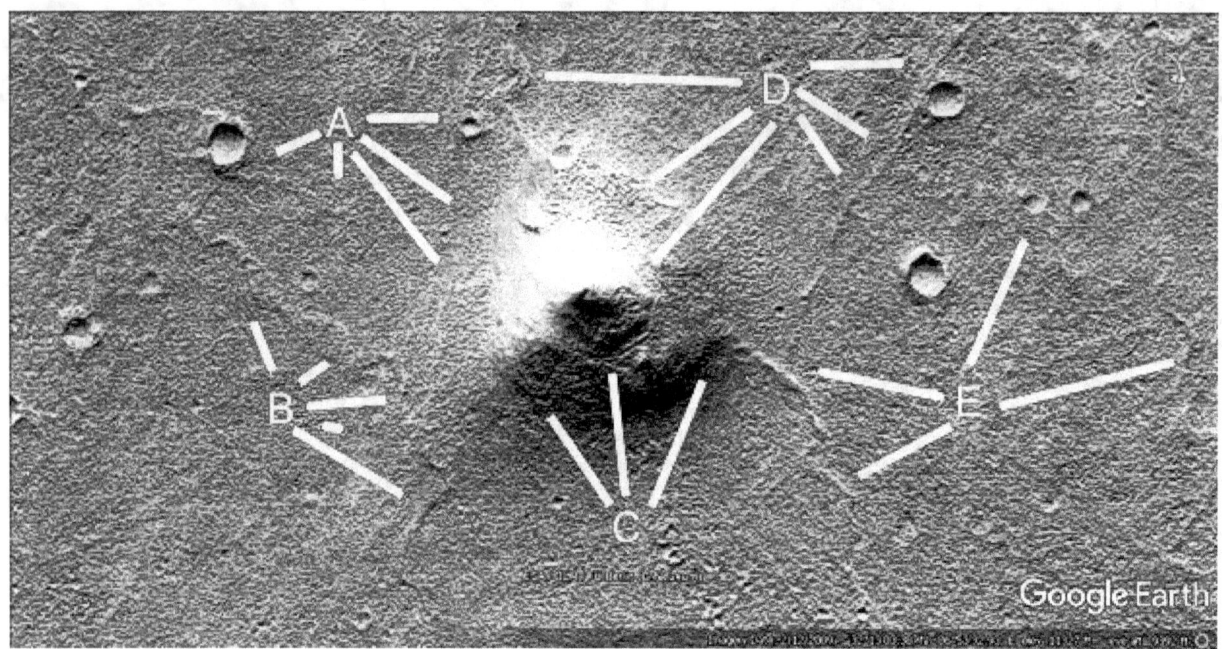

Prhh751

Hypothesis

A, B, and D show many small tubes connecting to craters. C at 10 o'clock shows a hill with tubes going into it, also on the larger tube. At 2 o'clock is a smaller tube. E shows a settled area on the roof at 2 o'clock, F and G show more tubes. H shows tubes connecting to a long hill at 11 o'clock, and another tube from 6 to 9 o'clock.

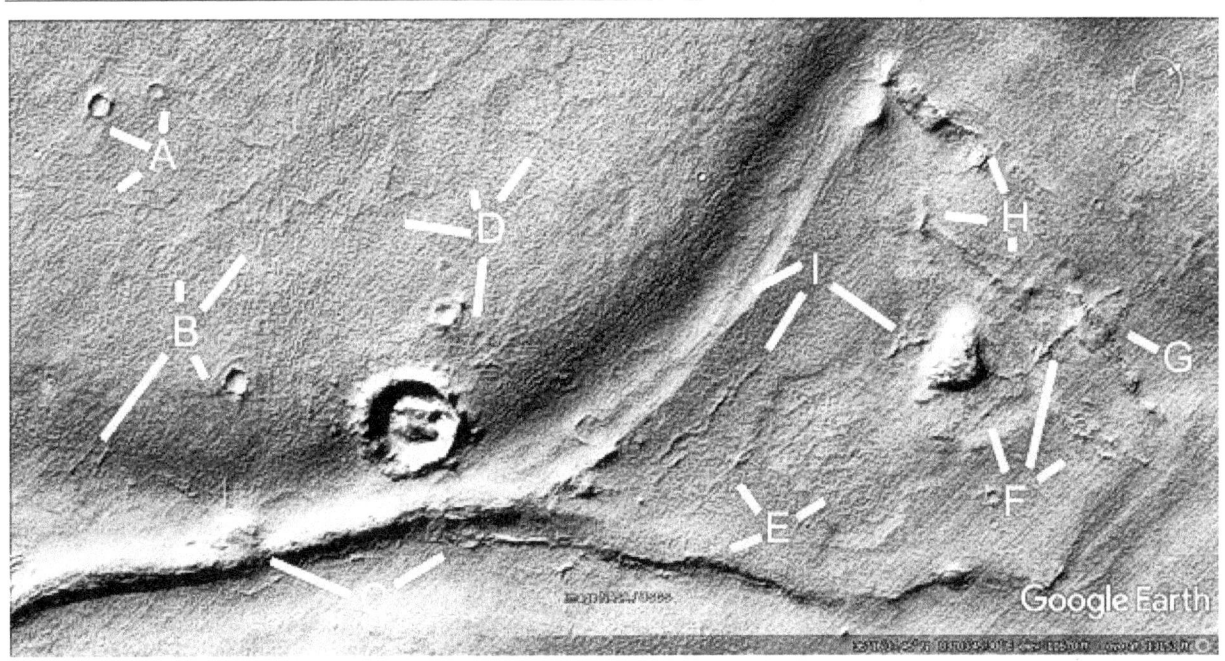

Prhh751a

Hypothesis

A parabola is shown, also the crater appears to form an interior tangent to the parabola.

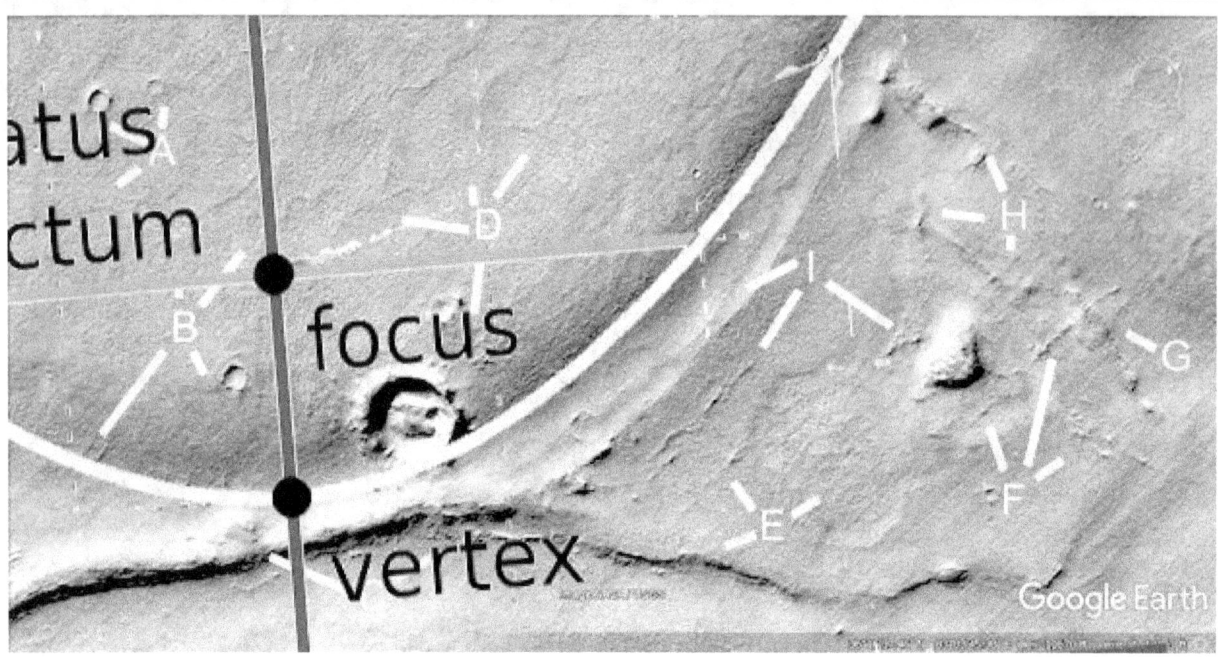

Prhh752

Hypothesis

A shows tubes going into the collapsing hill connecting at B at 12 o'clock, B shows a settled part of the roof at 1 o'clock, also at 2 o'clock. C shows tubes going into the hill. D shows an unusual shaped curved roof.

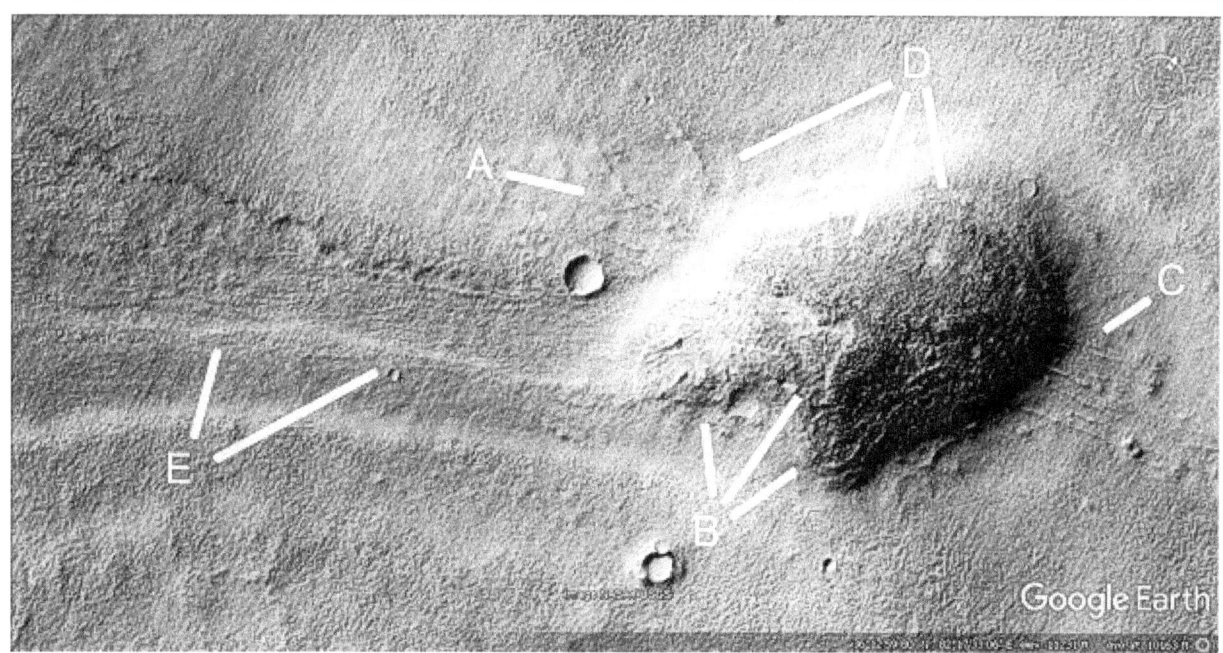

Prhh752a

Hypothesis

A parabola is shown.

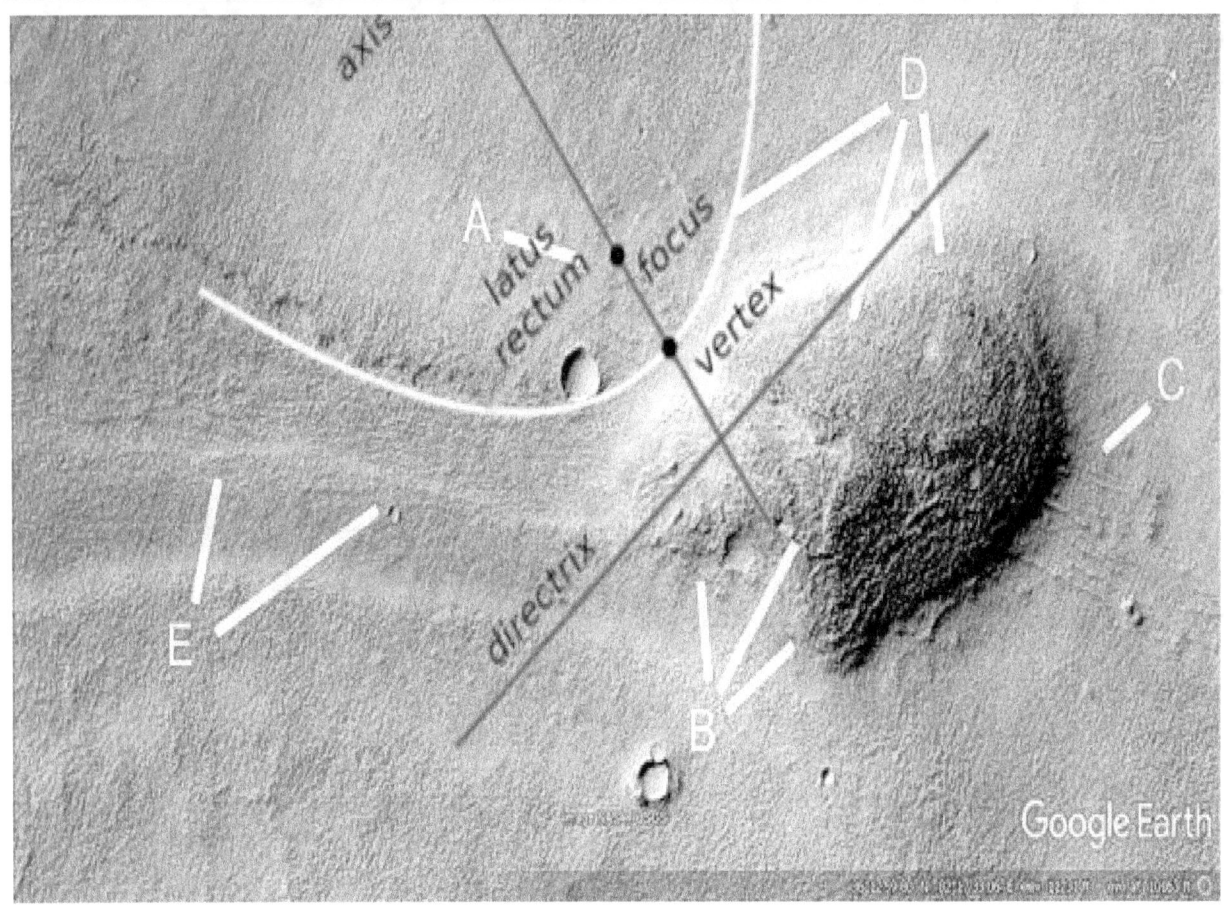

Prt753

Hypothesis

A shows many parallel tubes going through the long hill, continuing as E and E to the large nexus between E and F. This is a flat sheet like a roof in many areas. A at 5 o'clock and D at 7 o'clock show tubes crossing the parallel tubes so someone could have moved from one to another more easily. Above I there are nine parallel tubes going to the nexus, B shows about eight more parallel tubes. Under this is H with a grid or mesh of tubes, this continues on through C with more meshed tubes to the nexus. F shows about six more parallel tubes from 8 to 11 going to the nexus, between E and F there are about twelve more tubes going into the nexus. Between F and G there are about seven more tubes going to the nexus, many more of these form a tube mesh as well.

Prt756

Hypothesis

A shows many tubes though some have eroded, there are smaller tube segments remaining intact. Some have small hills connected to tubes. B shows a forked tube at 12 o'clock, this goes to a small hill at 2 o'clock, then to a T intersection under 4 o'clock to the main nexus between C, D, and F. At B at 5 o'clock many tubes cross each other. At C at 10 o'clock about six parallel tubes go into the nexus. From above C over to D there can be as many as twenty more tubes going into the nexus. E shows a small hill at 11 o'clock connected to a tube, at 7 o'clock three tubes interleave with each other so each goes over the tube clockwise to it. A large tube at 1 o'clock at F goes into the nexus, another at 11 o'clock. At 7 o'clock a small crater has many tubes going to it.

Prt762

Hypothesis

Many tubes come out of the hill, it has a settled area on its roof at 8 o'clock, three tubes come out at 6 o'clock, there is a small forked tube at 4 o'clock. B shows a tube parallel to the side of the hill at 9 o'clock, this connects at right angles to another tube at 7 o'clock. At 2 o'clock many tubes become parallel to go into the nexus, these also connect to the crater at C at 4 o'clock and beyond to D. C at 3 and 9 o'clock shows two parallel tubes. E at 10 o'clock shows the centre of the nexus, at 4 to 7 'clock are many more parallel tubes. Between F and G the parallel tubes are more eroded with many gaps. G shows more eroded tube segments. The dark lines at right angles are a HiRise image, a closeup is examined.

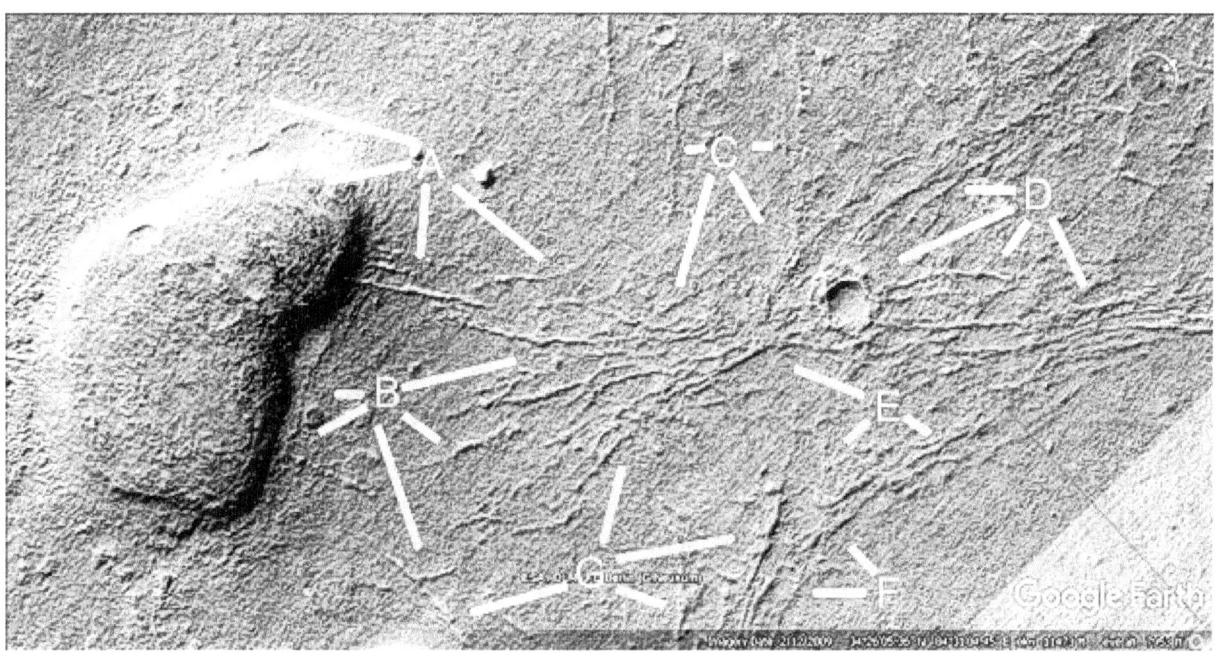

Prt762a

Hypothesis

A shows where the tubes are collapsing, at 10 o'clock there is a hole like the roof has collapsed onto the floor, the skin is breaking off at 1 o'clock. At 4 o'clock the wall of the tube has split off and points under the tube, away from the intersection with the other tube. B at 8 o'clock shows where two tubes connect to a larger segment like a room, but the roof skin is peeling off. At 10 o'clock the roof has collapsed. At 4 o'clock there is an intact flat roof like in the nexus nearby. This extends down to C and E, the walls are higher in some areas than the roof which may mean it is settling. C at 8 o'clock to the left is a long line of degraded skin on the roof.

Prt772

Hypothesis

A shows a tube at 7 o'clock from the larger to one of the small hills. At 9 o'clock there may be a tube on the roof. B shows a bend in the tube at 11 o'clock, some holes in the roof at 2 o'clock. At 3 o'clock is another tube going into the hill. A tube under B goes into the angled hill at C.

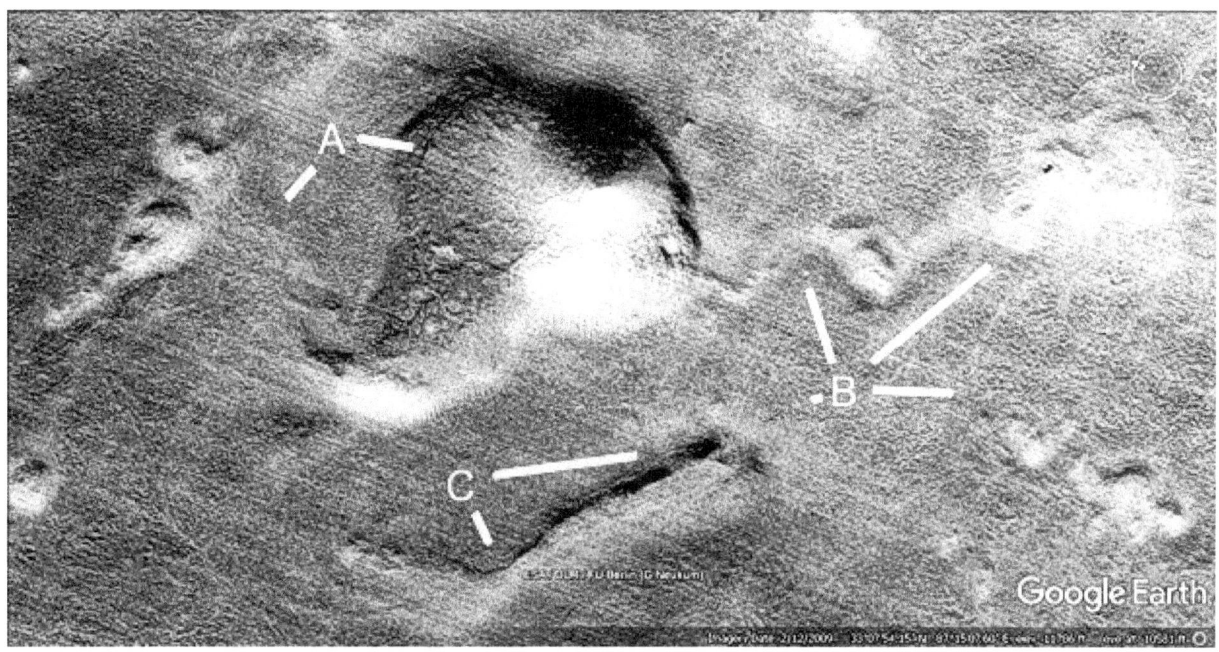

Prt774

Hypothesis

A at 6 o'clock has tubes coming out of the hill. B shows a collapsed hill at 8 o'clock, another at 11 o'clock. At 6 o'clock there is a tube nexus, also some parallel tubes at 3 o'clock going into a collapsed hill. C shows a tube free area but surrounded by a ring of tubes, they go into the collapsed hill at 1 o'clock. D shows more tubes coming to a nexus at 6 o'clock outside the hill. E shows another area clear of tubes, these may have been used for farming. F shows a tube nexus at 2 o'clock, a forked tube at 10 o'clock.

Prt774a

Hypothesis

A parabola is shown.

Prthh789

Hypothesis

A at 5 o'clock shows layers in the collapsing hill, at 3 o'clock the hill segment has probably collapsed. B shows where a collapsed tube goes to a crater, just the trench in the ground remains. C at 8 o'clock is probably a small collapsed tube or tunnel going to the crater. At 6 o'clock the tube appears to be degrading in some segments, at 4 o'clock is a triangular hill. At 12 o'clock there is a small tube going into the crater. D shows a faint tube going into the crater. F and D at 1 o'clock show tubes going into a collapsed hill.

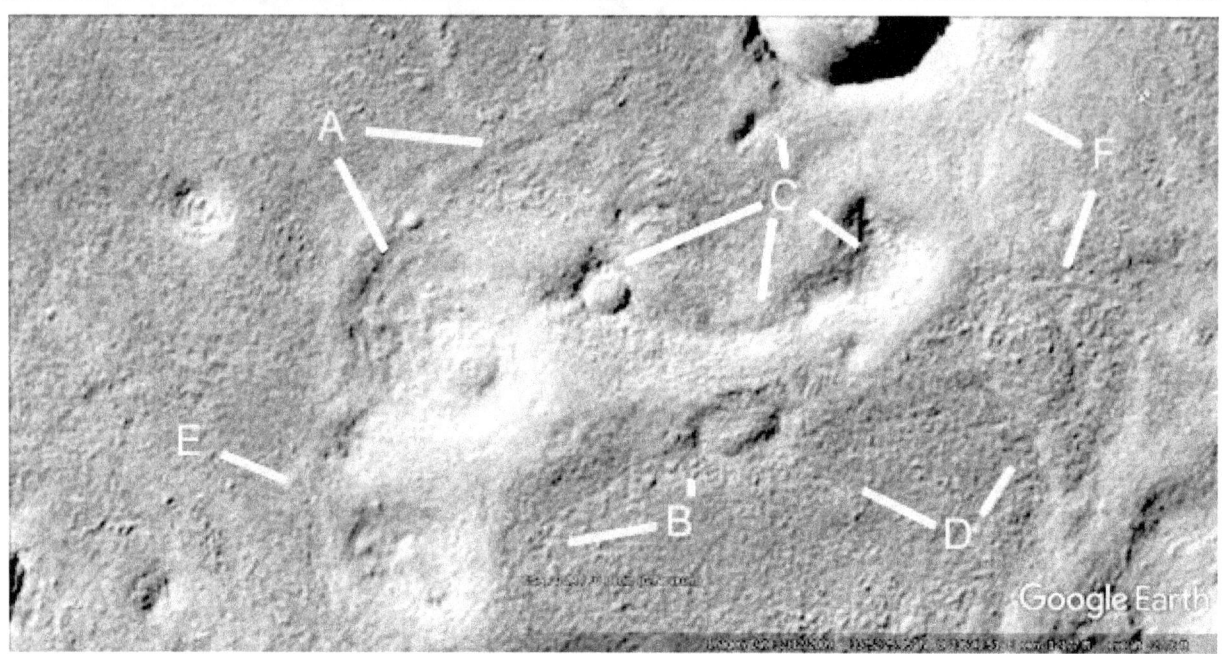

Prt792

Hypothesis

A shows many collapsed hills, B shows a tube going from B at 9 o'clock to a crater with an angled turn. It continues over to C at 6 and 7 o'clock, over to C at 3 o'clock and a pair of parallel tubes going up the image. D shows a straight tube at 7 o'clock, more tubes at 6 o'clock and a collapsed hill at 5 o'clock. E shows a collapsed tube, this continues on to F. G shows another collapsed tube.

Prt798

Hypothesis

A shows a hollow hill with cavities in the roof, it connects to a wider part of the hill at 6 o'clock. This has a twisted shape like a rope, it continues on through the twisted tube at B to connect to a collapsing hill at 2 o'clock. At 8 o'clock there is another tube. At 3 o'clock the roof has collapsed. D shows another tube going into the hill at 8 o'clock, this connects to the tube at 5 o'clock. This in turn connects to the hill above D with tubes at right angles to it. E shows a collapsed roof at 10 o'clock, at 11 o'clock is a tube. Bat E at 12 o'clock up to F at 6 o'clock is a symmetrical wall.

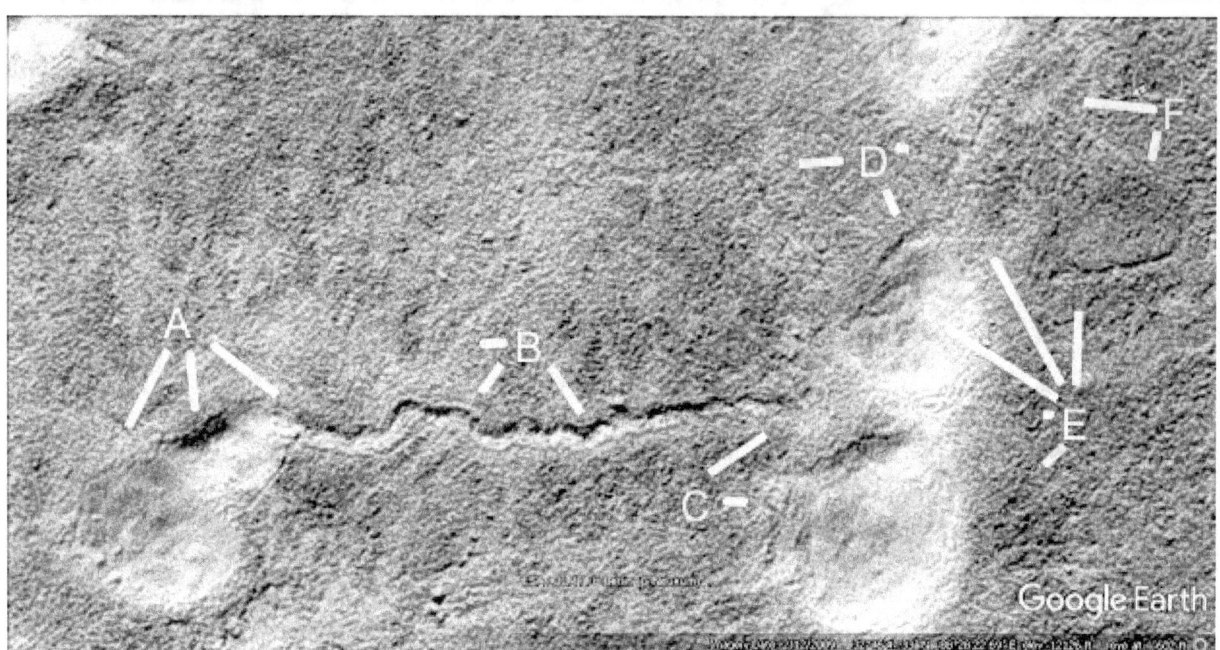

Prt804

Hypothesis

A shows more tubes between collapsed hills. B shows layers in the hill at 2 o'clock like a Cobler Dome. At 11 o'clock the tube from the chain of hills enters the hollow hill. At 3 o'clock is a thicker tube connected to a small hill. C at 8 o'clock shows the circular roof of the hill, it contains two parabolas, at 4 o'clock a tube goes into a small hill with a cavity on the roof. From 11 to 3 o'clock are other tubes. D at 5 o'clock shows the edge of this circular roof, the rest of D shows other tubes. E shows an arc of tubes connected to some collapsing hills.

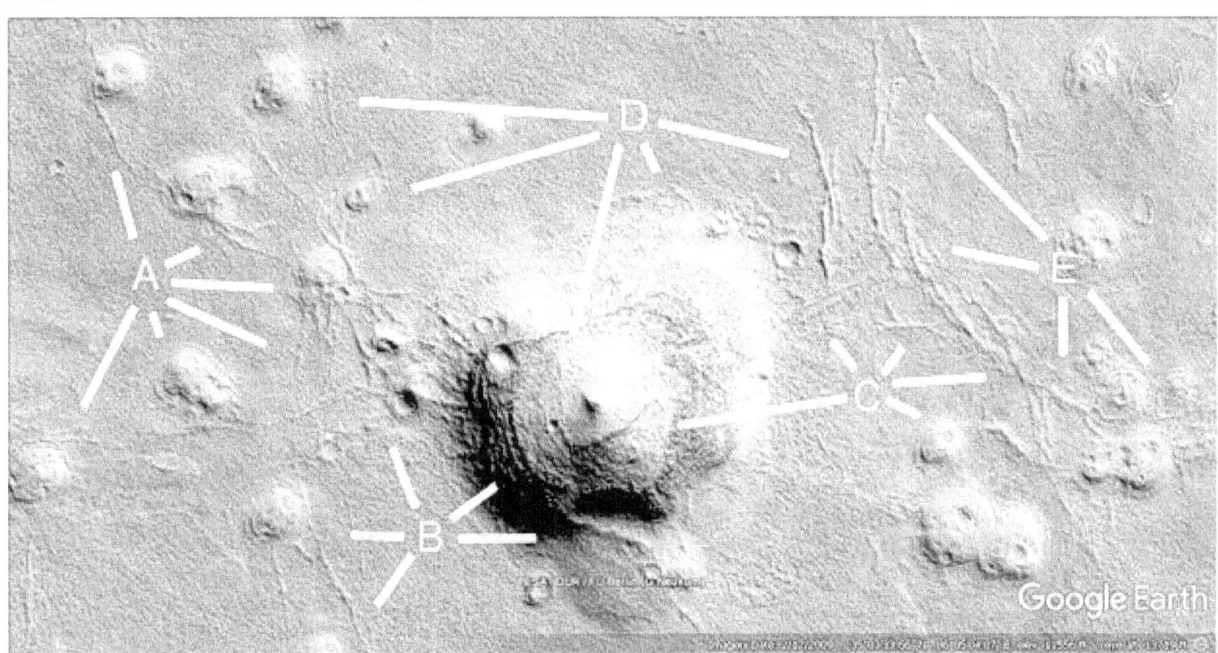

Prt804a

Hypothesis

The roof is close to a circle, here a circle is overlaid onto it. Also two parabolas are drawn onto the dark marks on the roof.

Prt805

Hypothesis

A shows a small nexus connecting the tubes from the hills. B shows a tube on the roof at 4 o'clock, at 7 o'clock is a tube going into a crater. C shows another tube ending in a small hill. D shows a settled roof at 10 o'clock, perhaps an interior support exposed at 12 o'clock, at 1 o'clock these may have been rooms in the hollow hill or walled fields outside it. E shows parallel tubes going to the crater. F shows a straight tube going to a collapsing hill.

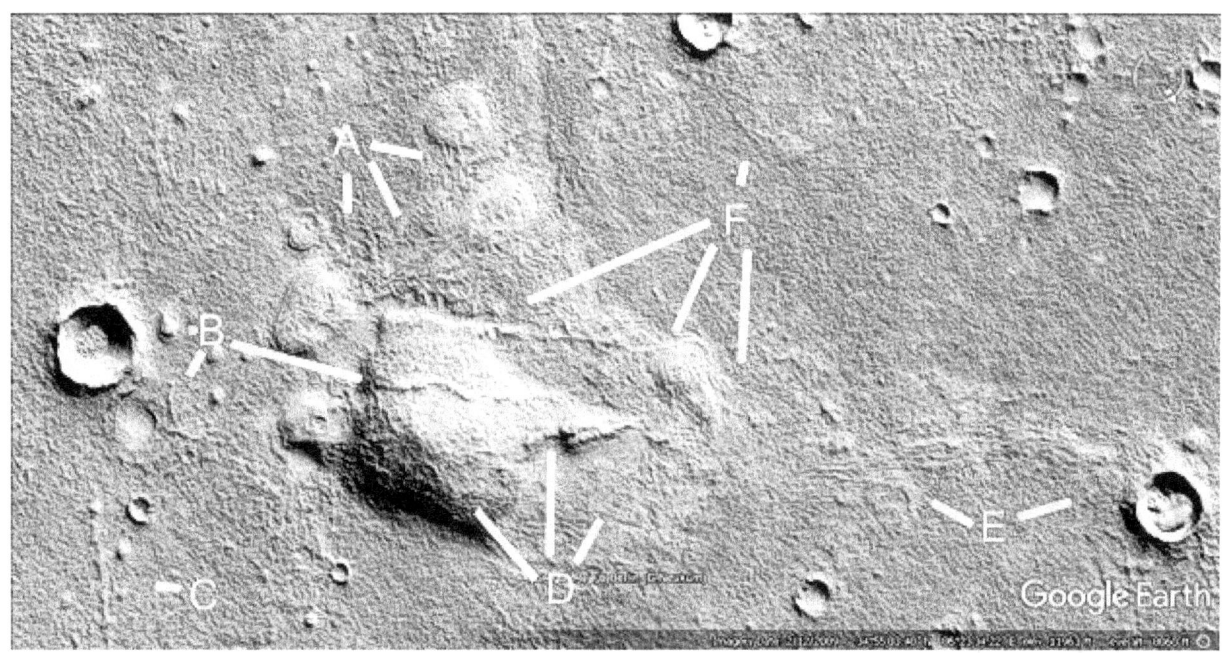

Prt812

Hypothesis

A, B, and C show many tubes connecting to hollow hills. These extend through C at 12 o'clock up to D and many more connections to hills. From 11 to 1 o'clock there are three parallel tubes coming out of the hill. E shows a tube at 12 o'clock broken into segments, these may have been held up by internal arches or pillars. From 6 to 11 o'clock are more tubes. From F to G are many more tubes.

Prt813

Hypothesis

A shows more tubes connecting hills and the two craters, parallel tubes at 7 o'clock. B shows an approximately straight tube from 10 o'clock over to C at 1 0 to 1 o'clock and to the right. It is a wavy tube, perhaps giving it strength with arch like bends in it. C at 5 o'clock shows a small crater connected to a hill. D shows a tube that may have collapsed down the image making it look wider.

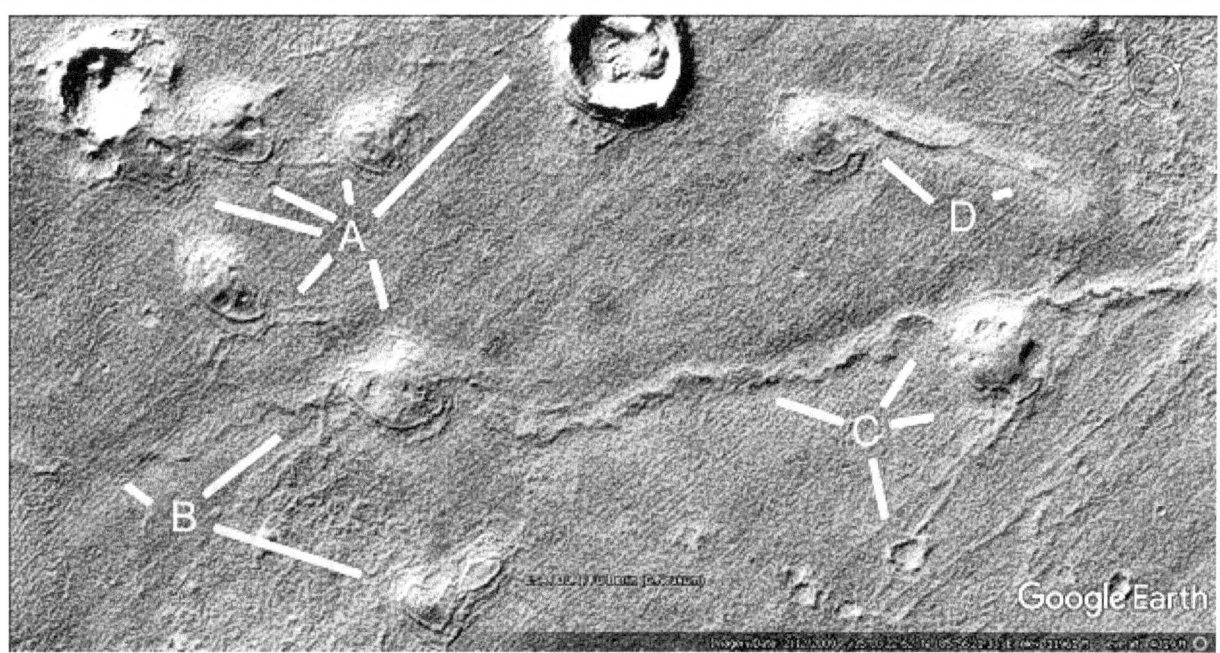

Prt813a

Hypothesis

The lines show how straight the tubes are, they seem to zig zag perhaps for strength.

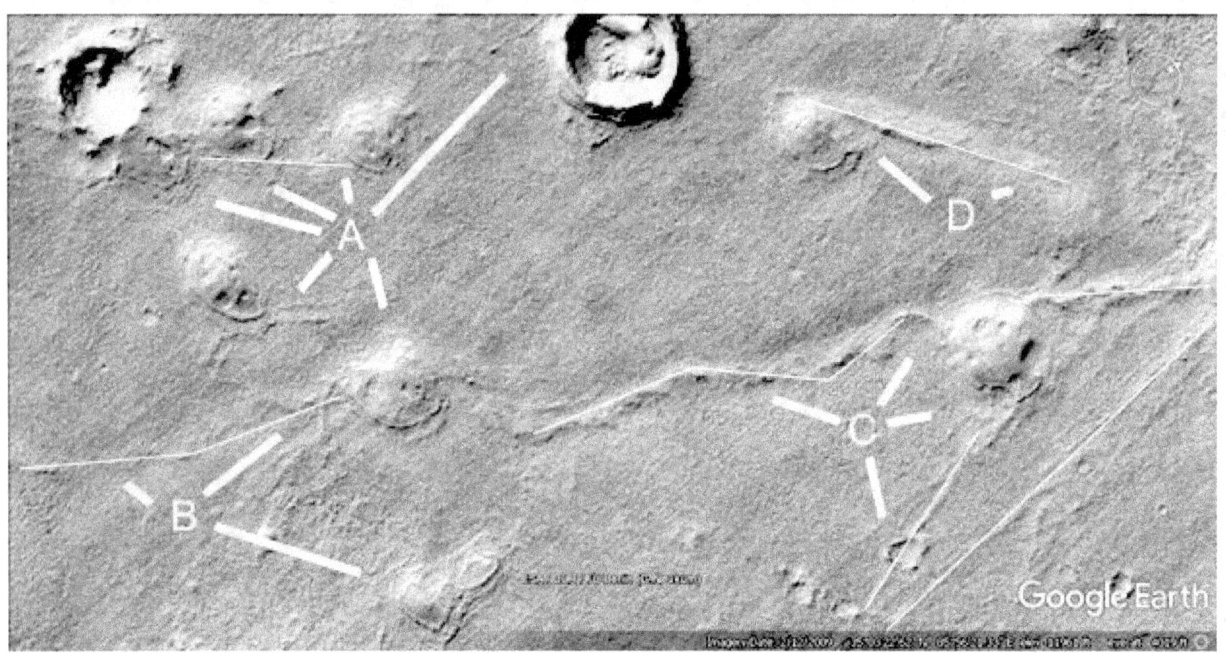

Prt814

Hypothesis

A from 5 to 7 o'clock shows two collapsed hills connected by a tube, the holes in the roof may have been rooms. At 8 o'clock is a tube. B at 10 o'clock shows a collapsed hill connected by a tube to A at 7 o'clock. B from 4 to 7 o'clock shows small hills connected by tubes, also some tubes go to the crater under it. C at 6 o'clock shows many tubes connected to the crater, at 7 o'clock a tube goes through a collapsed hill over to 4 o'clock and then up to the nexus at F at 1 o'clock. At 4 o'clock a forked tube comes out of a collapsed hill. C from 10 to 2 o'clock shows a tube coming out of the collapsed hill continuing over to the nexus. D and E show more tubes connecting to the hills and over to the crater at E at 4 o'clock.

Prt819

Hypothesis

A shows a tube from 5 to 1 o'clock going into a collapsed hill with three cavities in the roof. At 8 o'clock there is an imprint of a completely eroded hill and tube. B shows eroded tubes from 2 to 4 o'clock that shown segments, like the pillars or arches are all that remains. From t to 9 o'clock are more tubes going into the hill with three cavities. C shows a tube from 5 to 7 o'clock going into a crater, another at 6 o'clock, and a tube mesh around a hill at 12 o'clock. D shows more tubes connecting hills. E shows the other side of the tube mesh at 8 o'clock and highly eroded tubes at 3 and 4 o'clock. F shows a tube connected to a small crater at 4 o'clock going down to a forked tube at 5 o'clock.

Prt822

Hypothesis

A from 4 to 7 o'clock shows a straight tube from the hill to the crater, B from 11 to 2 o'clock is two straight tubes at right angles. Over to 10 o'clock is a degraded tube going into the crater and traversing it to the opposite rim. C shows a walled area with a collapsed hill at 5 o'clock, one at 12 o'clock , and at 7 o'clock. At 8 o'clock is a T intersection with a small tube going to the hill, another intersection is at 2 o'clock. D shows a tube between two craters at 4 o'clock second leg, the first leg shows parallel tubes. E shows a degraded tube going into the crater at 1 o'clock, another at 6 o'clock.

Prt827

Hypothesis

A shows a tube from two small hills at 10 o'clock, it goes down to G at 10 o'clock and then over to the symmetrical walls at A at 5 o'clock. At 4 o'clock tubes come out of this symmetrical formation. B at 7 o'clock shows a thicker wall or tube, at 6 o'clock there is a nexus also at E at 10 o'clock. At 2 o'clock E shows a collapsed hill with the holes as rooms, another at 6 o'clock. D shows a tube coming out of the hill at 10 o'clock, a grid of tubes at 4 o'clock with a tube going from 5 o'clock to 7 o'clock into a crater. F shows an unnatural shape for a crater and where tubes connect to it. G at 4 o'clock shows a tube going into a crater.

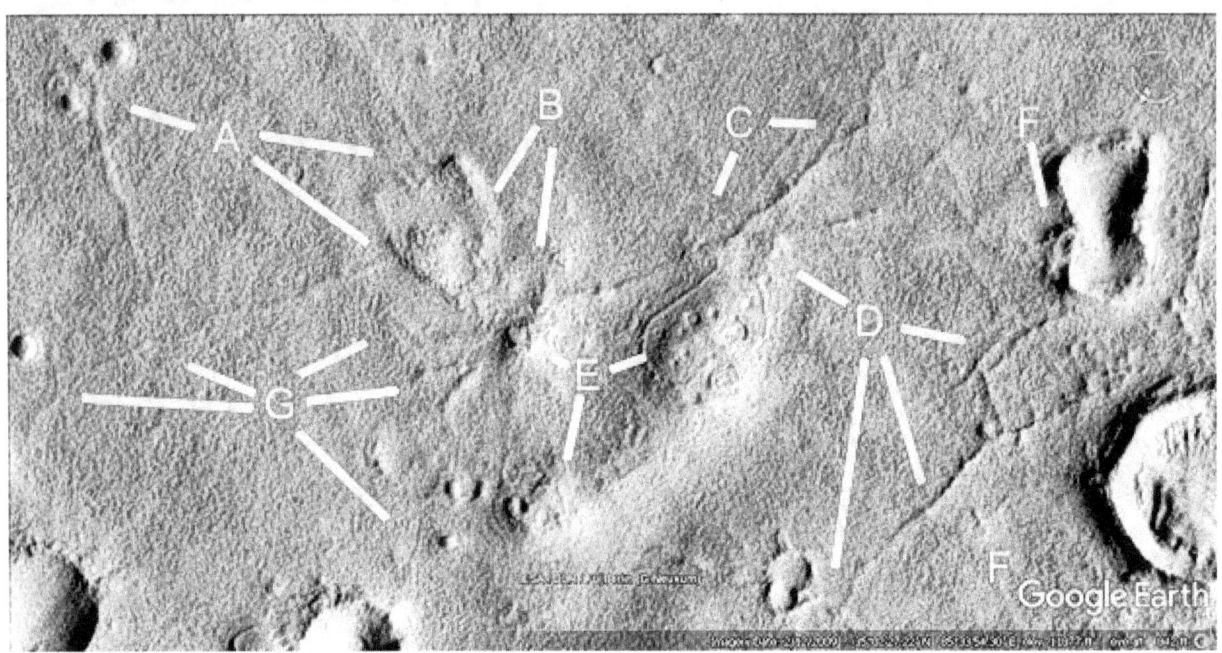

Prt838

Hypothesis

A shows a hill at 8 o'clock connecting to a tube, at 5 o'clock there are tubes on the roof, and at 6 o'clock a tube connects to the hill. At 4 o'clock the tubes connect to a collapsed hill with holes as rooms. B shows a tube from 12 in the hill to 2 o'clock as a collapsed hill. At 3 o'clock there is a straight tube. C shows a tube going into the crater at 11 o'clock, at 10 o'clock short leg there is a fork and the longer leg shows a small tube. D shows a collapsed hill at 10 o'clock, a tube at 9 o'clock goes into it and comes from an intersection at 8 o'clock. At 6 o'clock may be a nexus.

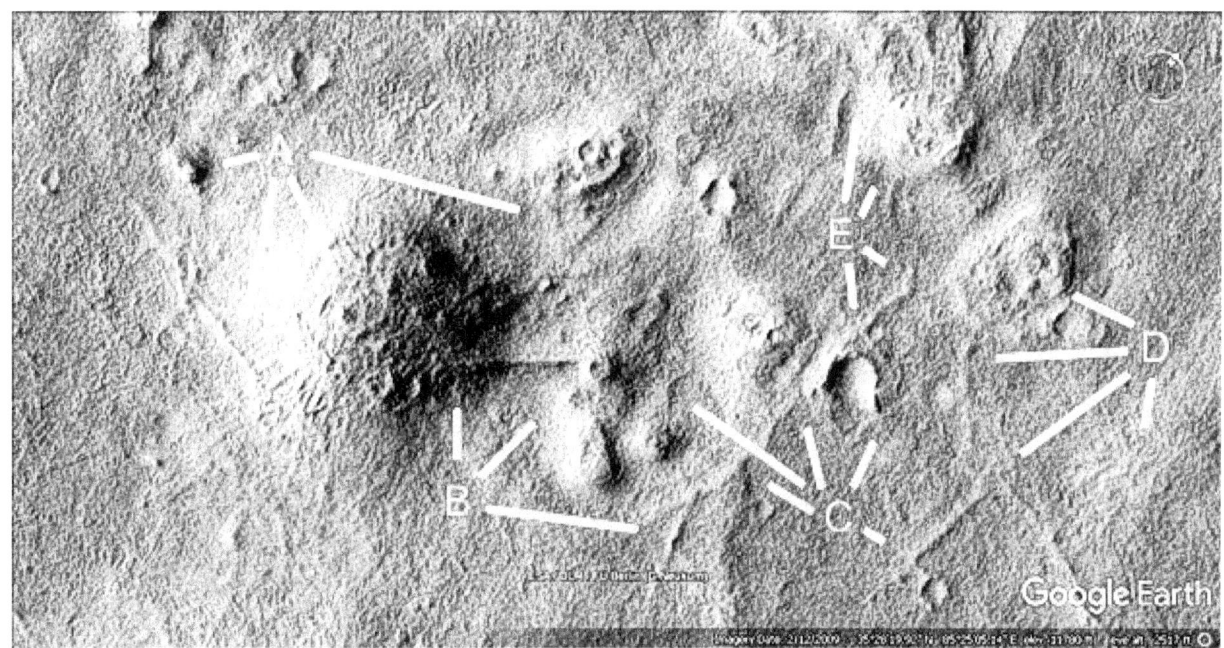

Prt841

Hypothesis

A shows a pair of tubes at 1 o'clock connecting at 2 o'clock then at 4 o'clock there may have been a hollow hill. This then goes to the right though a fork at B at 12 o'clock. The tubes connect to a pair of craters at A at 5 o'clock. B shows a tube ending in a small hill at 1 o'clock, at 8 o'clock a road connects to a crater. C shows tubes forking like a tree at 2 o'clock, more at 4 o'clock. D shows tubes going into the crater at 7 o'clock, parallel tubes at 6 o'clock. E shows many tubes coming off this road. F shows the other side of the road and more tubes as does G.

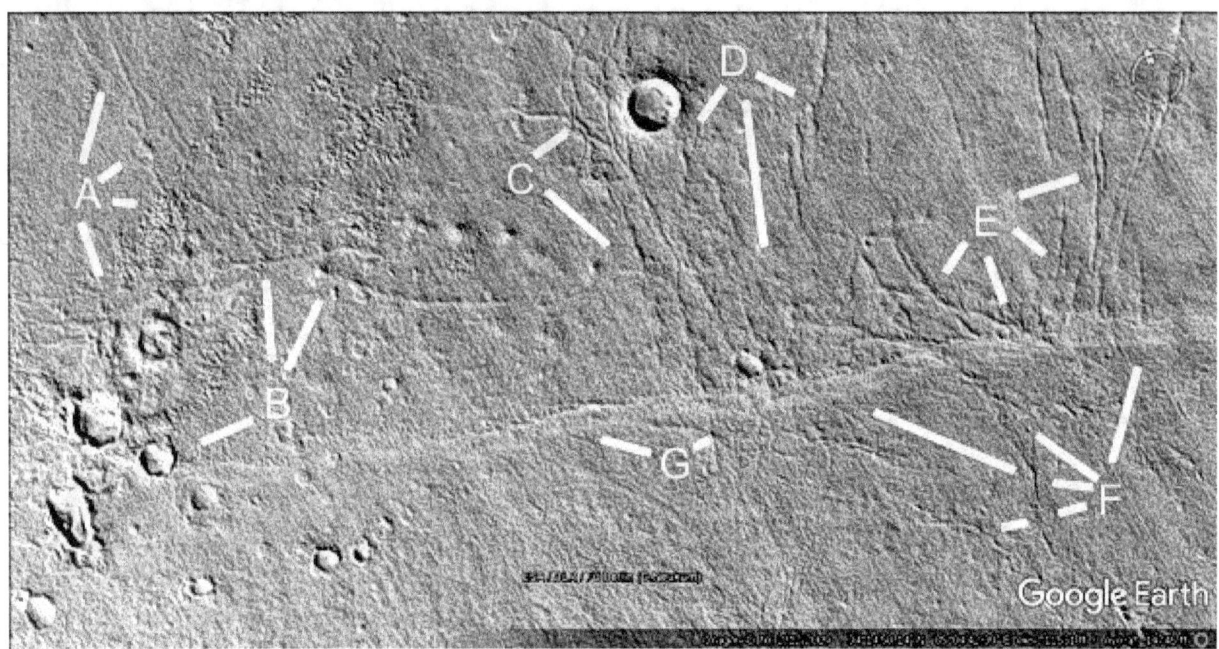

Prt842

Hypothesis

A, B, C, D, and G all show clear areas surrounded by many parallel tubes. E from 11 to 1 o'clock shows collapsed hills connected by tubes. From 5 to 9 o'clock many tubes go into the crater and come out at F going over to G. The tubes must have come after the crater because the impact must happen in a random location. The impression here is then the crater would have to form precisely where these tubes were aiming at. The road from 4 to 8 o'clock also has some small hills along it, the tubes connect to this road and generally stop inside it. So even though the road is a different kind of formation the tubes seem to be intended to go there.

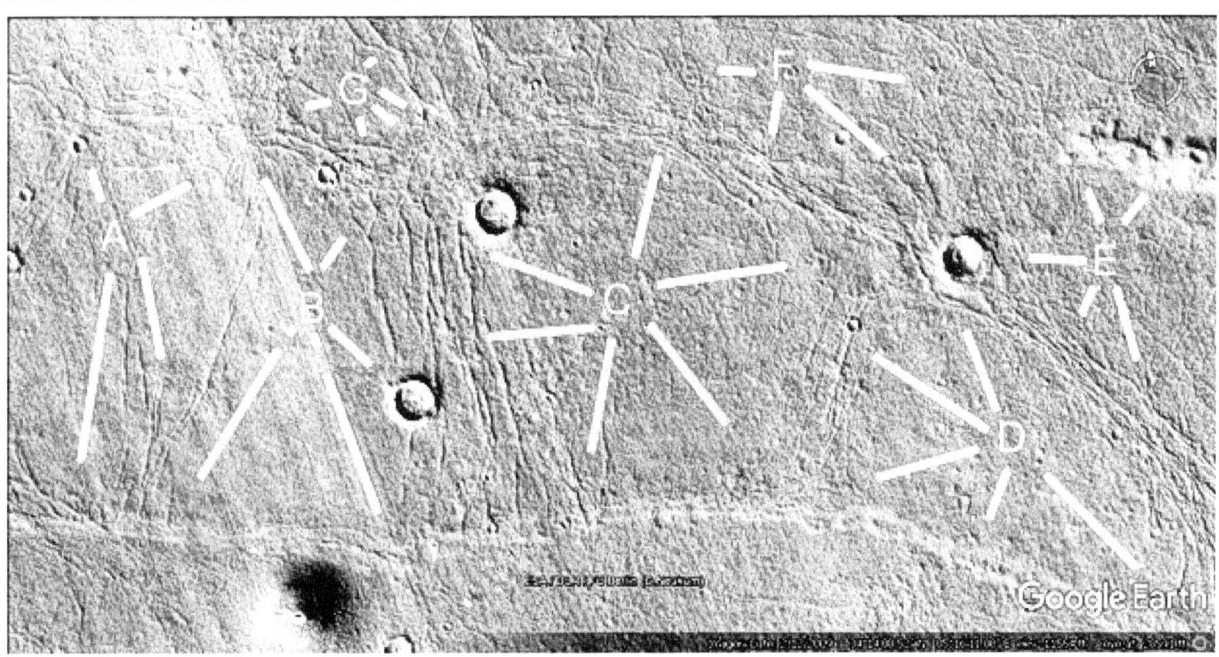

Prt844

Hypothesis

A, B, C, and D show many tubes going into the hollow hill at C at 5 o'clock. Some connect to the crater at B at 11 o'clock. The road at H connects to the side of the crater at G at 11 o'clock, this has an unnatural shape with a straight left side. At C at 7 o'clock there are 6 tubes in parallel going into the hill. E shows how the road went through the hill and now flows out the right side. There is a line down the centre, perhaps a tube. Directly under E there is a hollow like a collapsed tube. It may have changed from a road on the left to a tube on the right. F shows an eroded tube at 6 o'clock going into the craters and then up to F at 10 o'clock and into the hill.

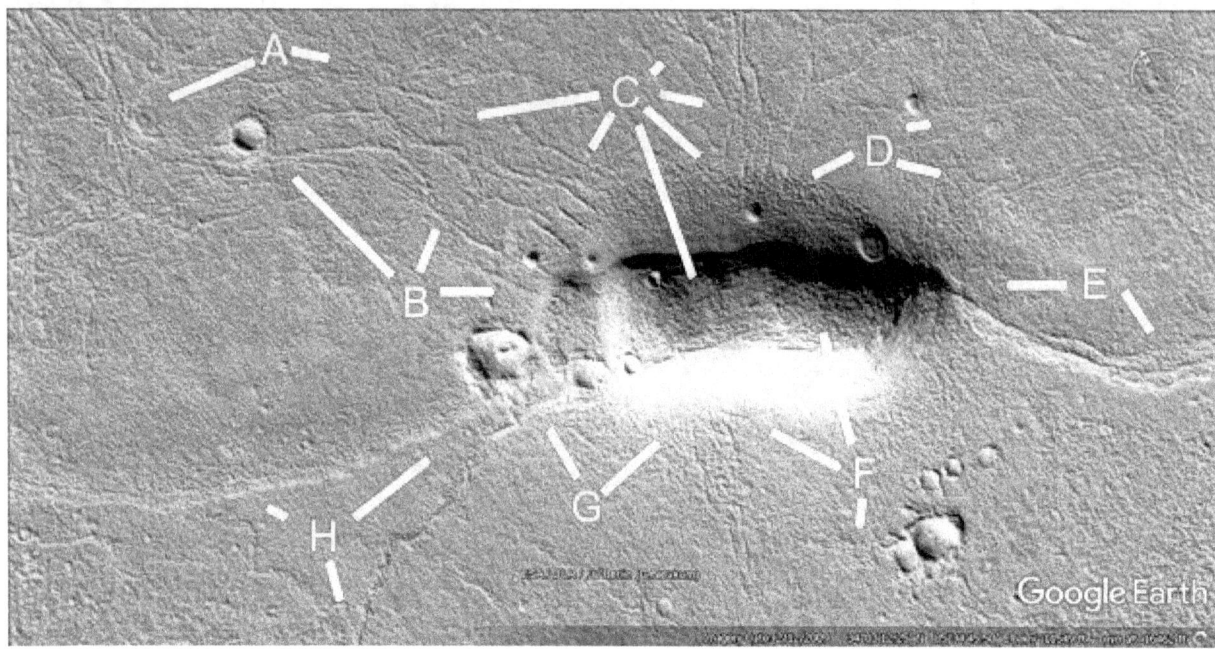

Prt849

Hypothesis

A shows tubes from the crater and collapsed hill going to the larger amphitheatre at C at 8 o'clock. It has concentric rings on its roof like a Cobler Dome, called an amphitheatre here because of the shape. A shows a small hill with two large tubes coming out of it at 4 o'clock. B shows tubes going into the hill at 11 and 12 o'clock though there are dozens of them overall. C shows more collapsed hills at 4 o'clock.

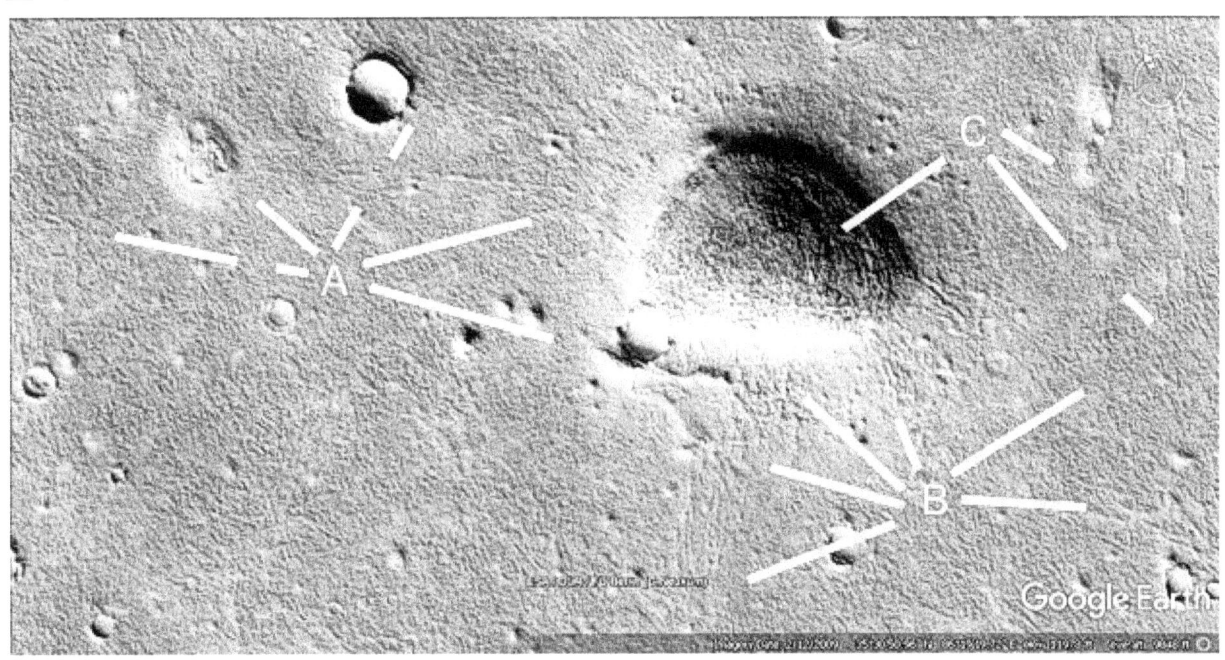

Prt849a

Hypothesis

A parabola is shown.

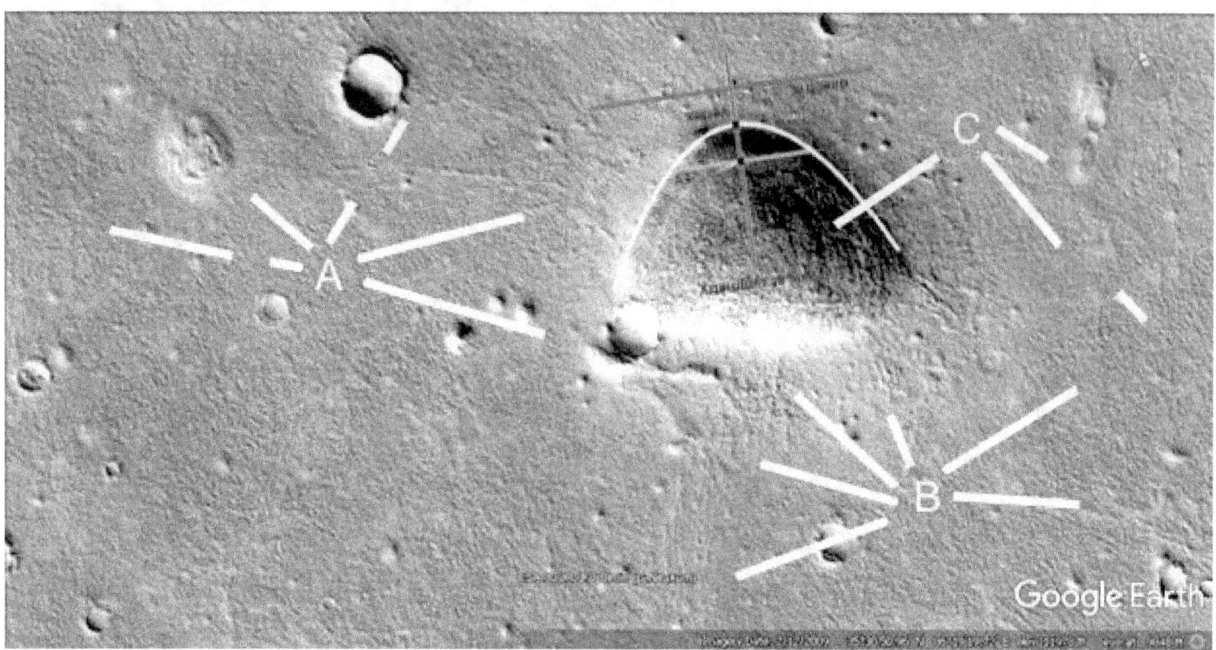

Prt857

Hypothesis

A, B, and C show many parallel tubes inside this farming area. Some connect to the craters at A at 7 o'clock. Between A and B there are about six parallel tubes, between B and C there are about four. B from 2 to 4 o'clock shows a tube going into the crater. D shows where many of these tubes converge, there may have been a hollow hill here. E at 7 o'clock shows a small hill and a straight tube extends up the image.

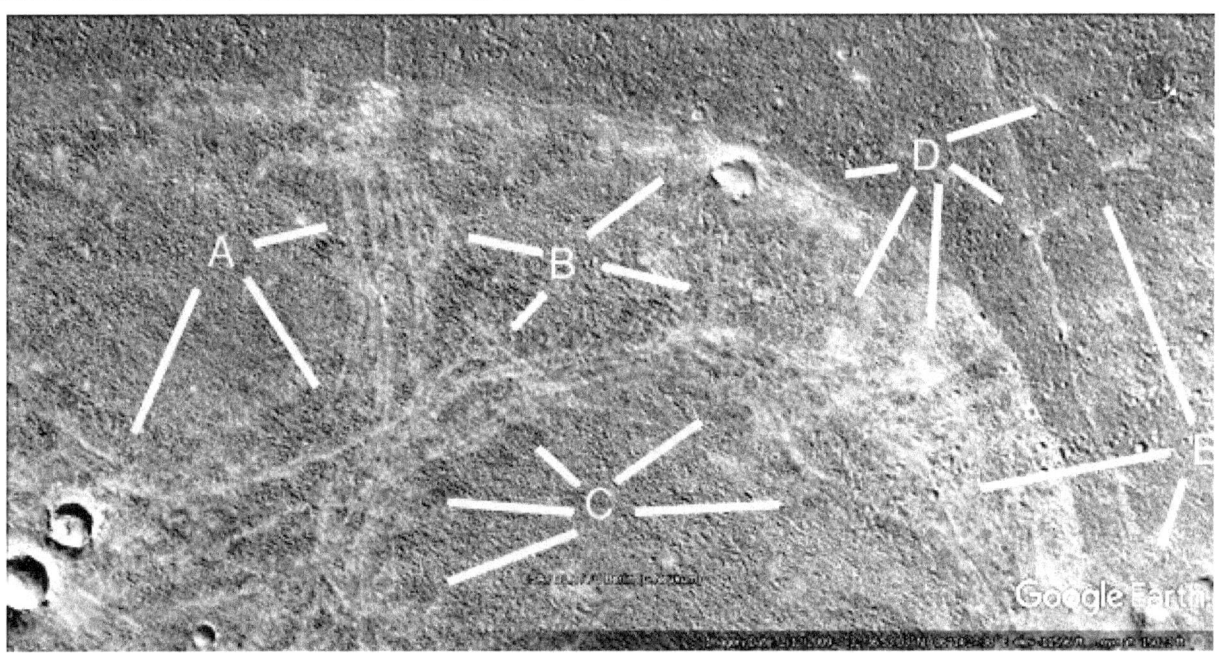

Prt857a

Hypothesis

A parabola is shown. Also the line shows how straight the long tube is.

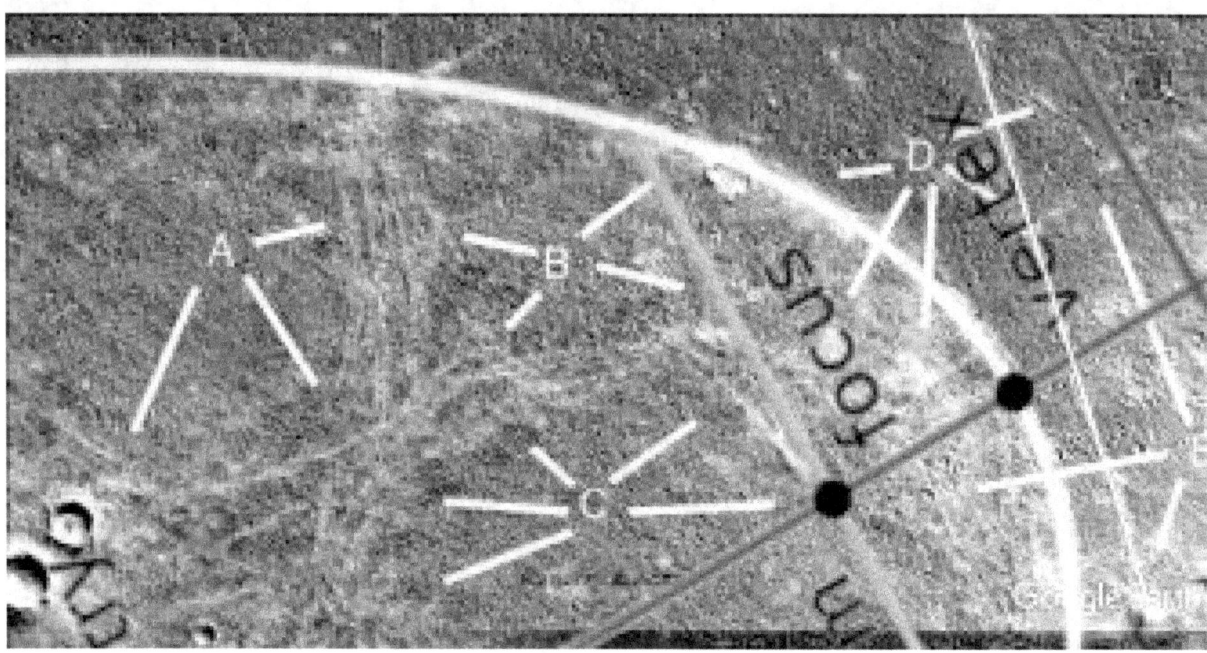

Prt860

Hypothesis

A from 2 to 5 o'clock shows a long tube going into a hill, it crosses the horizontal tube at 2 o'clock. At 4 o'clock is a small collapsed hill with a central wall. B shows a wavy tube going into an elliptical hill, C shows this is a knotted tube around 2 o'clock. Some of the tube is likely to be intact around arches and pillars, then collapsed between them. At 3 and 6 o'clock there is a parabolic tube. D shows another parabola, E shows other tubes going into the hill.

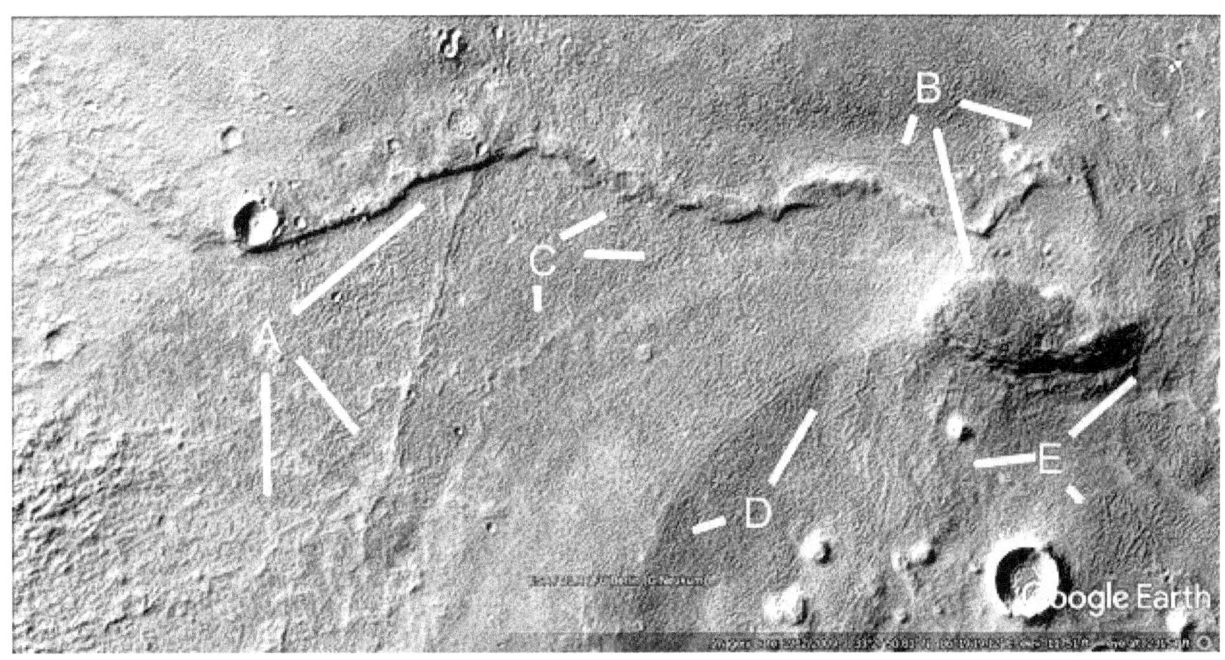

Prt860a

Hypothesis

Three parabolas are shown. Also the line shows how straight the wall is.

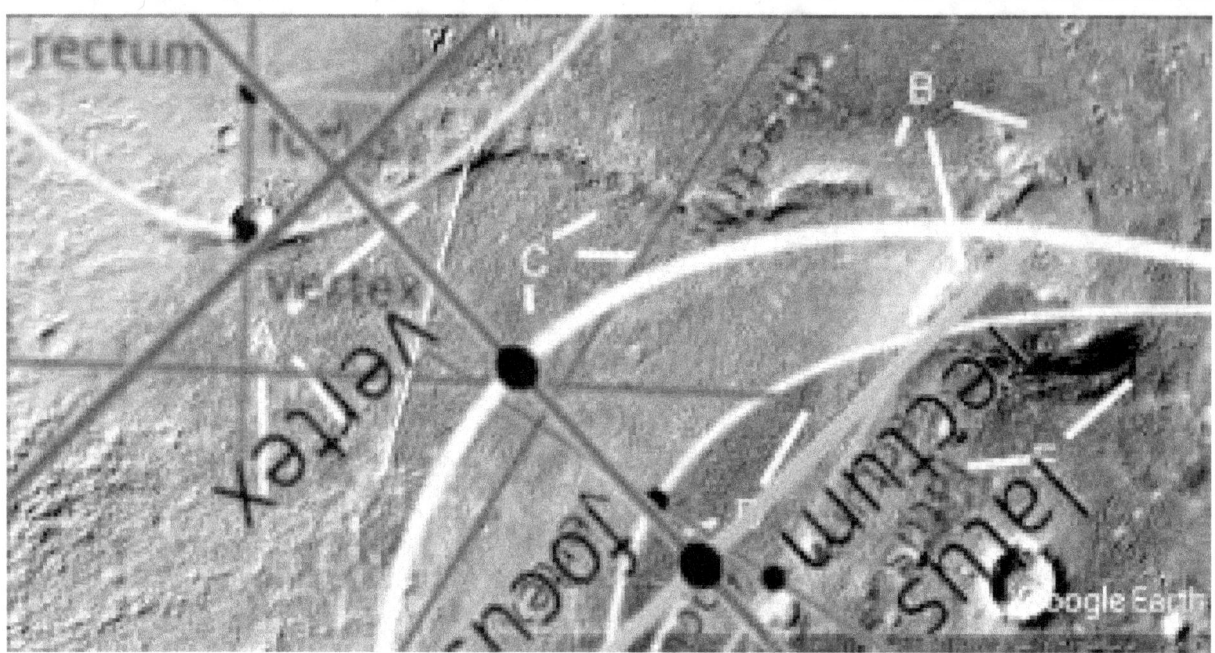

Prt863

Hypothesis

A, B, and D show many tubes in varying condition, at A at 10 and 12 o'clock they are more eroded. At 7 o'clock the tube is breaking up into segments with the arches inside them. At 3 o'clock is a forked tube. B at 7 o'clock shows a fork and a small hill. D shows more tube fragments. E shows more segments where the tubes have broken up. To the right of C there are many collapsed hollow hills near the crater.

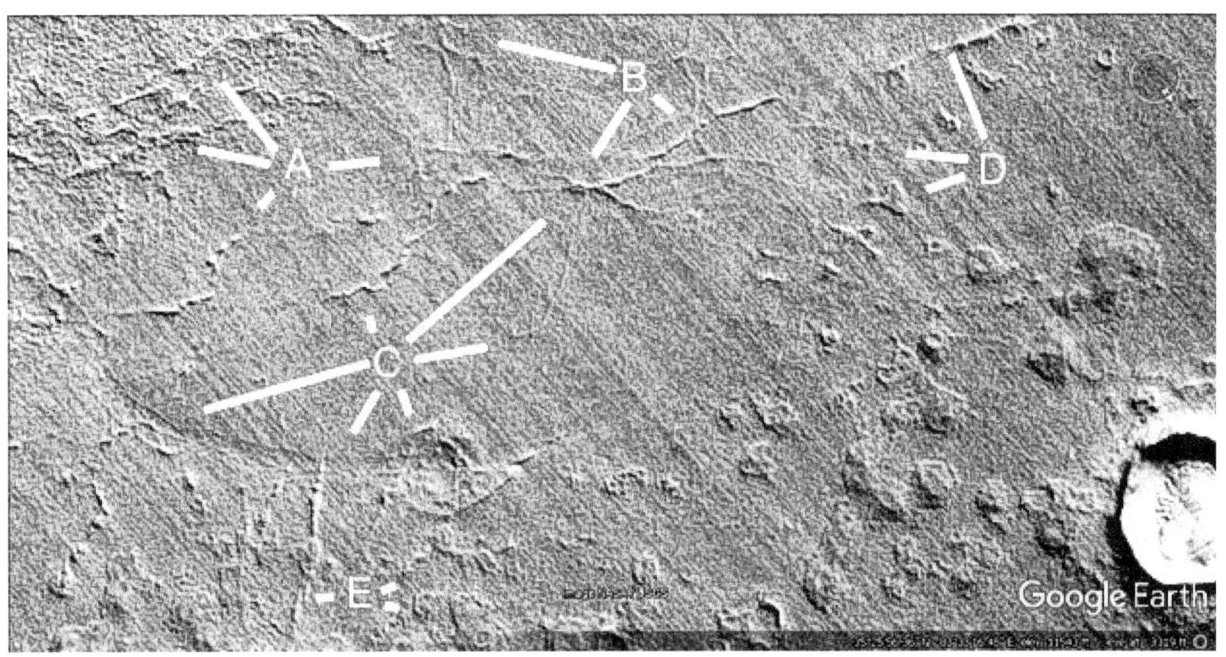

Prt863a

Hypothesis

A parabola is shown.

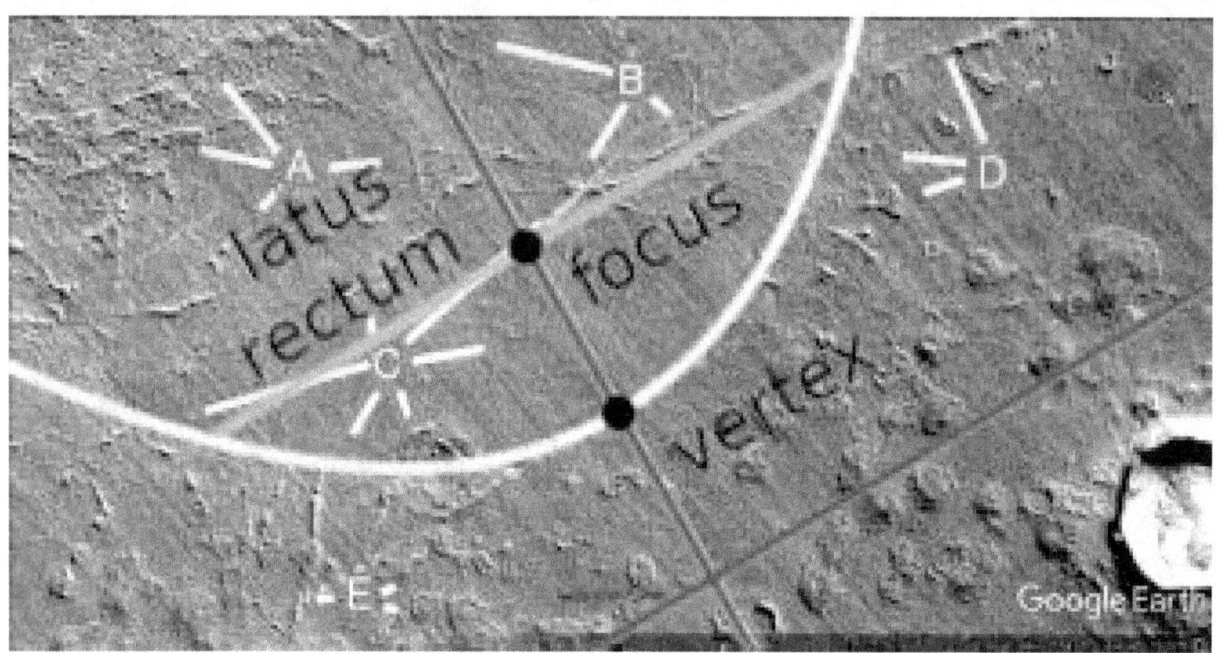

Prt870d

Hypothesis

This might look natural but it contains four parabolas, A shows the dam wall at 1 and 3 o'clock is degrading into two walls. This is seen on many dams, the construction technique appears to make them hollow or containing soil. To the left of 1 o'clock the wall is in better condition with some holes. B shows an intact segment at 11 o'clock.

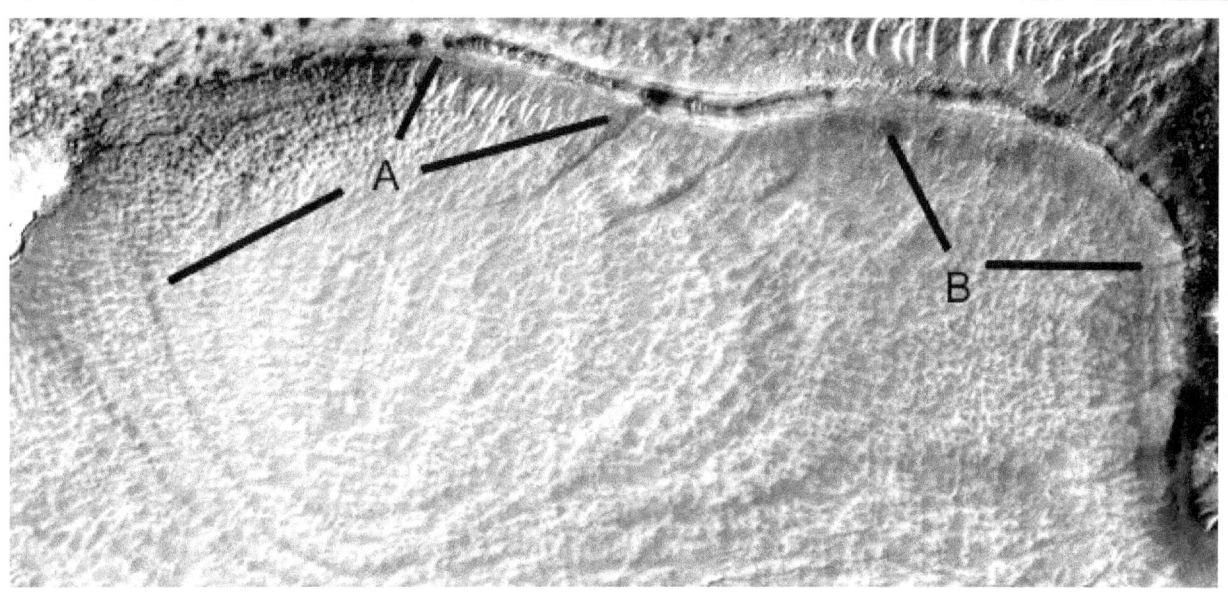

Prt870d2

Four parabolas are shown.

Hypothesis

Four parabolas are shown.

Prt870f

Hypothesis

A and B show where the dam wall is breaking up, at 3 and 4 o'clock at A there are regular shapes like tiles. A at 1 o'clock shows the crack going right through the wall but this has resisted breaking as much as the floor underneath it. C shows very regular undulations in the top of the wall, these are likely to be pillars used in its construction.

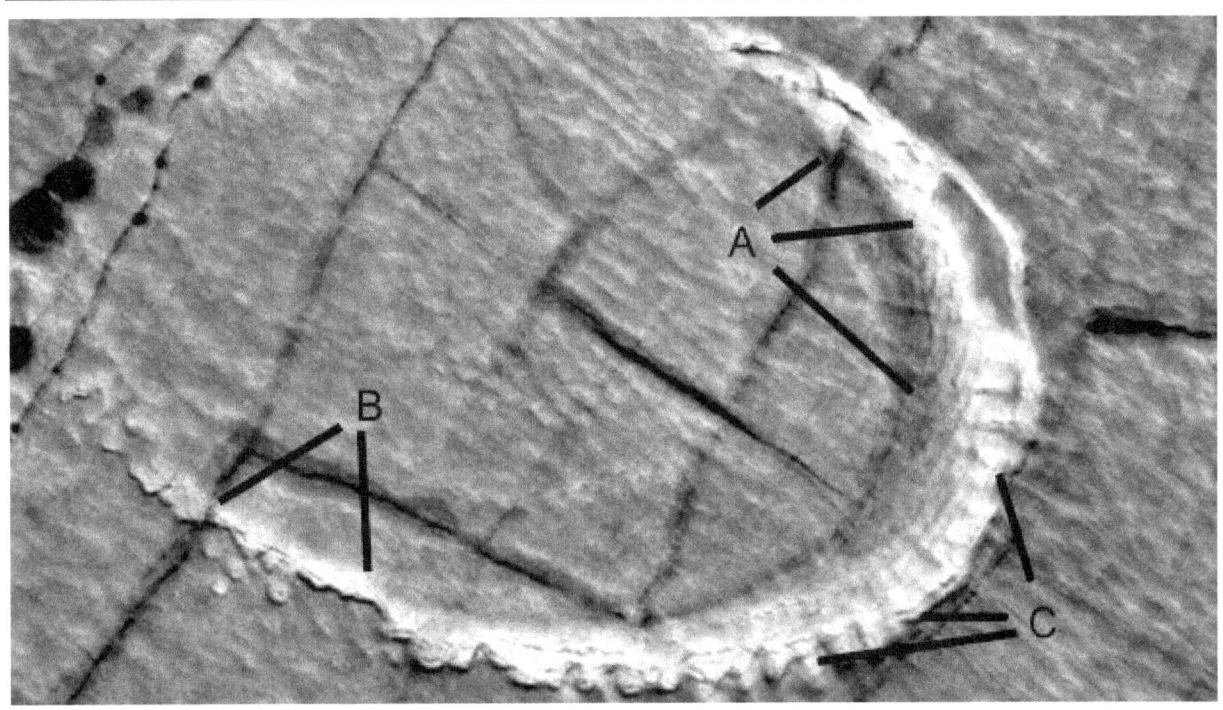

Prt870f2

Hypothesis

A parabola is shown.

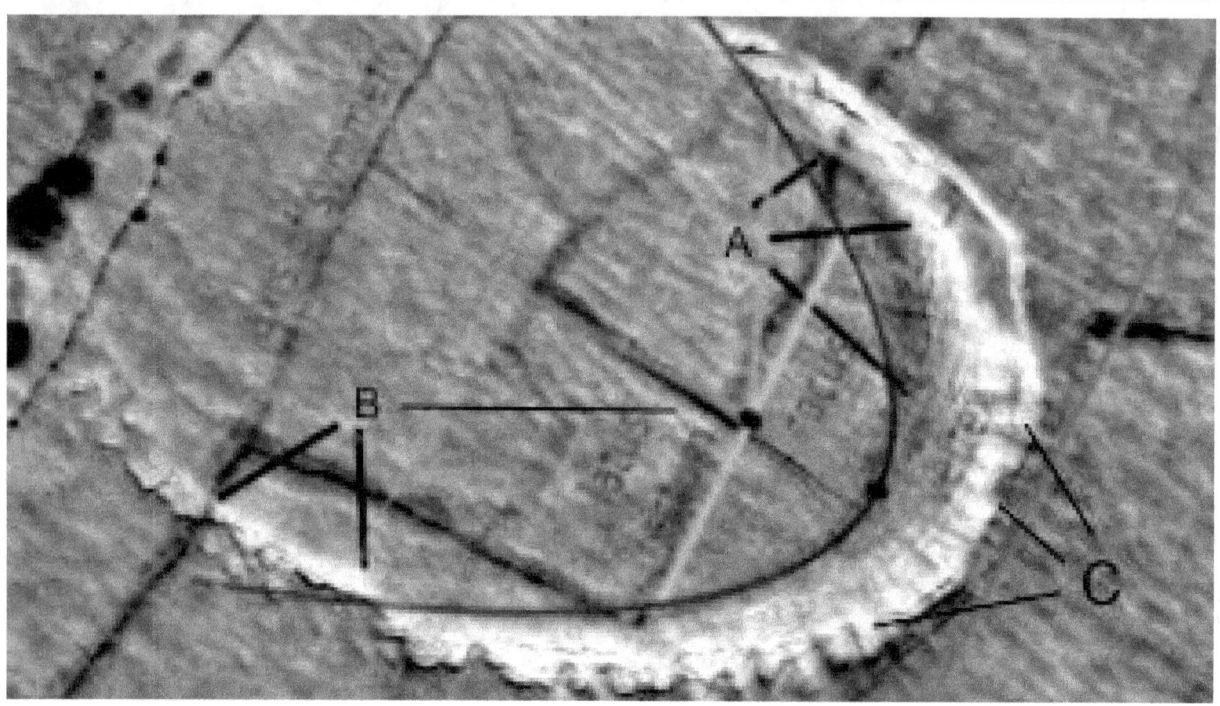

Prd879c

Hypothesis

A shows a sharp wall intersection at 9 o'clock, another at 2 o'clock. At 4 o'clock it appears to be much more eroded. B shows a sharp wall at 10 o'clock, at 2 o'clock it is starting to separate into a double wall. There is also a cavity in the wall at 5 o'clock. C shows two double walls forming.

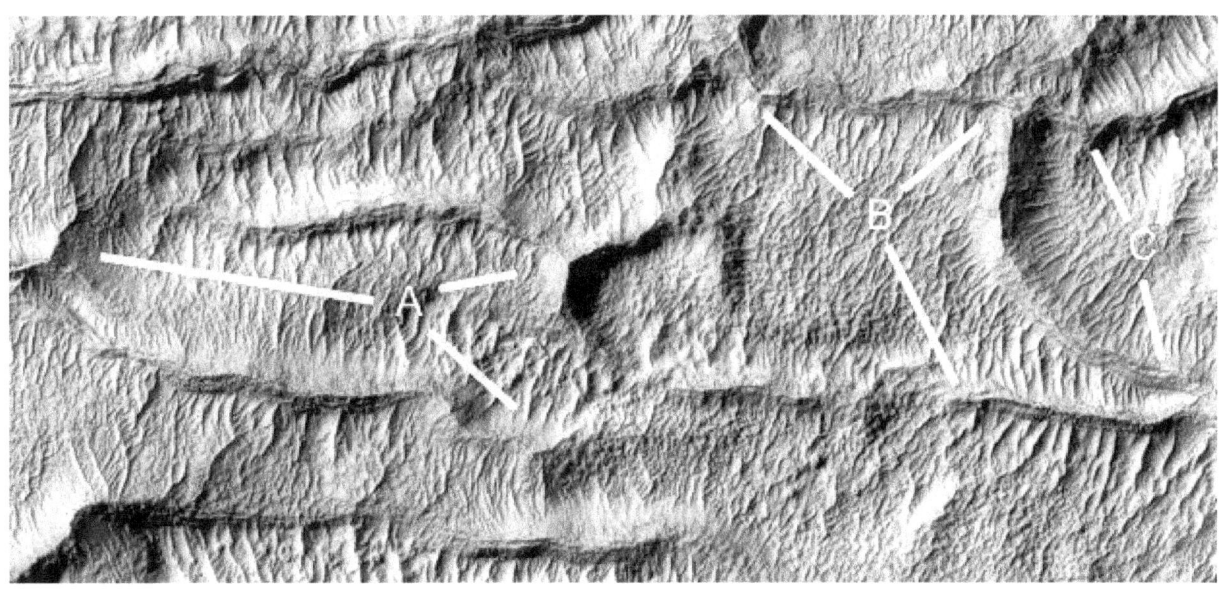

Prd879d

Hypothesis

A at 11 o'clock shows a double wall forming, also at 7 o'clock. At 10 o'clock the wall is in better condition. At 4 o'clock a double wall is forming, the wall may have been repaired with this rounded shape.

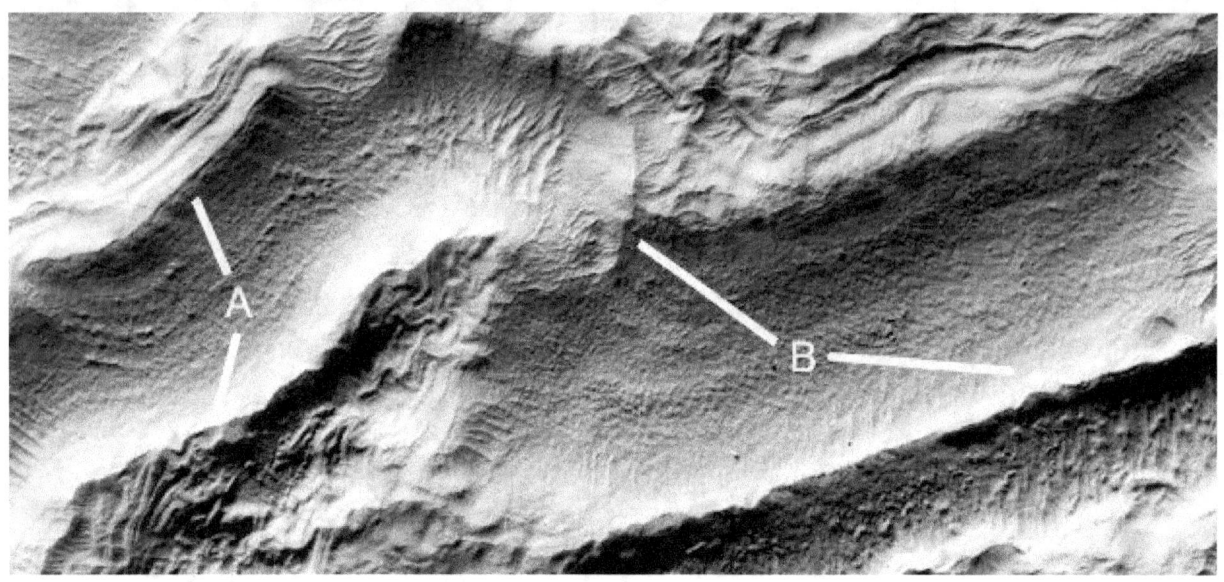

Prd882

Hypothesis

These lip dams, named because they look like lips on a face, often have parabolic sides. A shows possible hollow hills, at 2 o'clock may be a wall in a dam. B shows another hollow hill. C shows a small pit dam at 9 o'clock. D shows this lip dam.

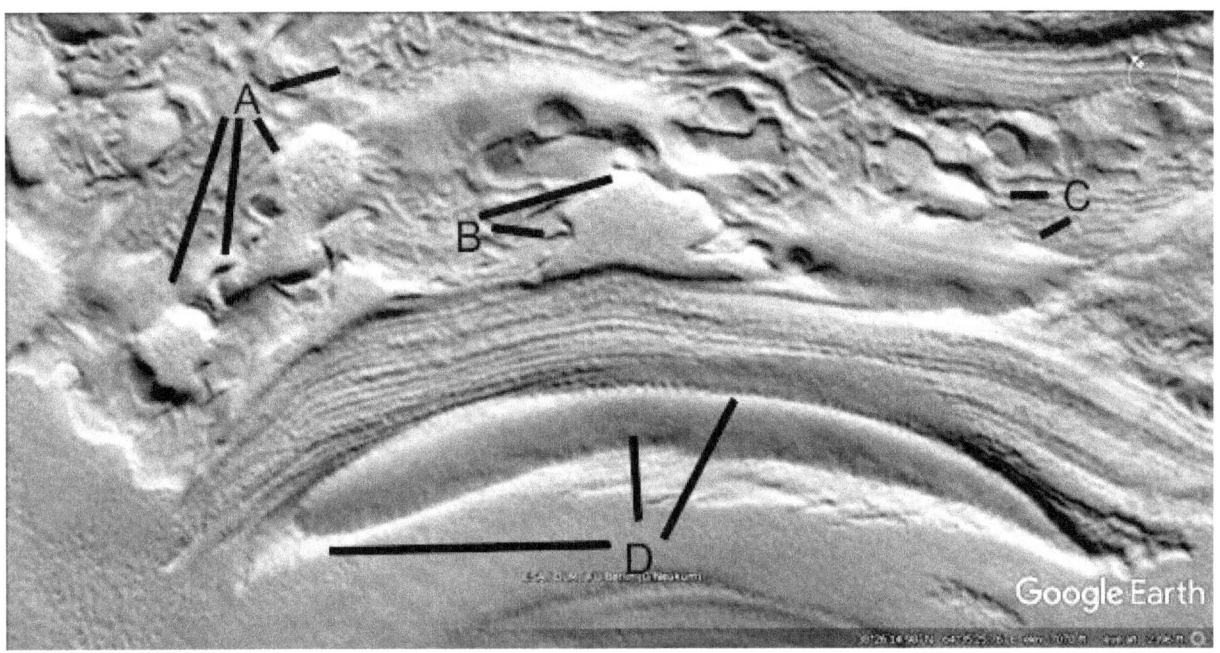

Prt882a

Hypothesis

Two parabolas are shown.

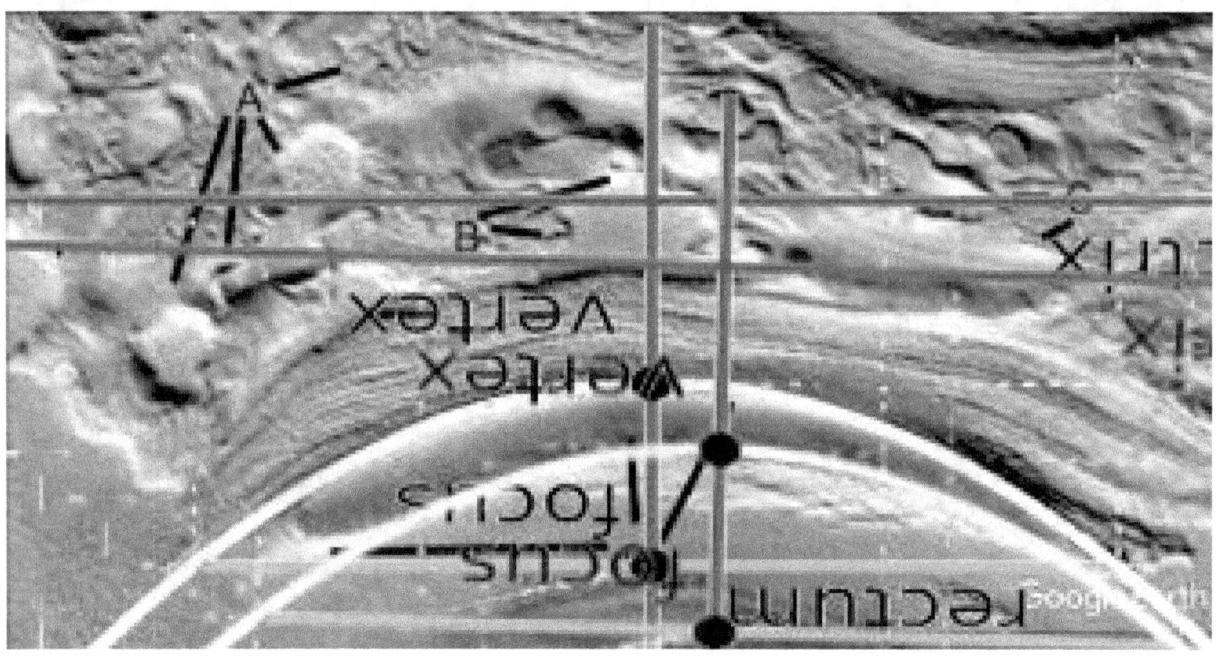

Prd886c

Hypothesis

A shows the double walls of this dam at 0 o'clock, also a small cavity in the wall at 8 o'clock. This connects to a star shaped wall from 7 o'clock to 3 o'clock. B shows this dam wall is intact at 10 o'clock, there is a wavy wall like some tubes at 7 o'clock. At 8 o'clock one of the walls is much shorter. C shows this double dam wall continuing at 5 and 9 o'clock, the wall at 12 o'clock has broken up into segments on its end. D shows another walled segment of the dam, below 10 o'clock the wall is more eroded. At 4 o'clock there is a small entrance between the walls.

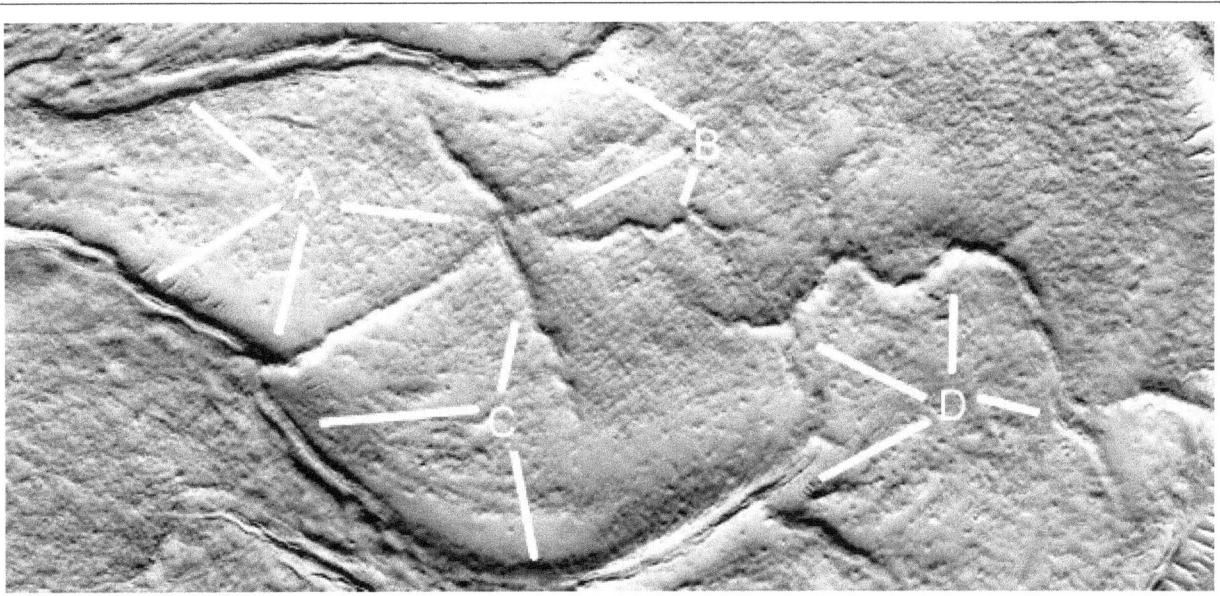

Prd886c2

Hypothesis

A parabola is shown. The axis of symmetry goes approximately through the centre of the star. The focus is also in line with the dam wall between E and F, the latis rectum or line through the focus would then approximately be an extension of this wall. A line is drawn from E to F to illustrate this.

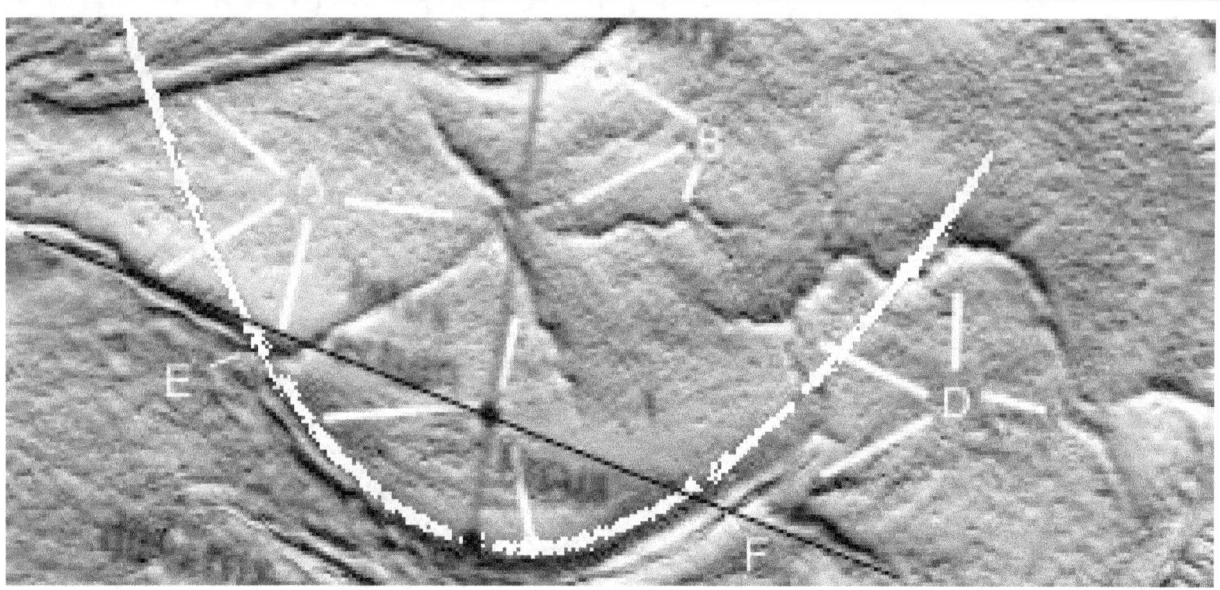

Prd886d

Hypothesis

More walls or tubes are shown, A has collapsed on its roof at 10 o'clock. At 1 o'clock there is an entrance to the next walled area. At 4 o'clock there is a double wall as the tube collapses, at 5 o'clock a hole in the tube. B at 4 and 7 o'clock shows a way line across the tube and layers under it, there appears to be layers of skin breaking off. At 2 o'clock there is a more intact segment, above and blow it the tube is more eroded. C shows two more double walls. D shows some erosion on the corner at 9 o'clock, a more intact top of the tube at 7 o'clock.

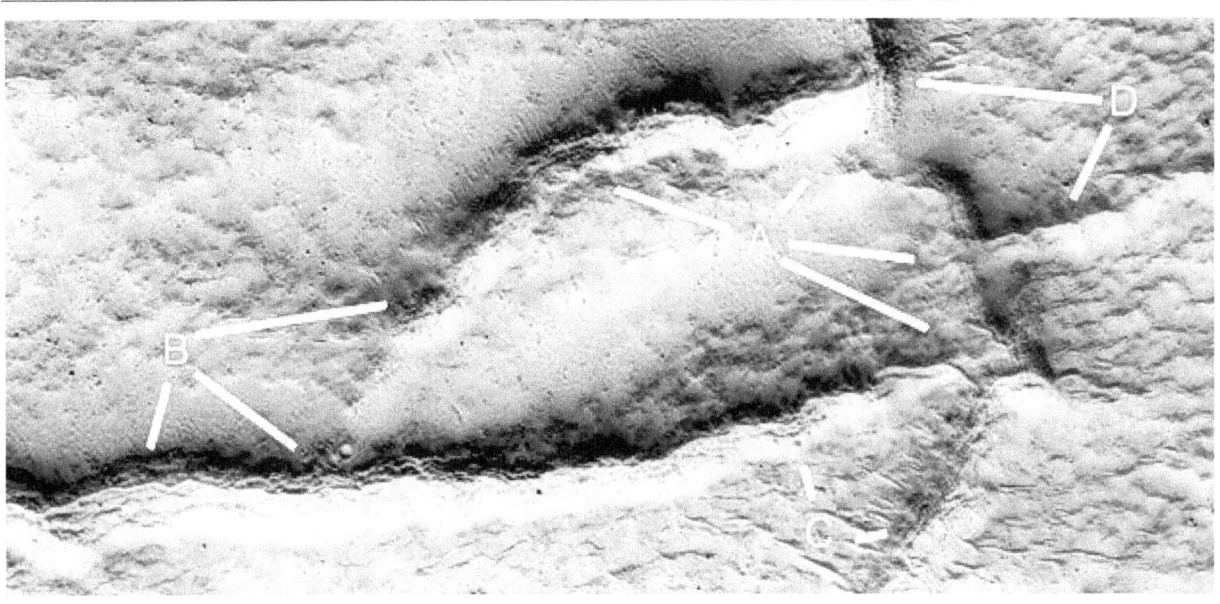

Prd886e

Hypothesis

A shows a pit dam, the walls have dark spots on them at 11 to 1 o'clock, like they are degrading. At 7 o'clock it appears to be a tube, another is to the left of A. B at 7 and 10 o'clock shows a double wall as the top of the dam wall is degrading, exposing the interior. At 6 o'clock the wall is flatter, this extends down to A at 8 o'clock, there is also a secondary wall at 7 o'clock which goes over to D at 12 o'clock. At 2 o'clock the wall bulges outwards in an arc.

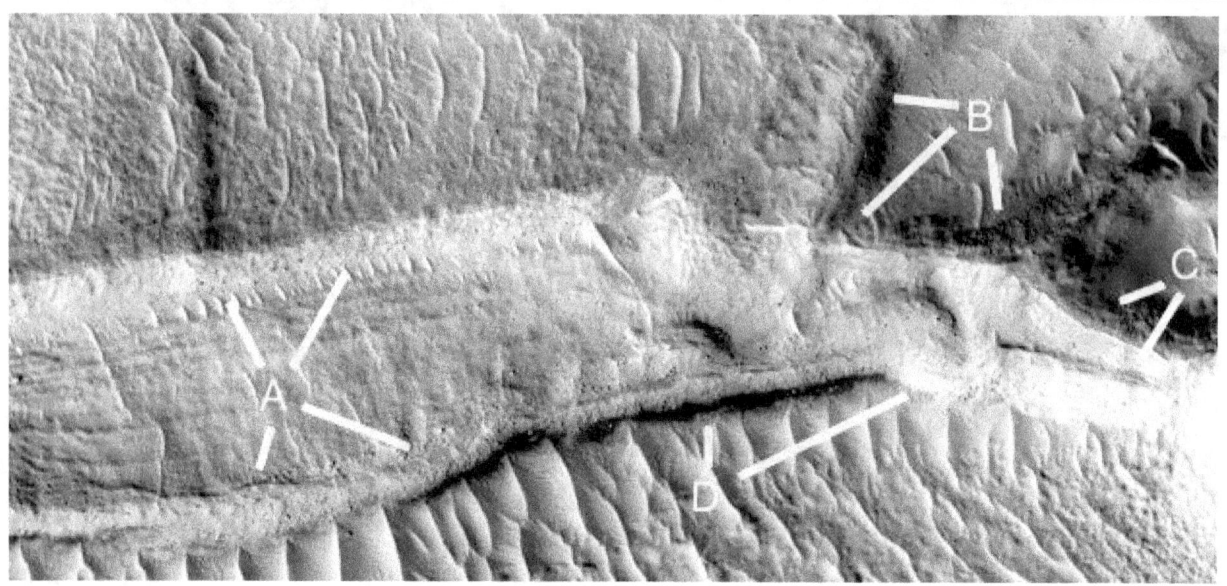

Prd891b

Hypothesis

A shows the top of the dam wall is breaking up at 10 o'clock, at 2 o'clock there are several parallel layers to the top of the dam wall, perhaps how it was constructed. There are also regular vertical pillars. B at 1 o'clock also shows regular segments like building blocks in the wall, these extend to C at 1 o'clock and further to the left. At 9 o'clock begins a chaotic exterior which continues to the left, to the right of this however the dam wall is flat like cement. This pattern might exist under the wall to give it strength like many arches. B has the skin breaking off at 5 and 6 o'clock, this extends at 8 o'clock to the left with a smoother hollow like an arch under C at 6 o'clock to the left.

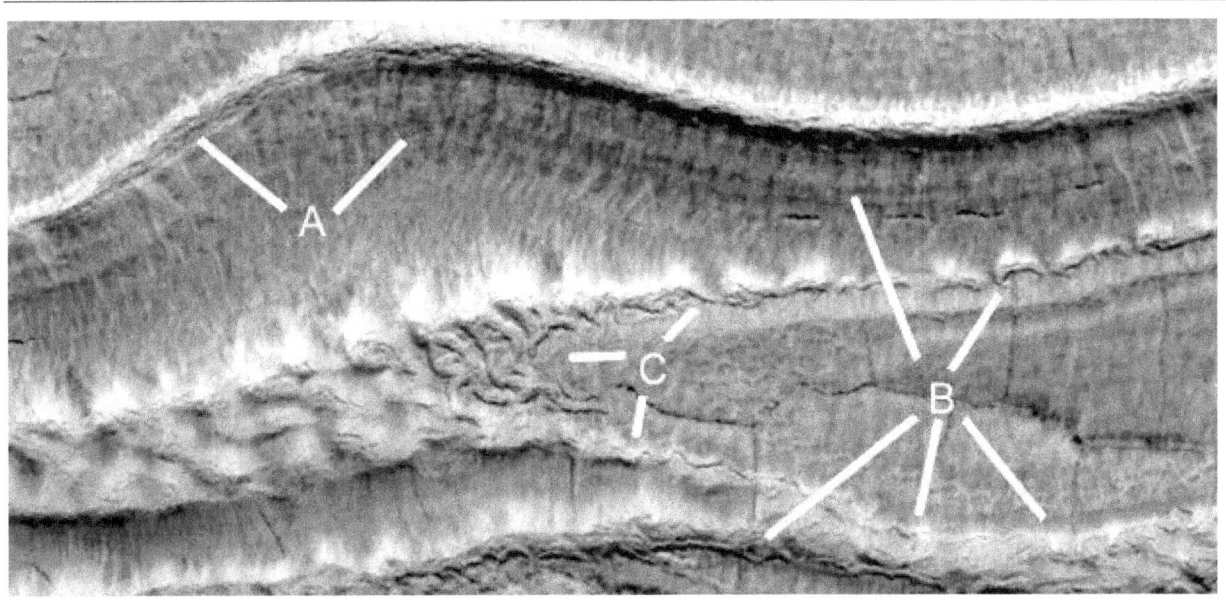

Prd891b2

Hypothesis

A parabola is shown.

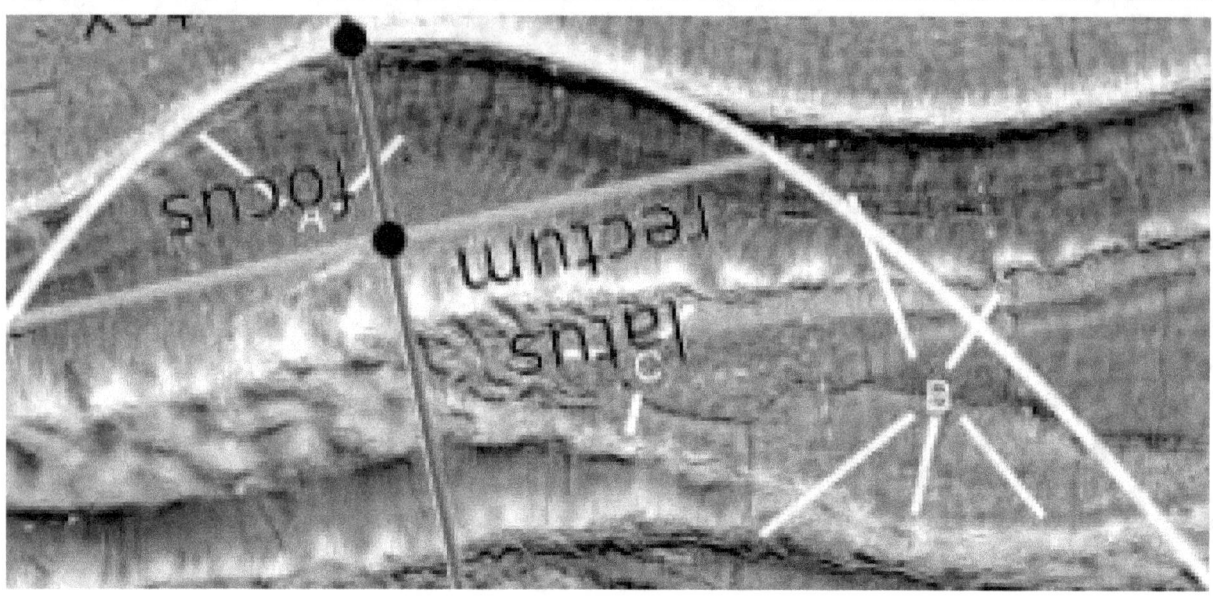

Prd891c

Hypothesis

A shows a wall breaking up, B shows cracks developing in the dam floor. These cracks appear to be small regular mounds, this might indicate the cracks have been sealed to keep the water in. To the right of this the cracks are not sealed. C also shows these seals at 4 and 7 o'clock, at 6 o'clock the dam wall has layers under it as a way of constructing it. D shows the top layer at 12 o'clock and 2 o'clock has regular bulges in it about the same size as on the sealed cracks. The hollow at D at 9 o'clock has a fairly constant width, also the slopes inside it are regular. This may have been a water channel. E shows a degraded wall at 7 o'clock, at 10 and 1 o'clock there is a wall or sealed crack, but the other cracks at E and under it are not sealed.

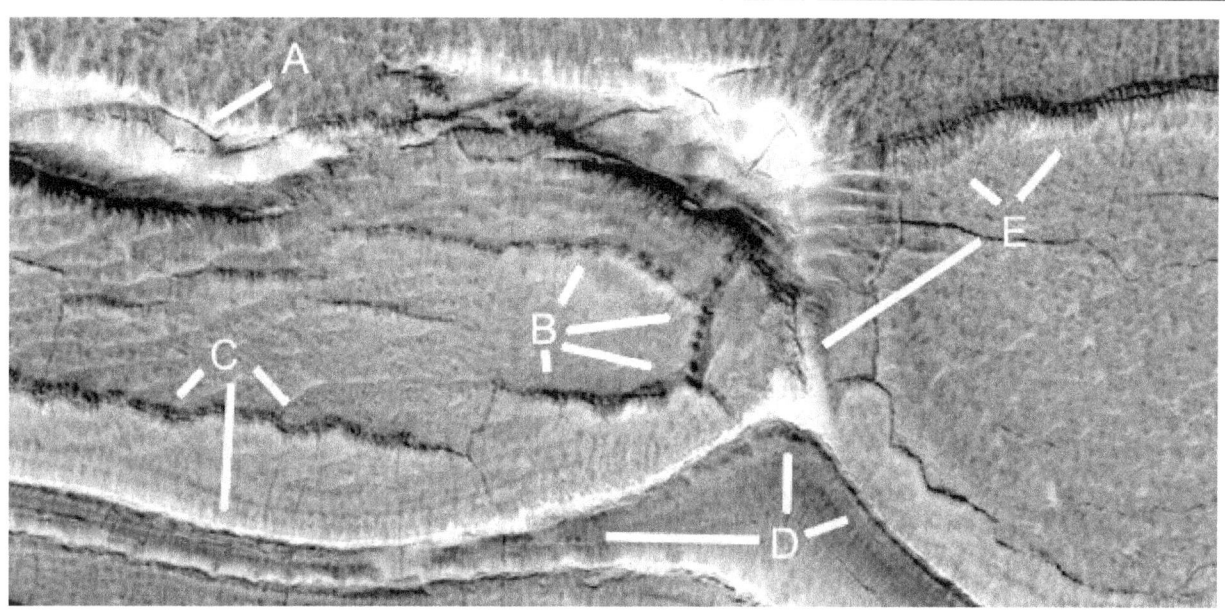

Prd891c2

Hypothesis

A parabola is shown.

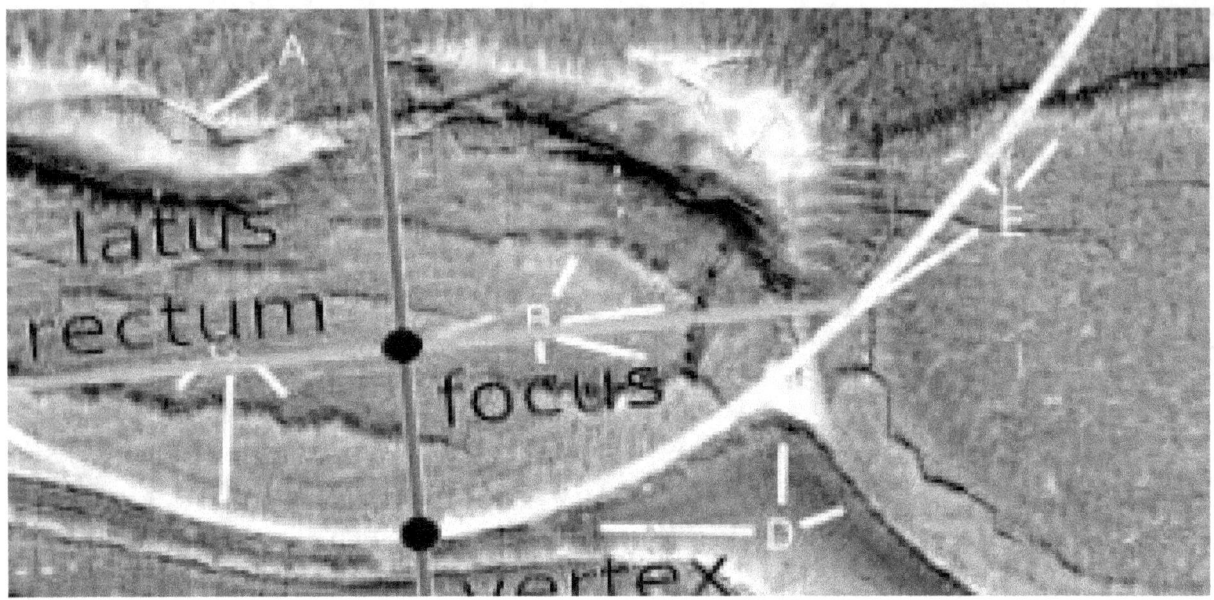

Prd901a

Hypothesis

This would seem natural except there are two parabolas in it. A shows the tops of the pit dam walls are cracking. B shows another double wall at 12 o'clock, at 3 o'clock may be a dam. C at 1 o'clock shows a smooth dam floor like cement, at 5, 6, and 12 o'clock there are smaller walls. D shows another pit dam at 6 o'clock, at 8 o'clock may be a water channel down to E.

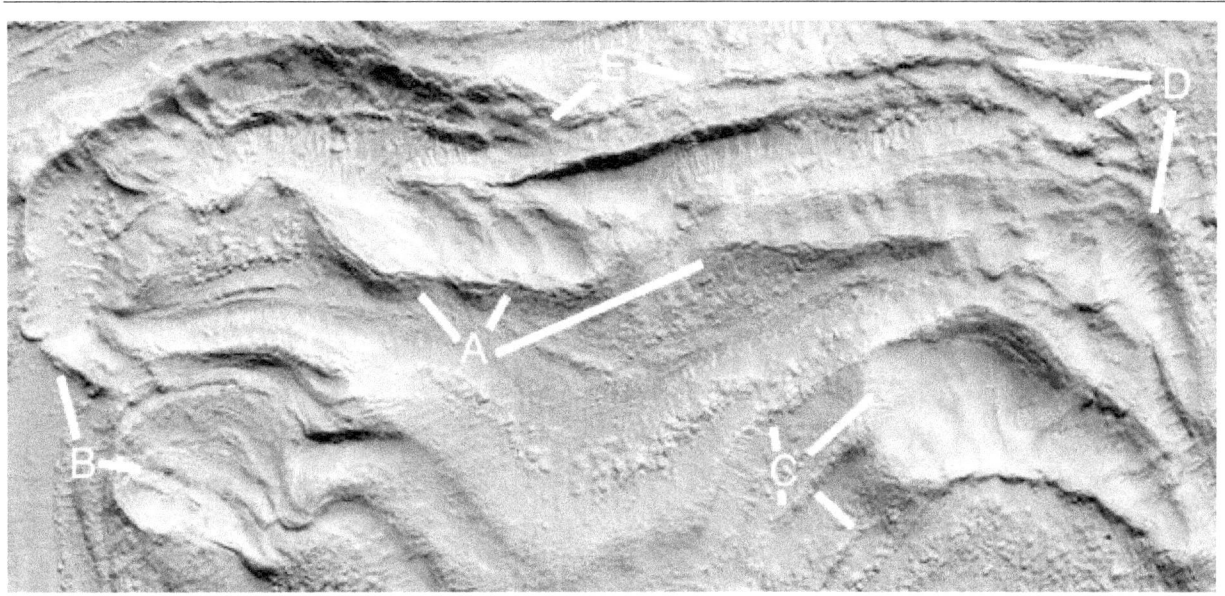

Prd901a2

Hypothesis

Two parabolas are shown.

Prd911b

Hypothesis

This shows more pit dams, A at 4 o'clock is smooth like cement and at 3 o'clock may be a water channel. B shows four other pit dams. C shows a degraded wall at 11 o'clock, a rougher interior between dams at 1 o'clock, and a dam wall in good condition at 2 o'clock. S shows a smoother contained area at 8 o'clock, the dam wall extends above this surface all around it. A similar texture is shown at 10 o'clock, at 4 o'clock the dam floor may be cement.

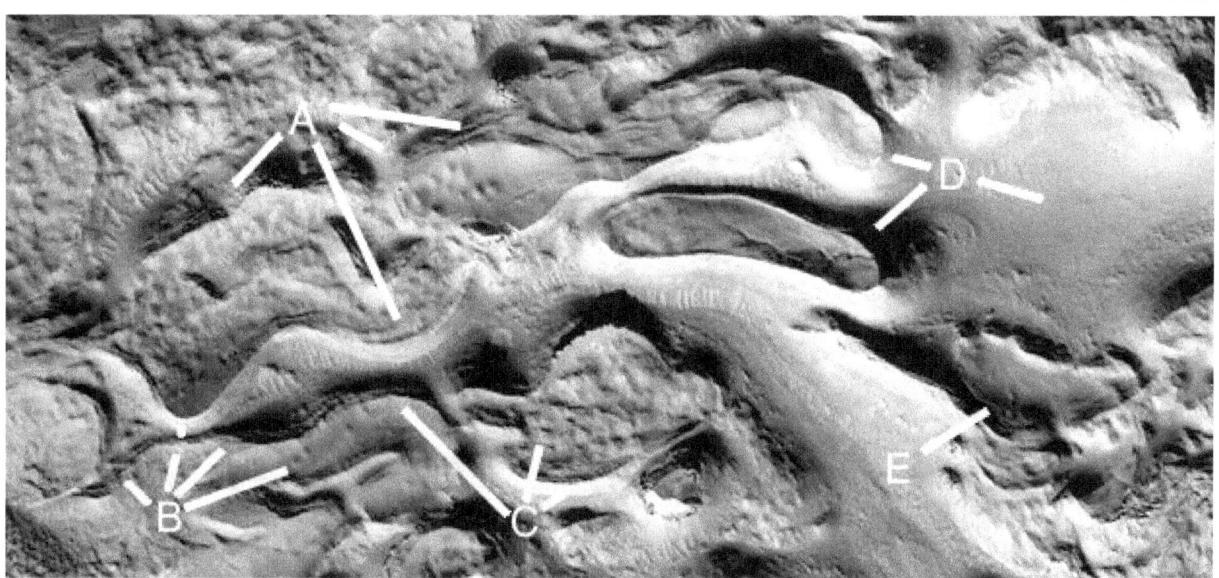

Prd911b2

Hypothesis

Eight parabolas are shown. This is a good example of how natural looking areas in a crater can be looked at more carefully. With a closeup there cold be even six more parabolas here.

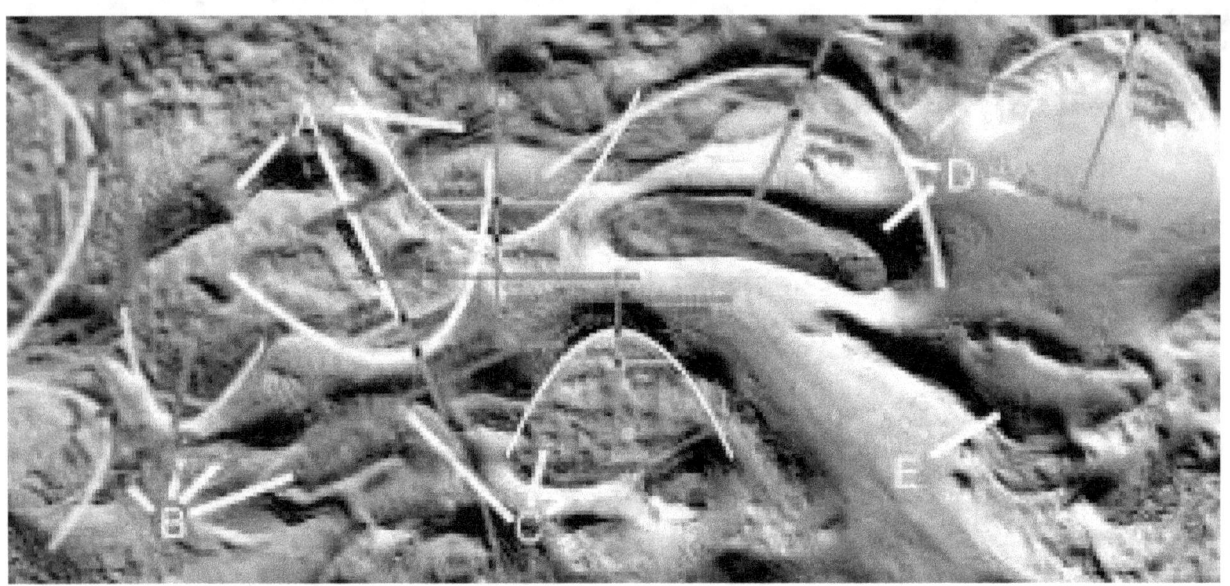

Prt912

Hypothesis

A shows tubes coming out of a hill. These look more like tree branches at 4 and 2 o'clock. At 8 o'clock they go into the hill. B shows two more branchings. C shows parallel tubes at 3 o'clock, a collapsed hill at 12 o'clock, and a small hill connected to the tube at 10 o'clock.

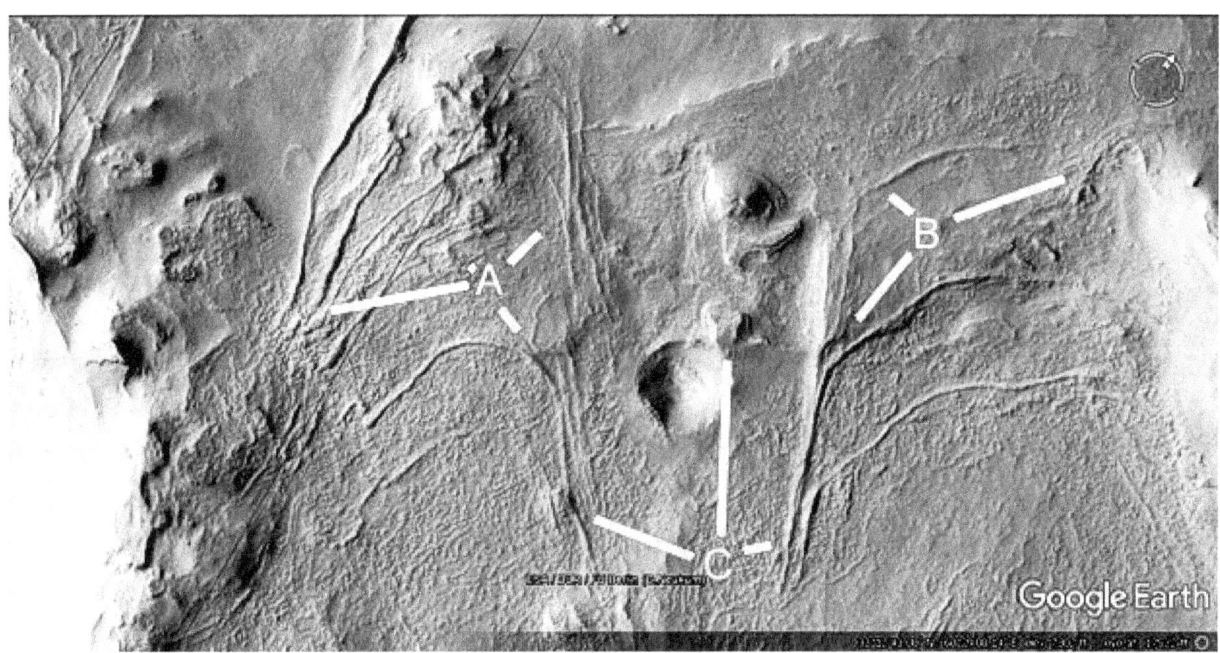

Prt912a

Hypothesis

A shows an eroded tube at 8 o'clock, to the left it becomes a double wall or collapsed tube. At 6 o'clock there is a faint wall parallel to the larger wall at B at 10 to 2 o'clock. This may be the other wall of a collapsed tube. At 5 o'clock there is a narrow tube. C shows another tube from 7 to 11 o'clock, some segments are missing from erosion. At 2 o'clock is a double wall or collapsed tube. This extends up to D at 9 o'clock into a small hill, at 12 o'clock is a collapsed hill with exposed rooms.

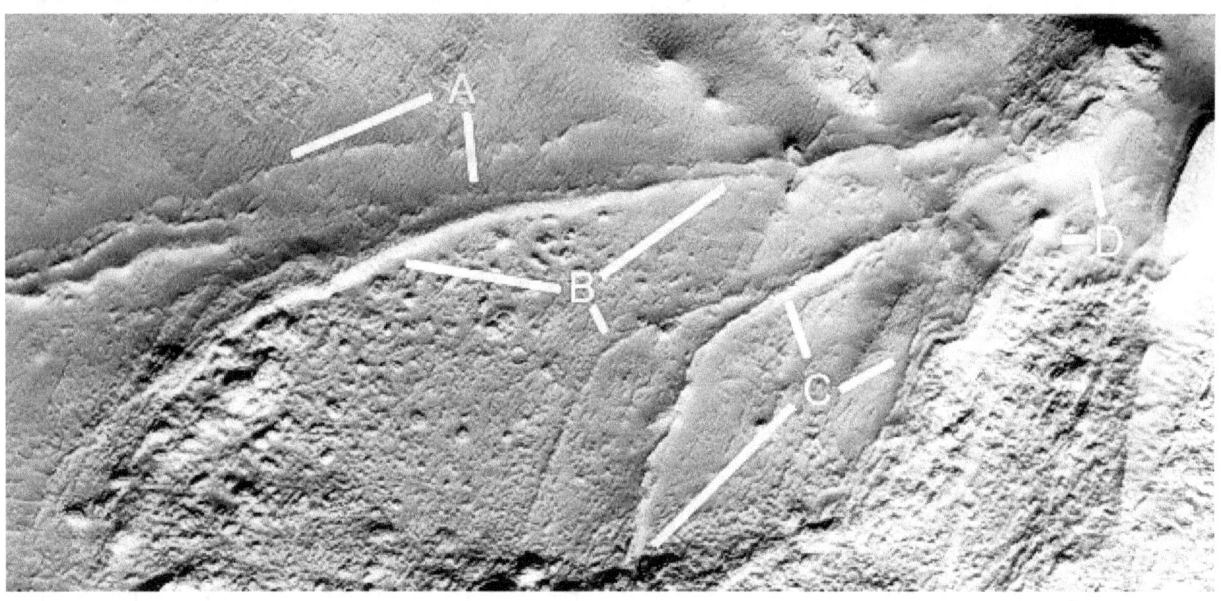

Prt912b

Hypothesis

A shows a collapsing tube breaking into segments like pillars, at 5 o'clock there is a cavity in the tube segment. The tube ends in a triangular hill at 4 o'clock, also at D at 10 o'clock with another hill at 8 o'clock. B shows the tube is in good condition at 11 o'clock, at 4 and 6 o'clock it has turned into a double wall as the roof collapsed. C shows another double wall or collapsed tube, also at E at 5 o'clock. There is a small tube at 12 o'clock.

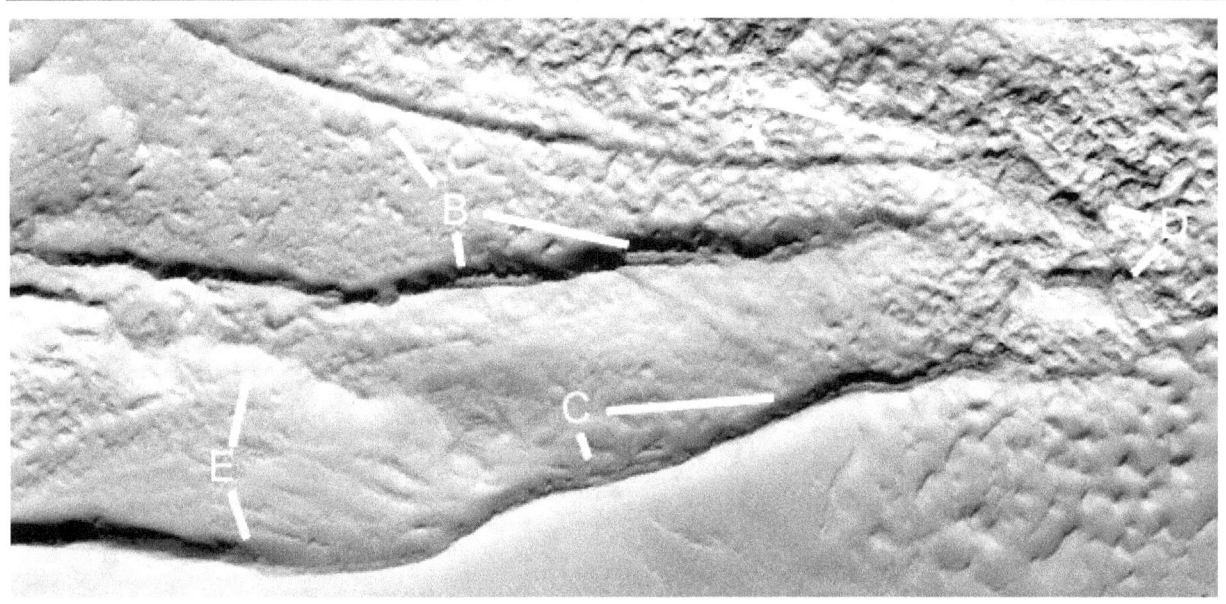

Prhh936

Hypothesis

The hollow hill has many parabolic arches on its roof for strength, at 9 o'clock at A it connects to another hill.

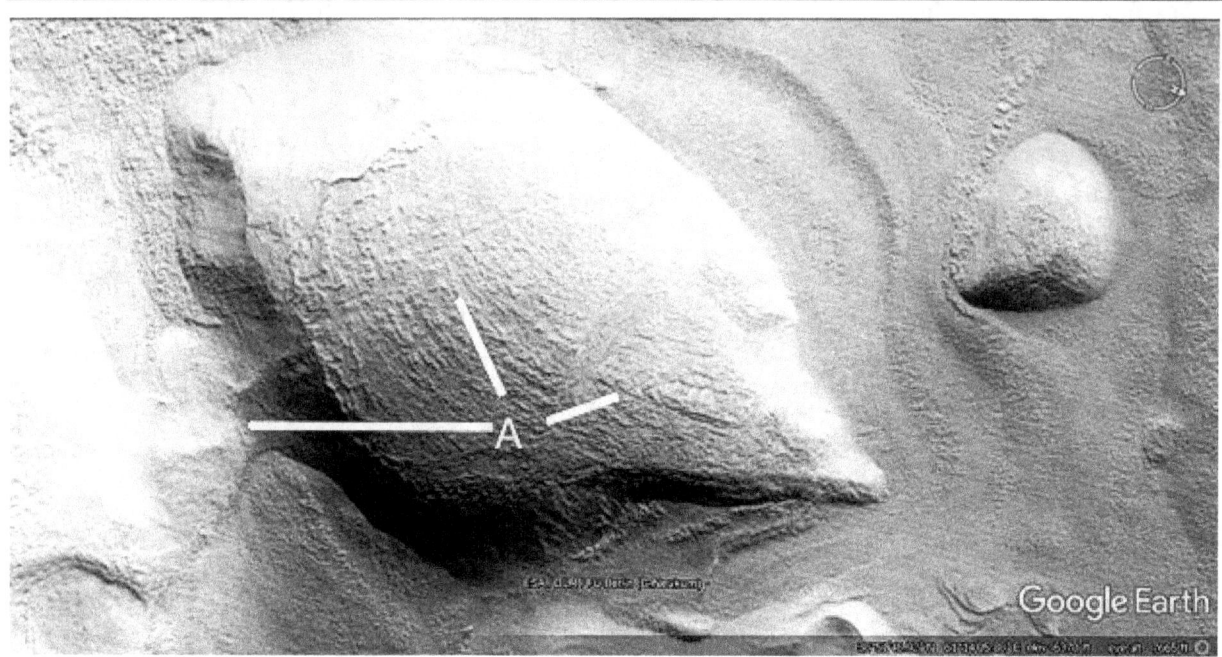

Prhh936a

Hypothesis

Three parabolas are shown, there would be several more but they are too faint.

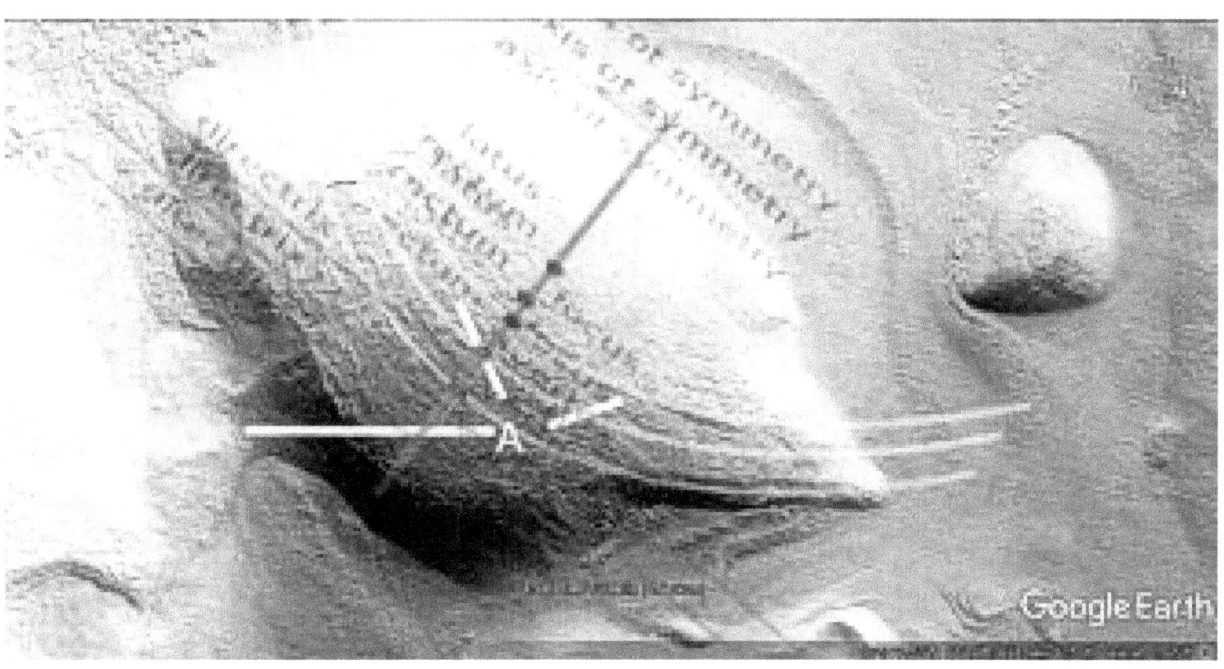

Prhh939

Hypothesis

A, B, and C show unusual connections between these hill like tubes.

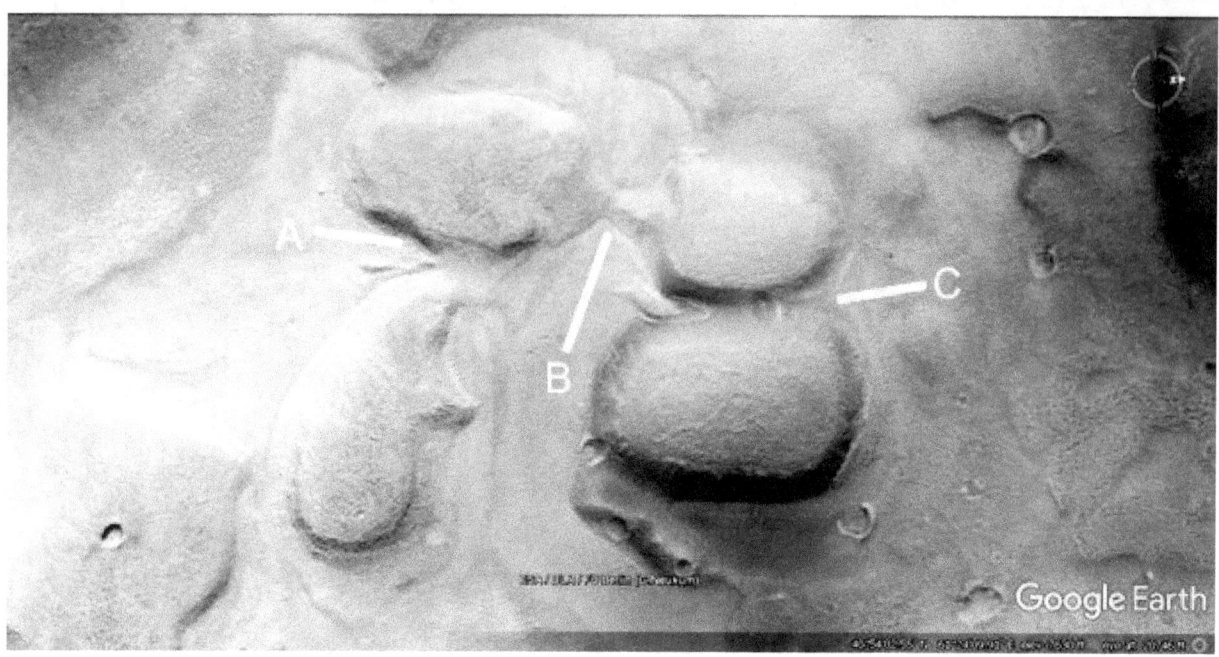

Prhh939a

Hypothesis

This hill appears to be a parabolic dome, which would probably be the strongest construction method. Only a few hills are analysed like this but there are many more which could be checked.

Prhh940

Hypothesis

These hills are approximately parabolic domes from ground level. A shows a collapsed segment. B shows a tube running up the hill side. C shows another tube running up to D.

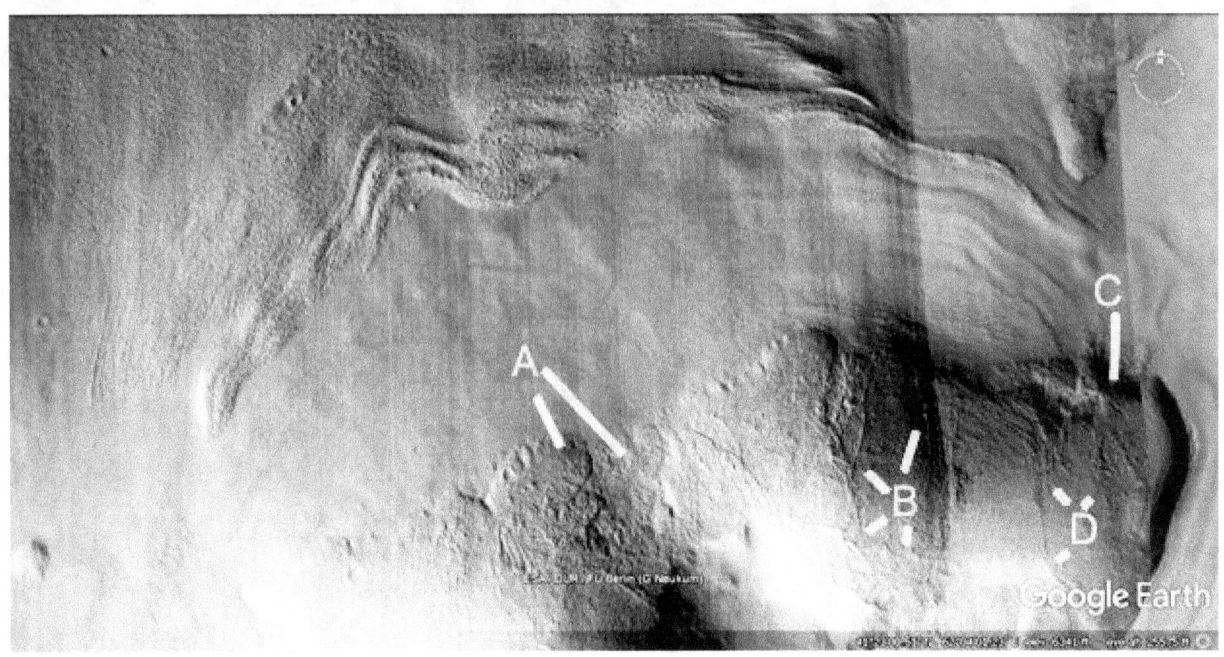

Prhh940a

Hypothesis

Two more parabolic domes are shown.

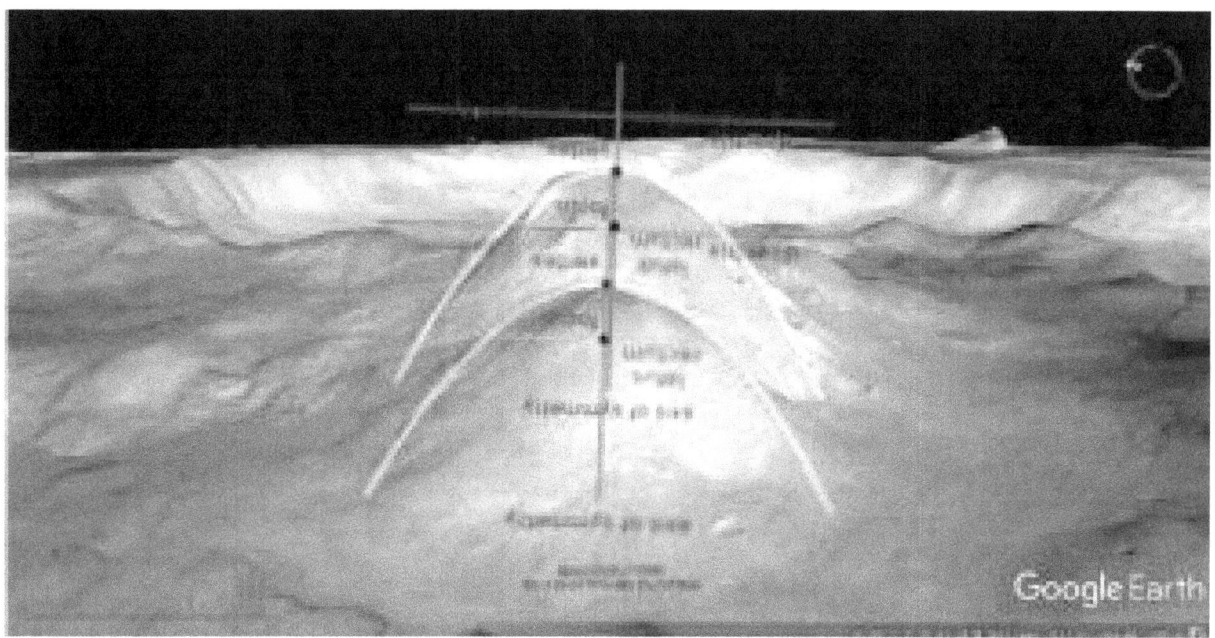

Prhh944a

Hypothesis

A shows smooth sides of the hill like cement, rougher on the roof at 12 o'clock with signs of the cement breaking off. At 9 o'clock a layer is shown also peeling off, this continues over to 3 o'clock and then down to 5 o'clock. B shows a parabolic segment at 1 o'clock then this degrading layer at 3 and 6 o'clock. It continues along to the right, it may also indicate the wall of the hill was built up with layers of different heights. At 5 o'clock is another layer with a small tube coming out of its apex.

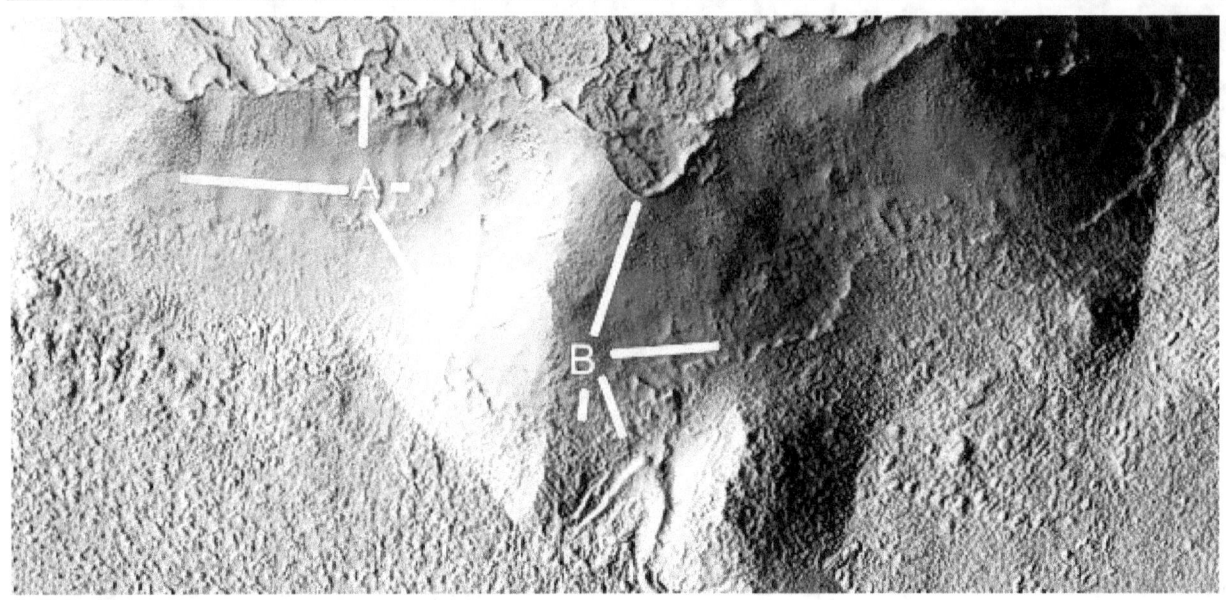

Prhh944b

Hypothesis

This is a parabola, indicating the layer might end here rather than be breaking off. There is also no sign of broken wall fragments next to the hill. A shows the thickness of the wall, B shows the edge has a raised lip. C shows a smoother area which may be from erosion.

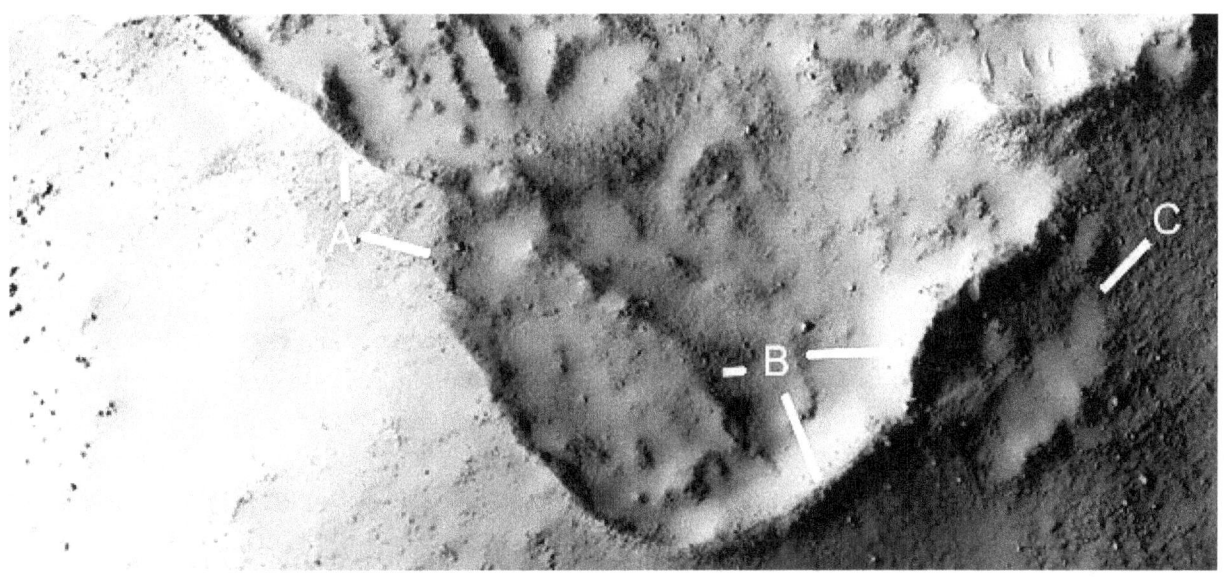

Prhh944ba

A parabola is shown. This is a standard shaped parabola, meaning it has not been narrowed or broadened to fit the formation. A parabola can be made proportionally narrower for example and still retain its load bearing properties.

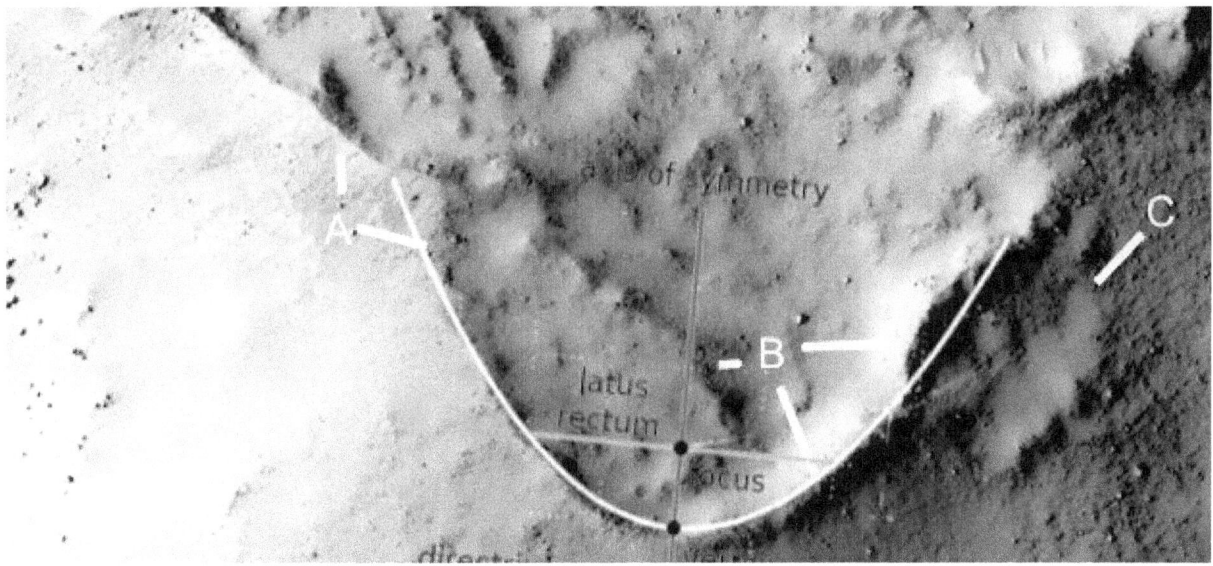

Prhh944c

Hypothesis

The top of the layer here is shown at A at 12 o'clock, at 10 o'clock is a tube. B shows multiple layers under it, this may be the construction technique. C shows a broken wall segment at 8 o'clock second leg, this may be two thinner layers broken together. At the first leg is a tube. At 9 o'clock second leg is another broken layer. At 6 o'clock the tube appears to come from here, this has a collapsed side and a gap between it and 8 o'clock first leg. At 12 o'clock the texture of the roof is different to the wall layers.

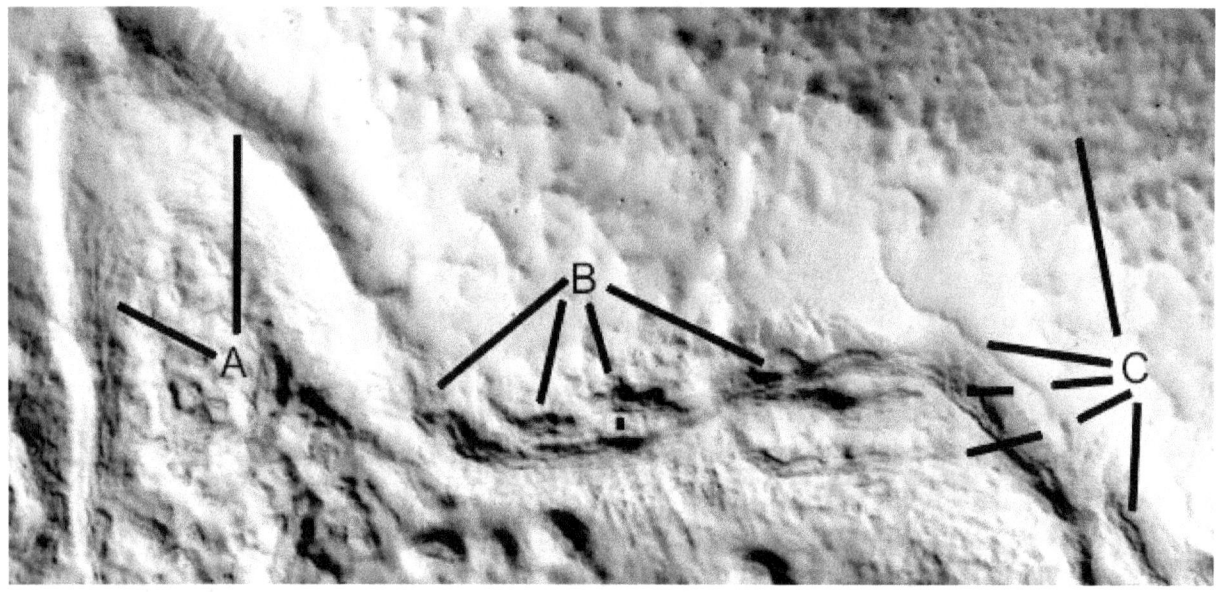

Prhh944c2

Hypothesis

Three parabolas are shown, like a parabolic wave. This can be an approximation to ocean waves which are elliptical.

Prhh944f

Hypothesis

A shows tubes or eroded segments on the roof. B shows contours which may have been used for strengthening the roof. C shows a settled area. D shows many parabolic arcs to strengthen the roof at 9 and 10 o'clock, at 2 o'clock there is an exposed grid perhaps used for reinforcing the roof.

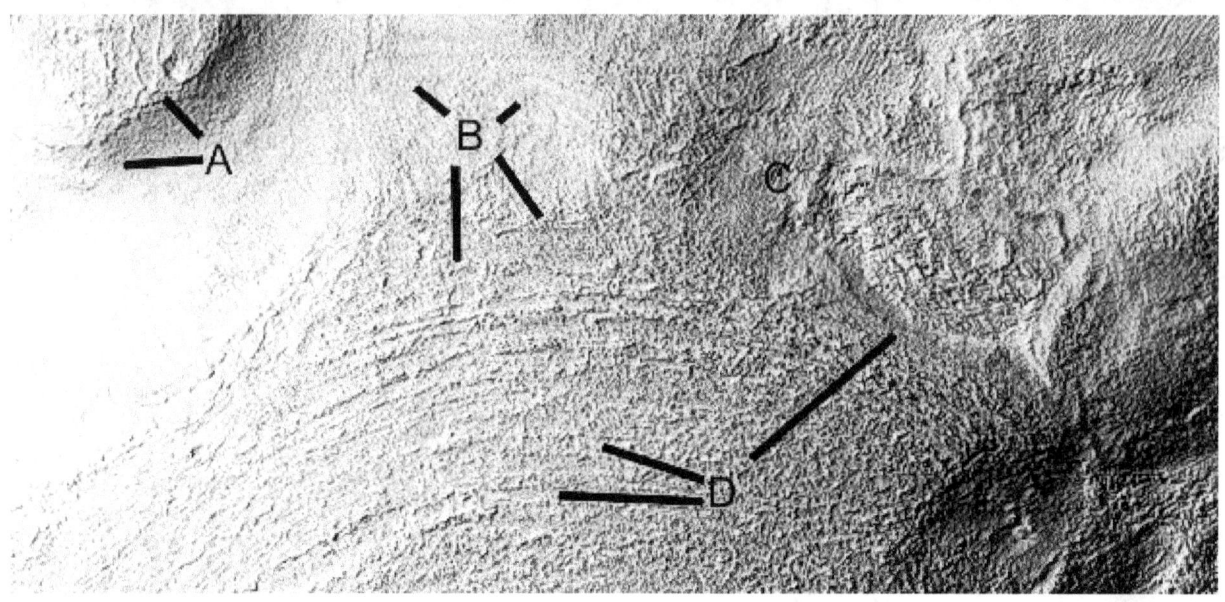

Prhh944f2

Hypothesis

Three parabolas are shown, there are several more but these are the clearest. The axis of symmetry of each is closely aligned but each parabola is smaller than the one surrounding it.

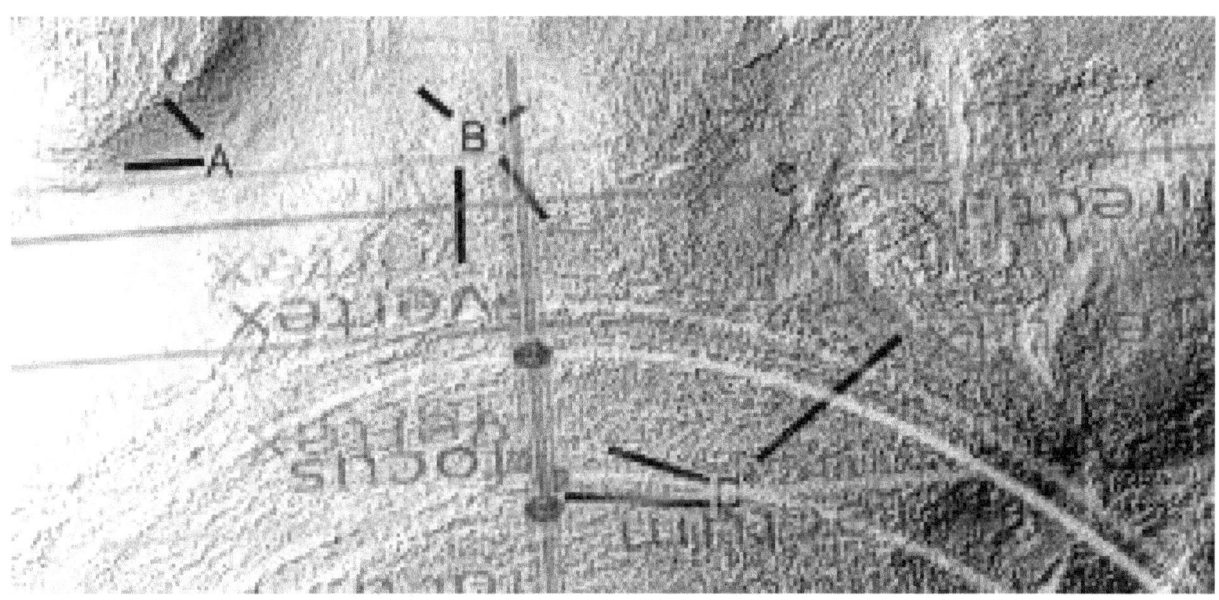

Prhh944j

Hypothesis

This may be a Cobler Dome where the parabolic layers of bricks are exposed. They are less visible at A at 10 o'clock, at 4 o'clock the top of the hill may be peeling off. B shows a smooth skin like cement that may have broken off on the upper side exposing the layers. C shows the parabolic layers, D shows two skins that have eroded away exposing the arcs.

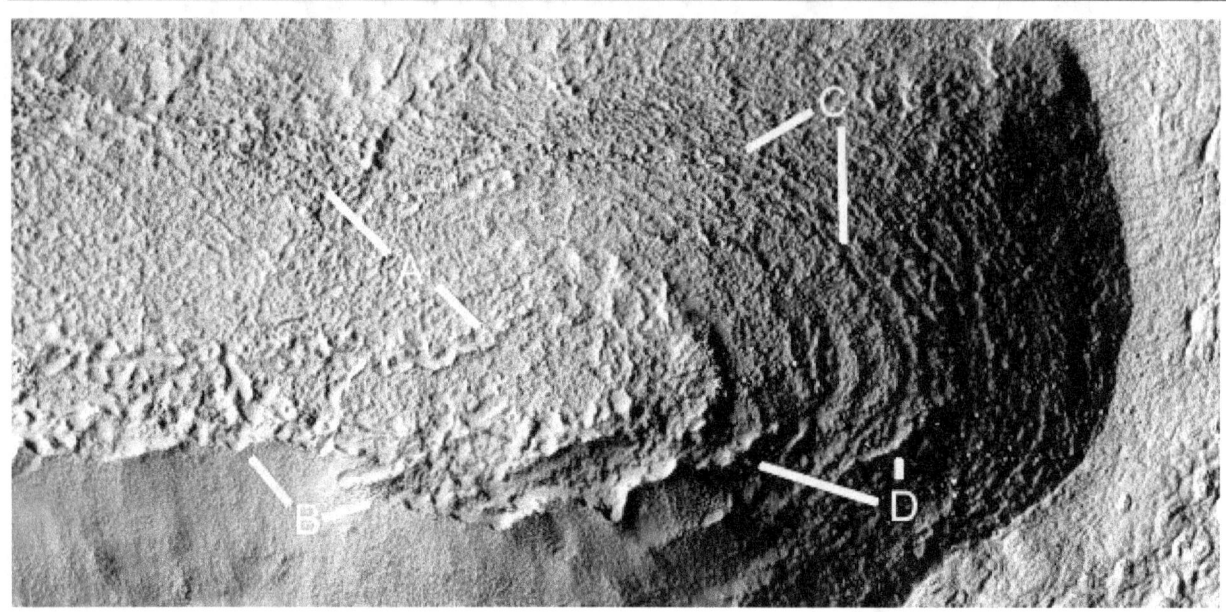

Prhh944j2

Hypothesis

Three parabolas are shown, there are several more which are too faint. Straight ridges are also overlaid by lines.

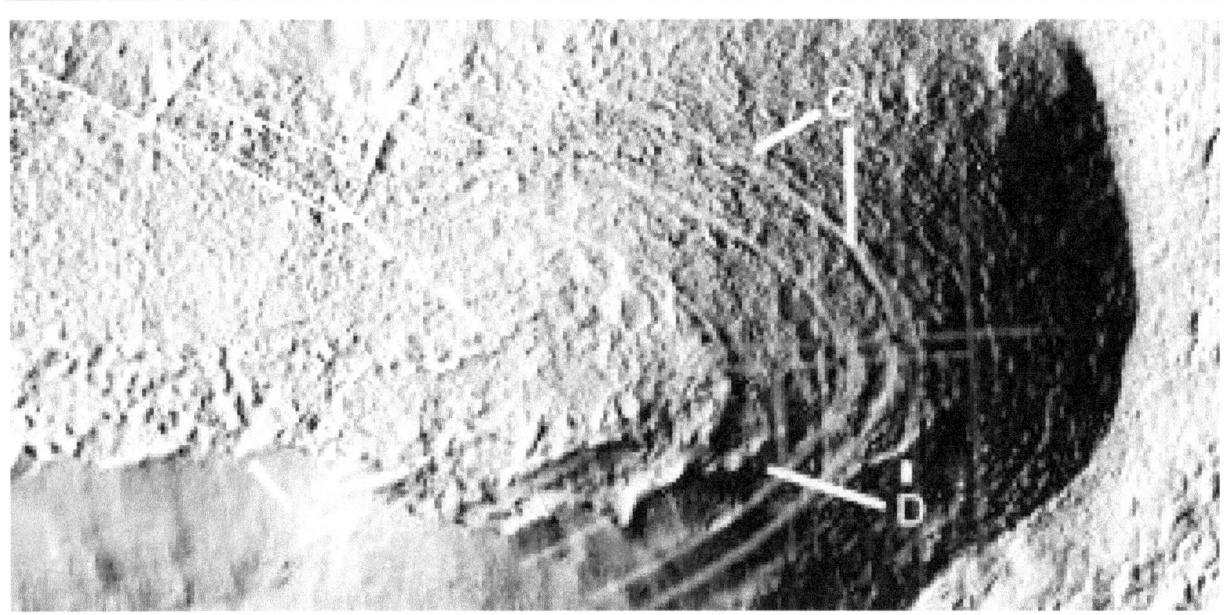

Prhh945

Hypothesis

The roof at A is approximately a rectangle at 8 o'clock, a second approximate rectangle is at 3 o'clock. B at 9 o'clock shows parallel ridges on the roof, a rectangular cavity is at 7 o'clock. Tubes are shown at 1 and 3 o'clock, also at 4 o'clock.

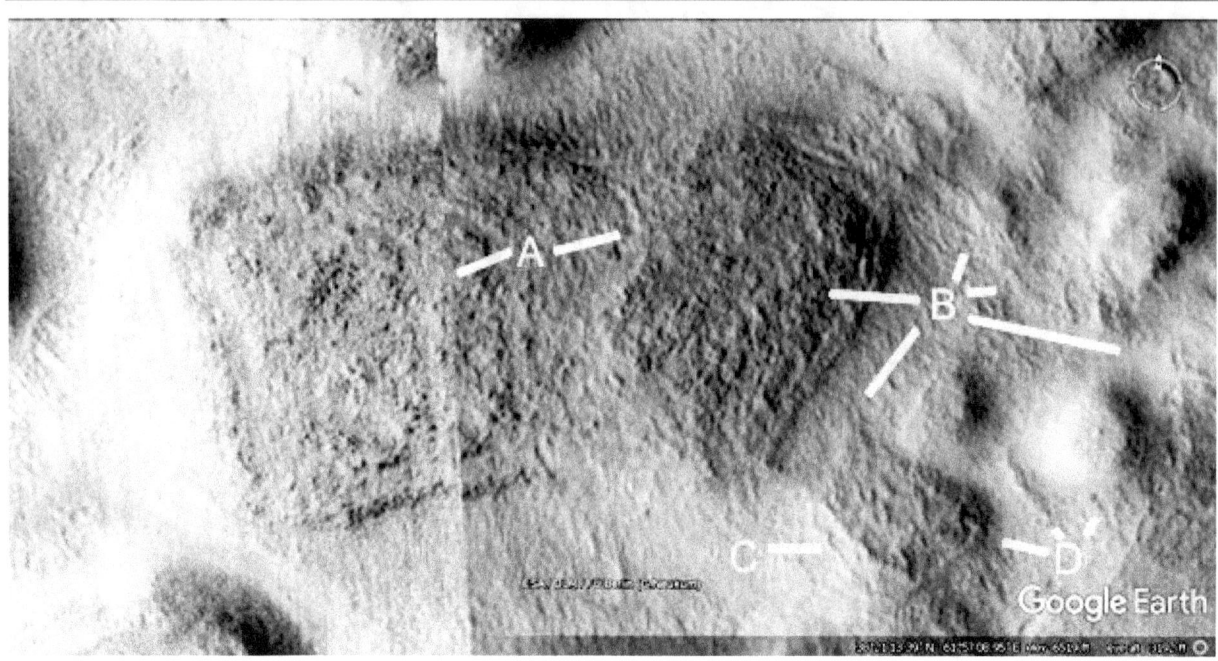

Prhh945a

Hypothesis

The lines show how straight parts of the formations are. The angles are approximate because there is some latitude in how the lines can be drawn. A parabola is shown connecting to two straight walls.

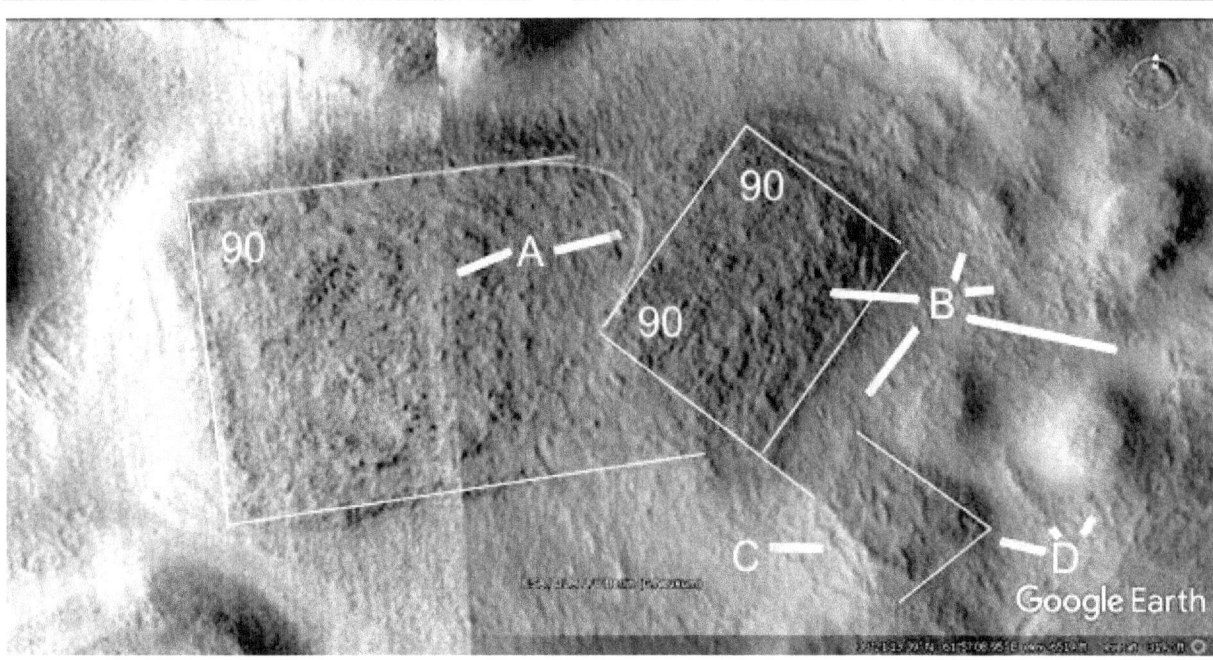

Prhh946

Hypothesis

A shows more exposed parabolic layers, B also shows layers in the side of the hill. C shows smaller hills connected by tubes.

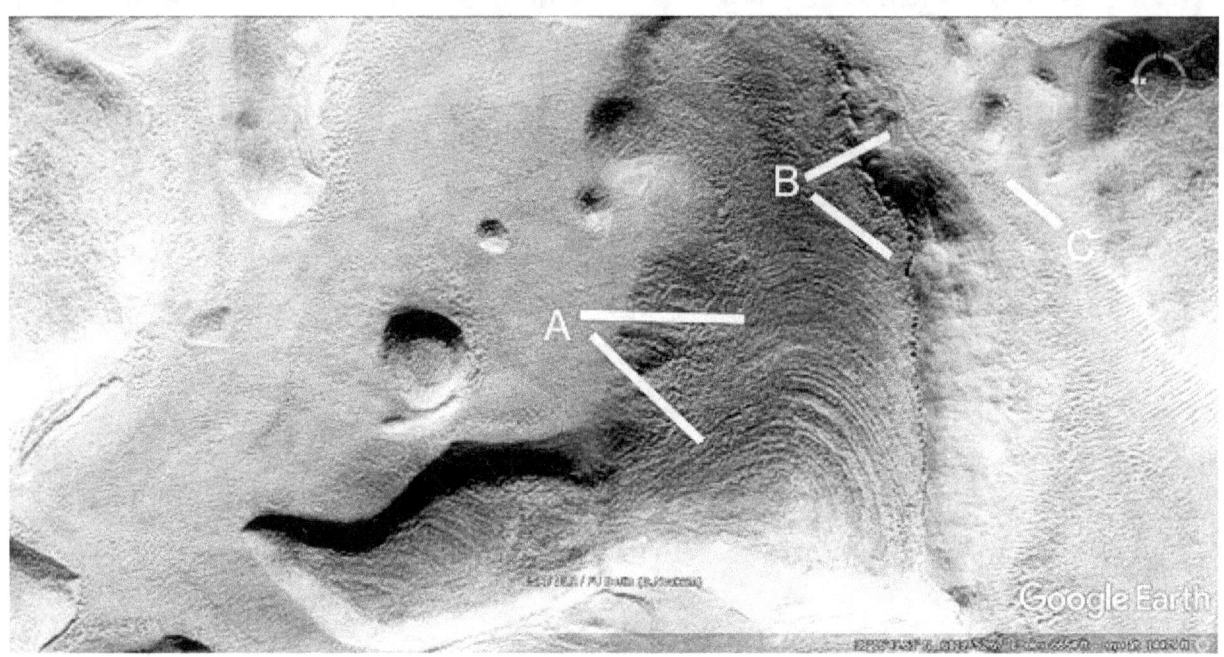

Prhh946a

Hypothesis

Five parabolas are shown, several more are too faint to include.

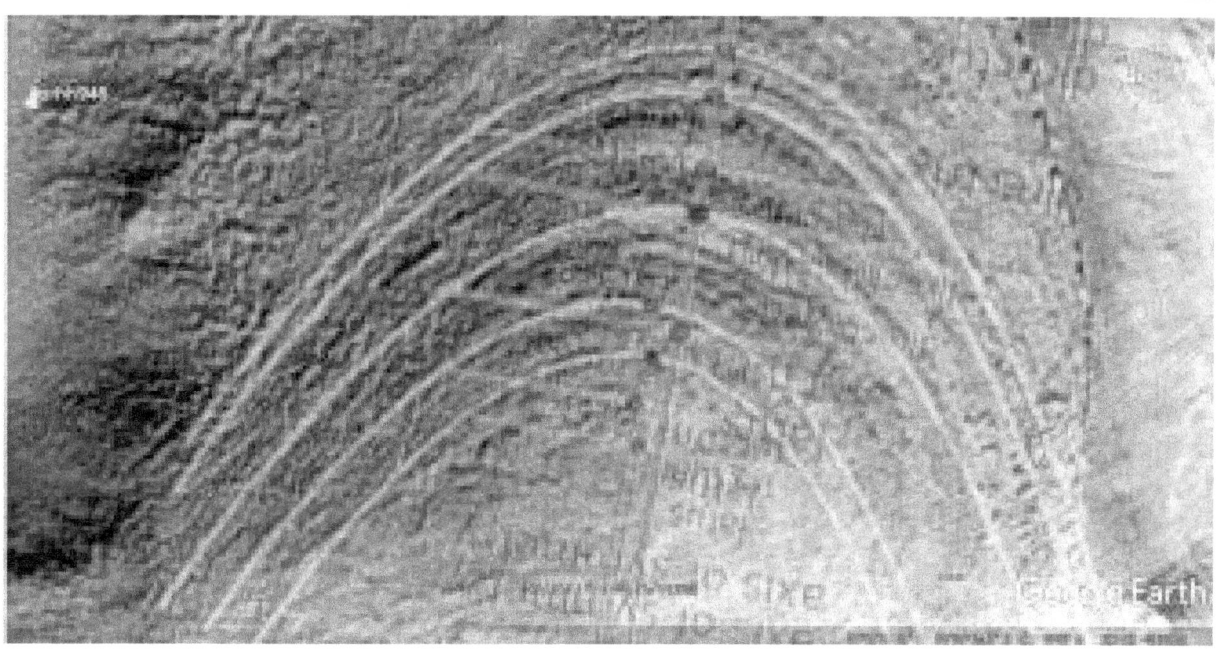

Prhh949

Hypothesis

An unusually shaped roof close to an ellipse. A at 4 o'clock shows a tube going from the hill below up to a tube mesh at 1 o'clock. At 8 o'clock the roof is settling. B shows arcs in the roof like a Cobler Dome or Amphitheatre seen in many other areas. C shows tubes going down the side of the hill. D shows rounded segments like tiles or patches from 7 to 4 o'clock, at to o'clock there is a tube. E shows another tube at 8 o'clock and a small hill connected by a tube at 7 o'clock.

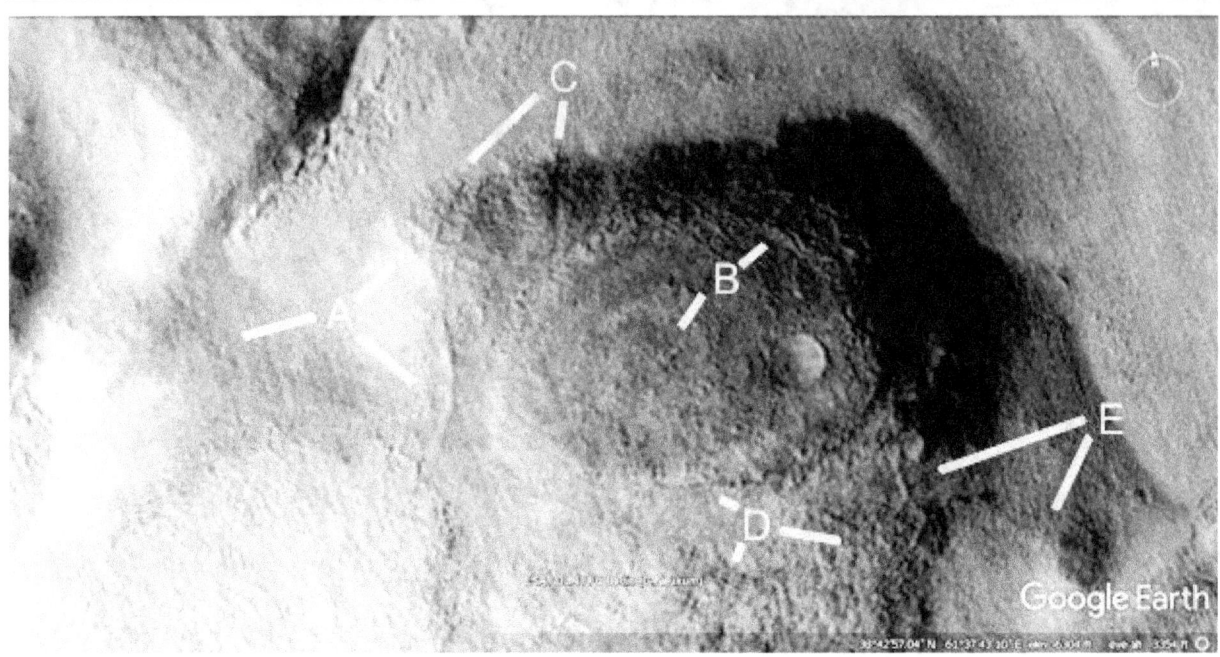

Prhh951

Hypothesis

A shows layers in the hill, B shows more like a Cobler Dome. The side of the hill is smooth like cement, perhaps the roof skin has peeled off. A at 6 o'clock shows these layers going into the crater implying these were repaired after the impact, or this is a dam not a crater.

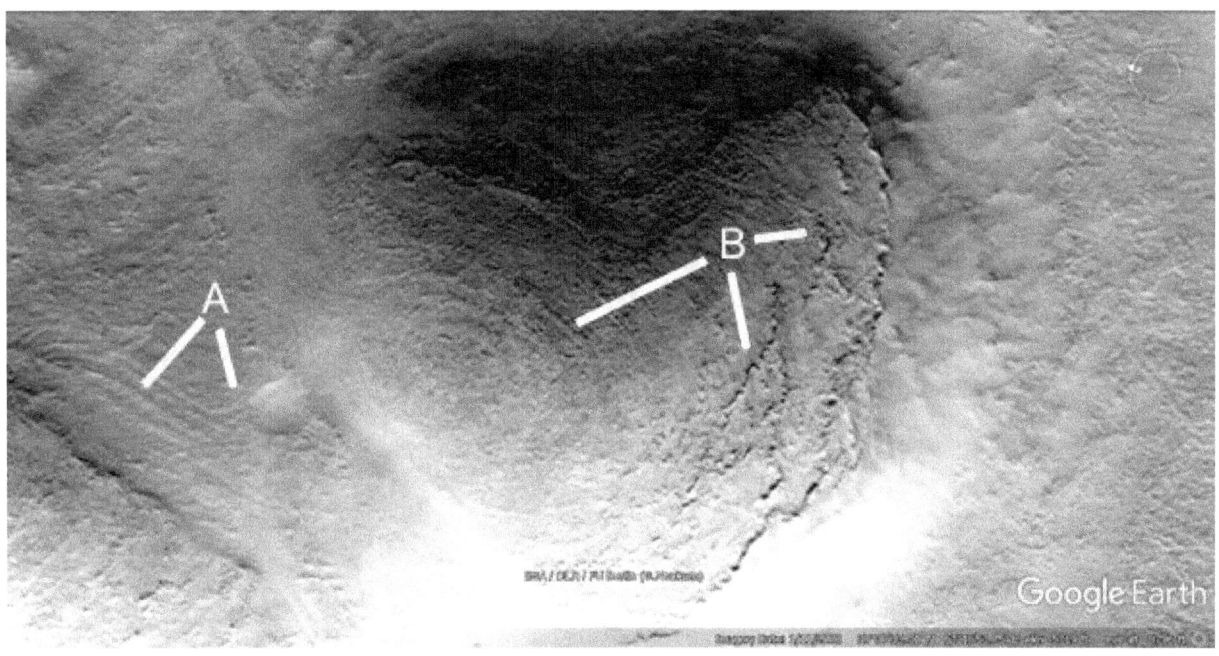

Prhh951a

Hypothesis

This shows 3 parabolas and 2 ellipses.

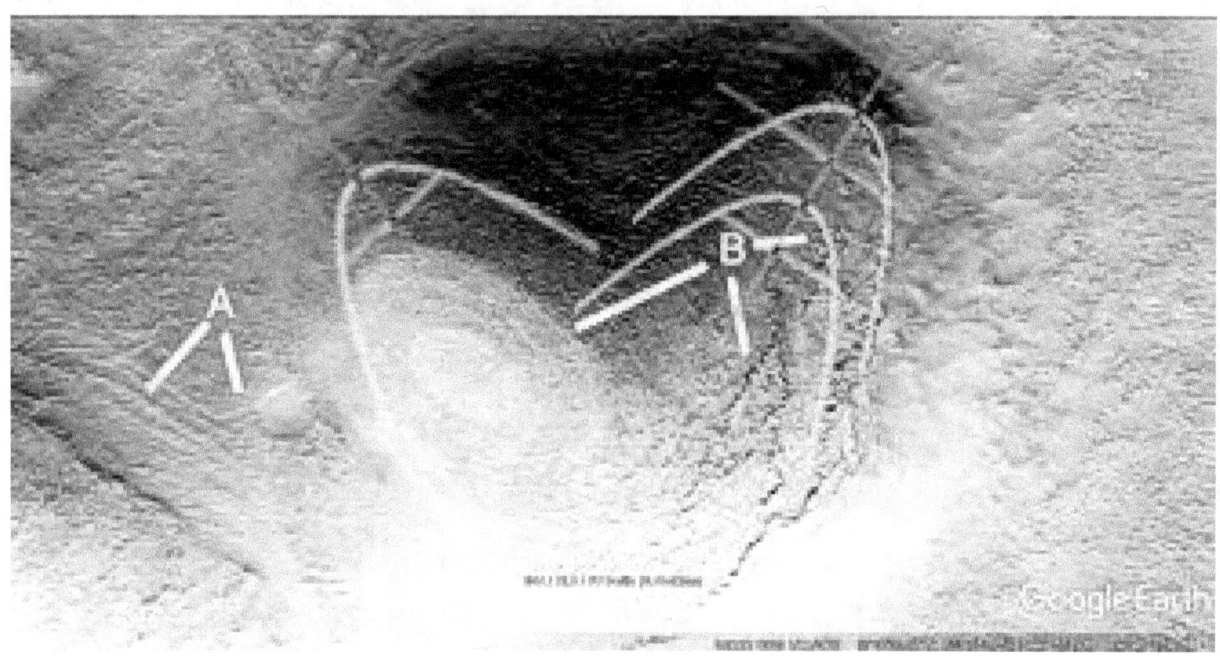

Prhh952

Hypothesis

The tube at B is quite straight going into a collapsed hill. A shows another straight tube at 4 to 7 o'clock.

Prhh956

Hypothesis

A shows two collapsed hills, B at 12 o'clock has its roof intact. At 5 and 3 o'clock the hills have collapsed, 3 o'clock might also be a rectangular dam. C at 8 o'clock shows a right angled walled formation. C at 4 and 5 o'clock shows a layer like a step in a dome. This might then be a Cobler Dome but with three layers. D then would be the upper layer at 3 o'clock, at 7 o'clock is a collapsed hill. E at 4 to 2 o'clock shows a smooth edge to this layer like cement. At 7 to 9 o'clock this is much rougher like the cement skin is peeling off.

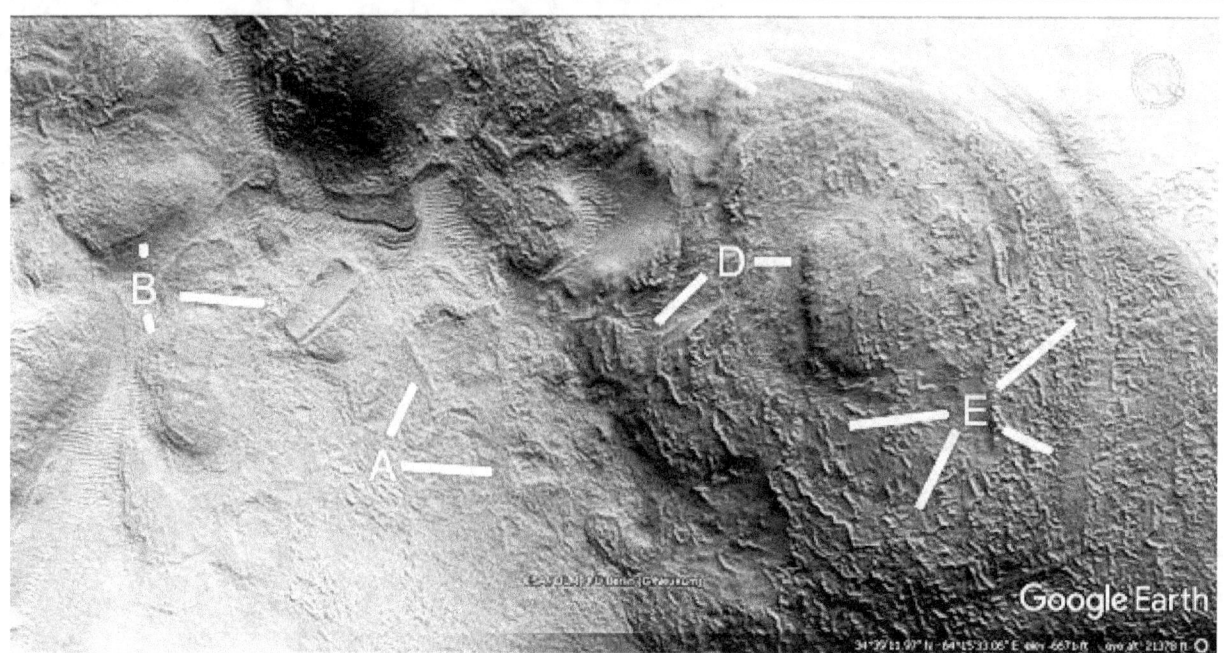

Prhh956a

Hypothesis

A parabola is shown.

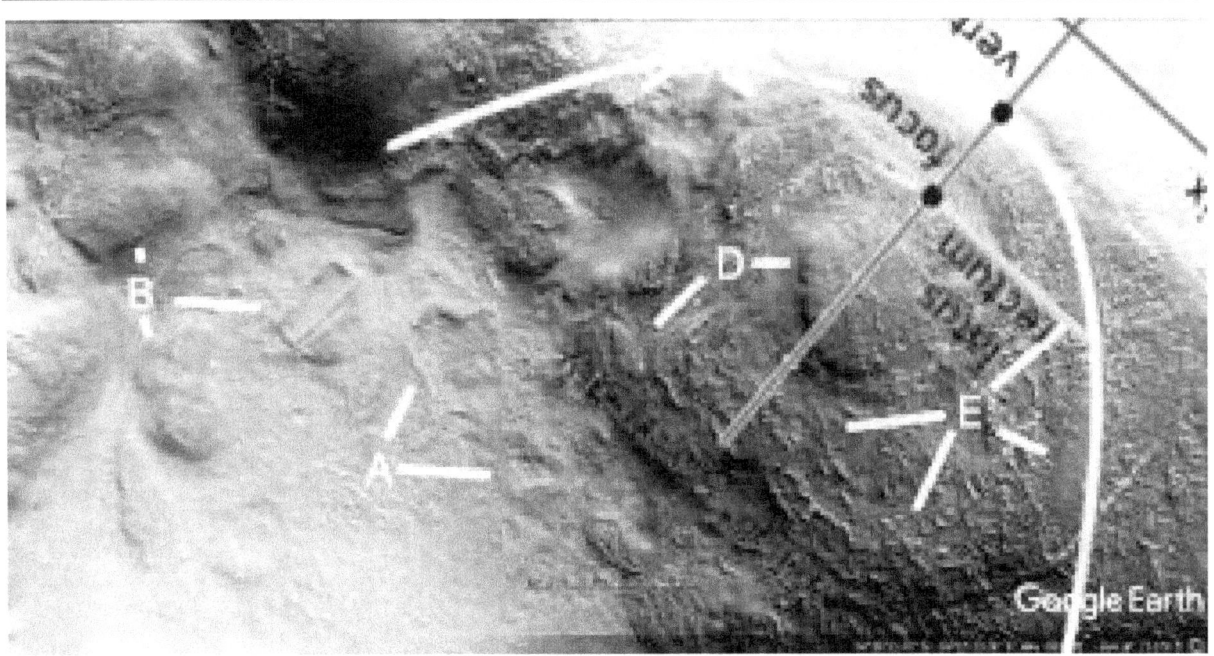

Prhh958b

Hypothesis

A shows where the hill roof is settling, also a wall like an interior support from 3 to 5 o'clock. B shows a settled roof at 7 o'clock, at 4 o'clock it is like a collapsed tube. C shows the edge of the smooth cement skin, also the edge of a patched roof segment at 12 o'clock second leg. D also shows a smooth cement side and a rougher segment of the roof.

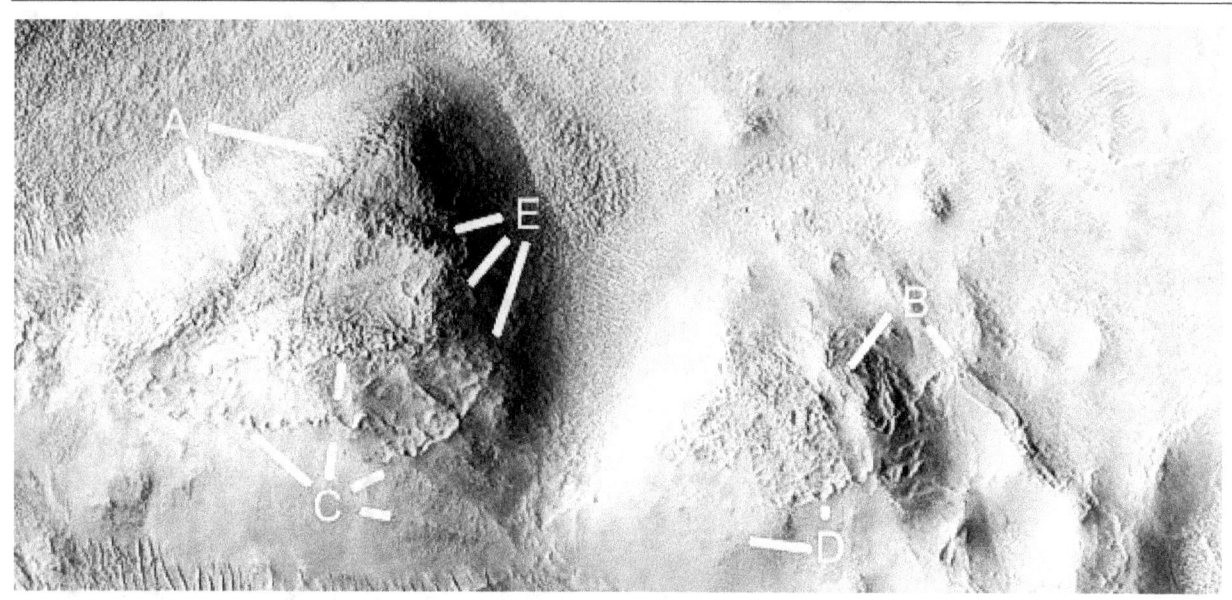

Prhh958b2

Hypothesis

A parabola is shown.

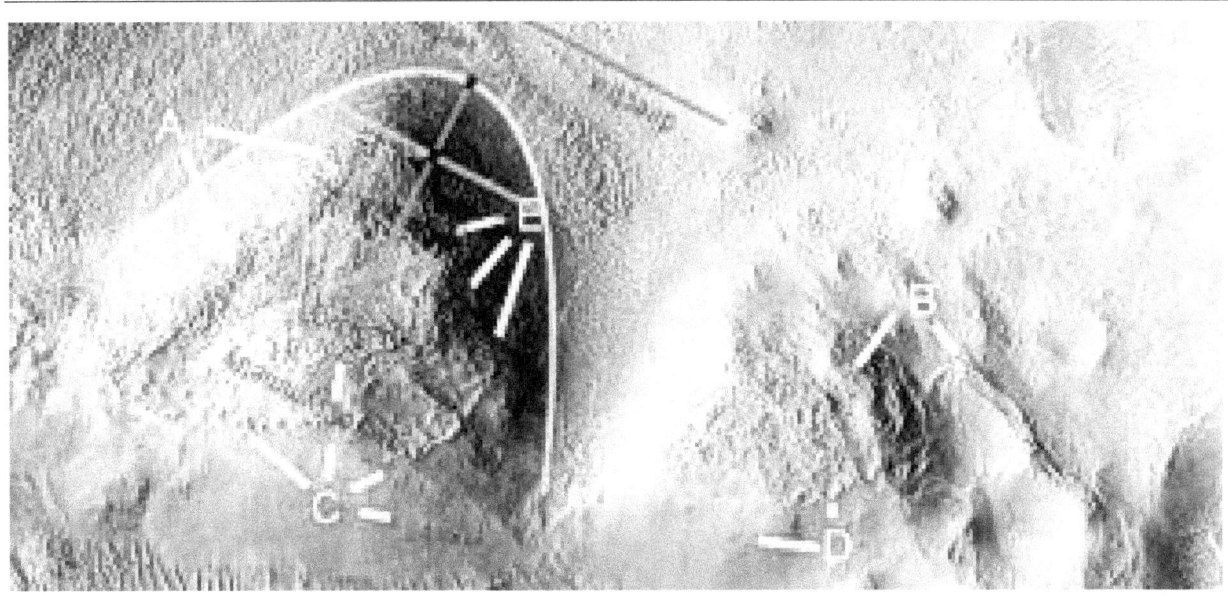

Prhh958c

Hypothesis

There is a squarish cavity between 4 and 7 o'clock at A, as if the roof has settled. From 7 to 8 o'clock there may be another squarish segment.

Prhh960

Hypothesis

A is a straight wall going into a hollow hill.

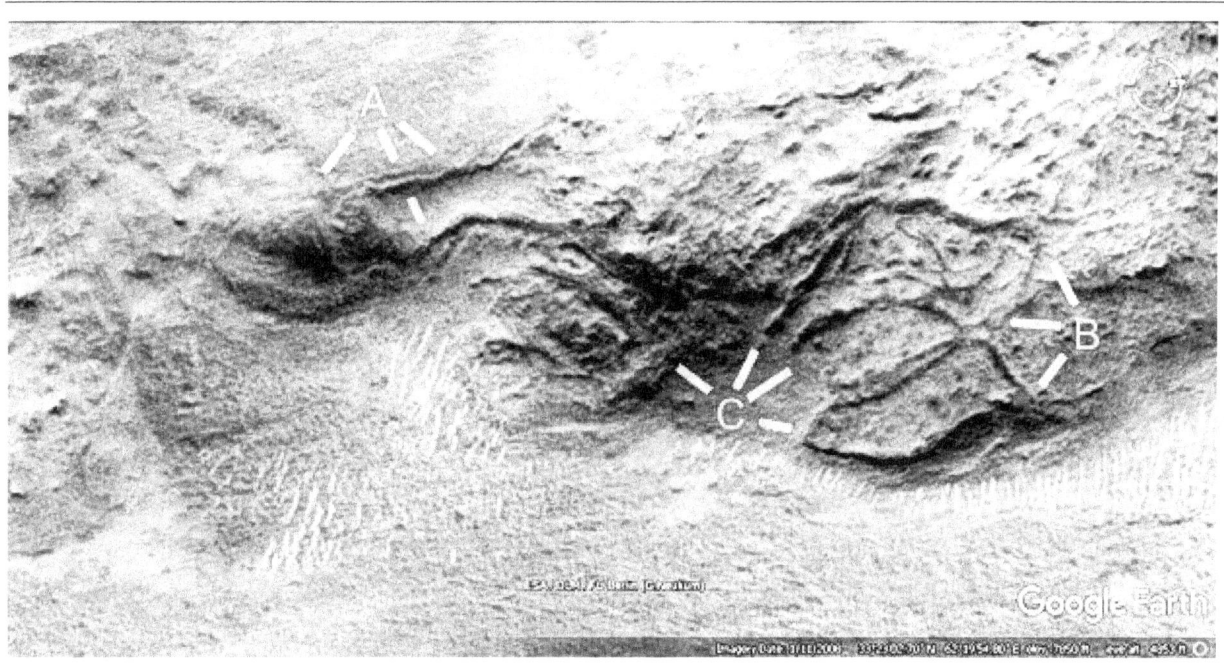

Prhh960a

Hypothesis

A parabola is shown between B and C, also 3 straight segments.

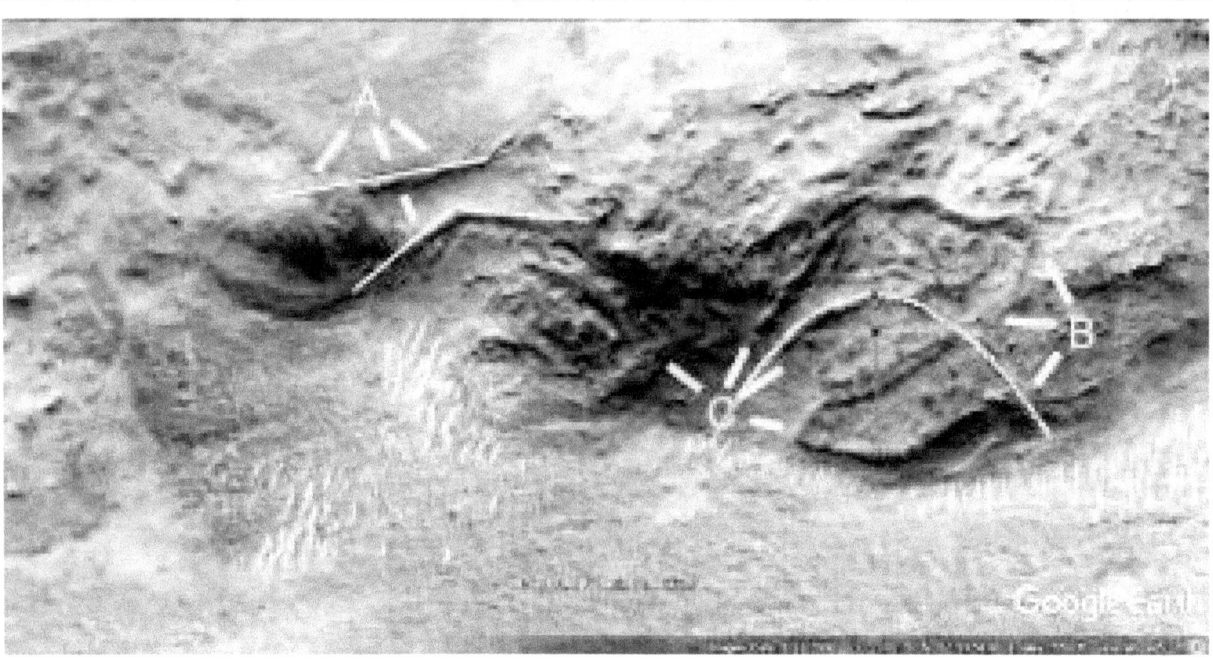

Prhh961

Hypothesis

A shows an exposed roof of a hollow hill at 9 o'clock, also at 3 o'clock, a collapsed roof at 5 o'clock, at 7 and 8 o'clock a hill may have collapsed with 8 o'clock being a remaining segment of the roof. B shows settled part of the roof of two hills. C, D, E, and F show more settled roofs.

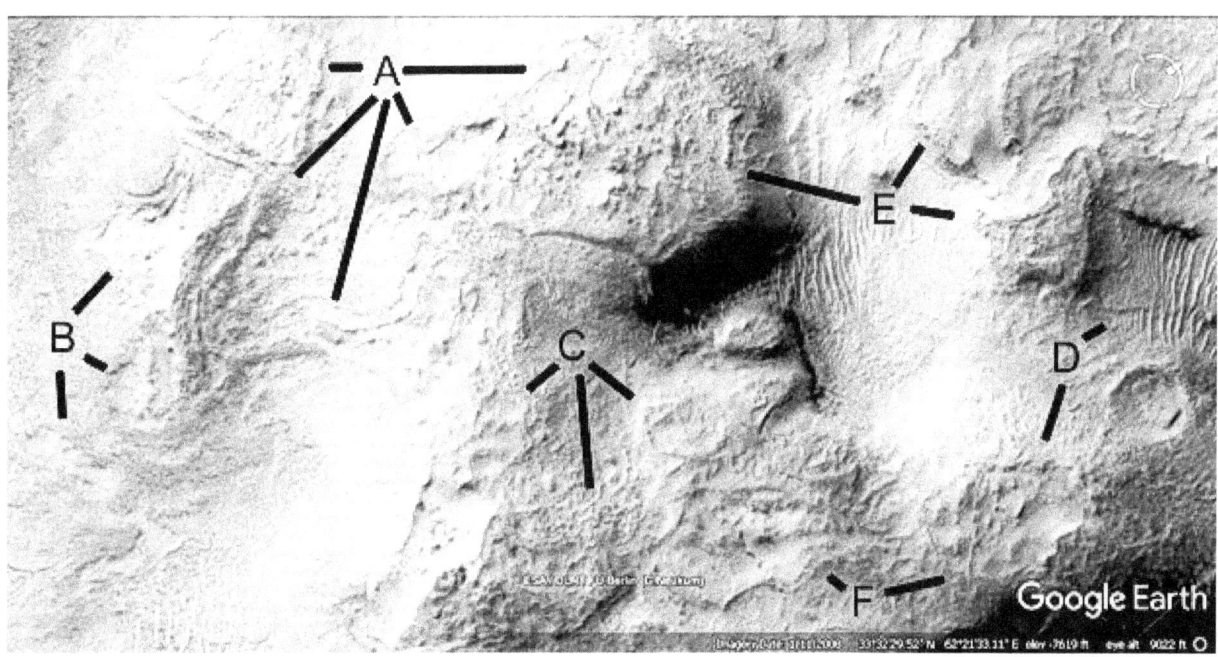

Prhh961a

Hypothesis

A parabola is shown.

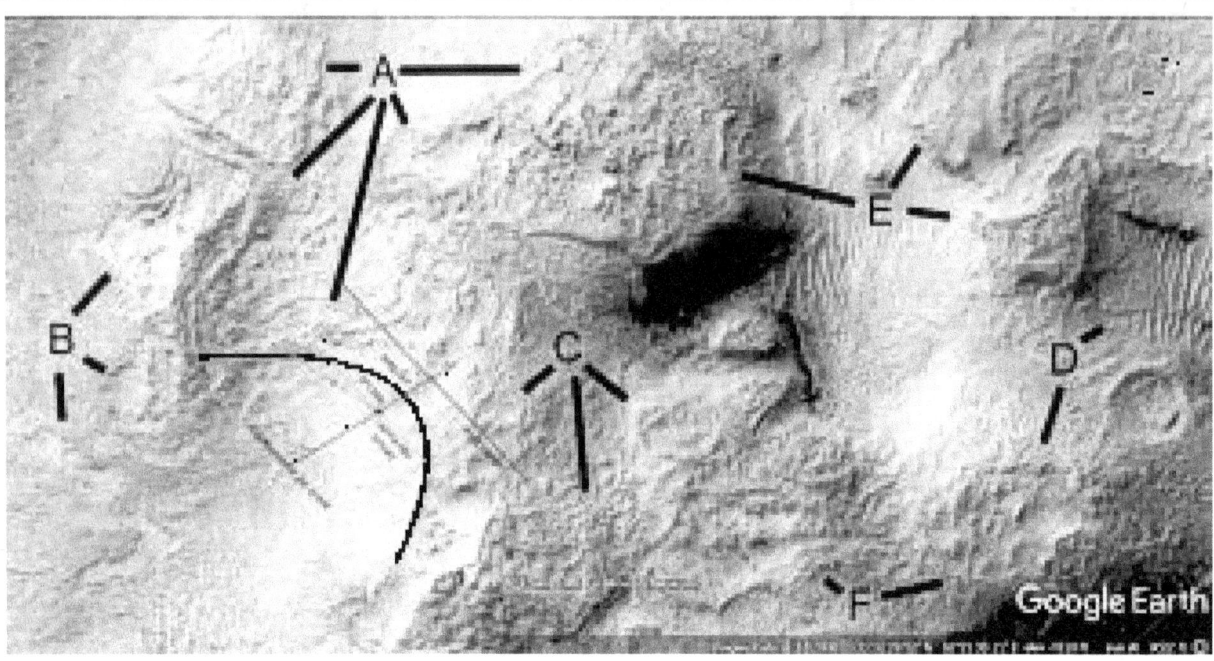

Prd965b

Hypothesis

A shows the double wall of a pit dam at 8 o'clock indicating it is degrading, at 4 o'clock it retains a sharp edge. B shows a cracked wall at 7 o'clock, another at 11 and 1 o'clock. C also shows signs of a cracked top of the wall.

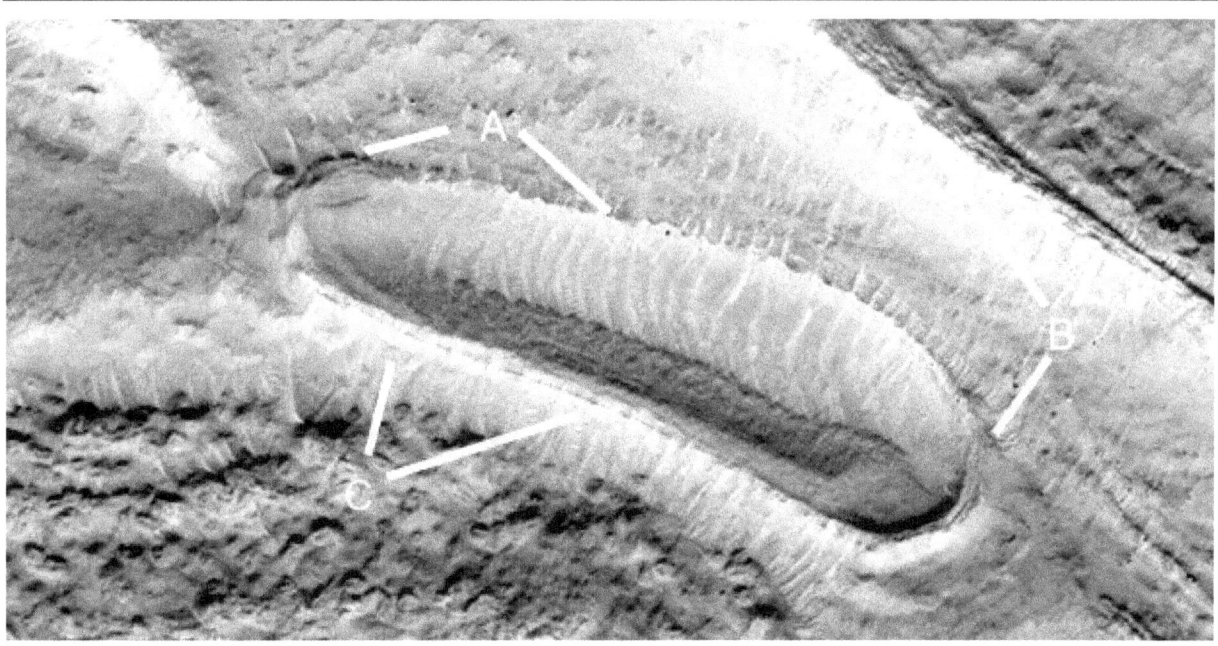

Prd965b2

Hypothesis

Two parabolas are shown as well as 3 straight sections of the pit dam wall.

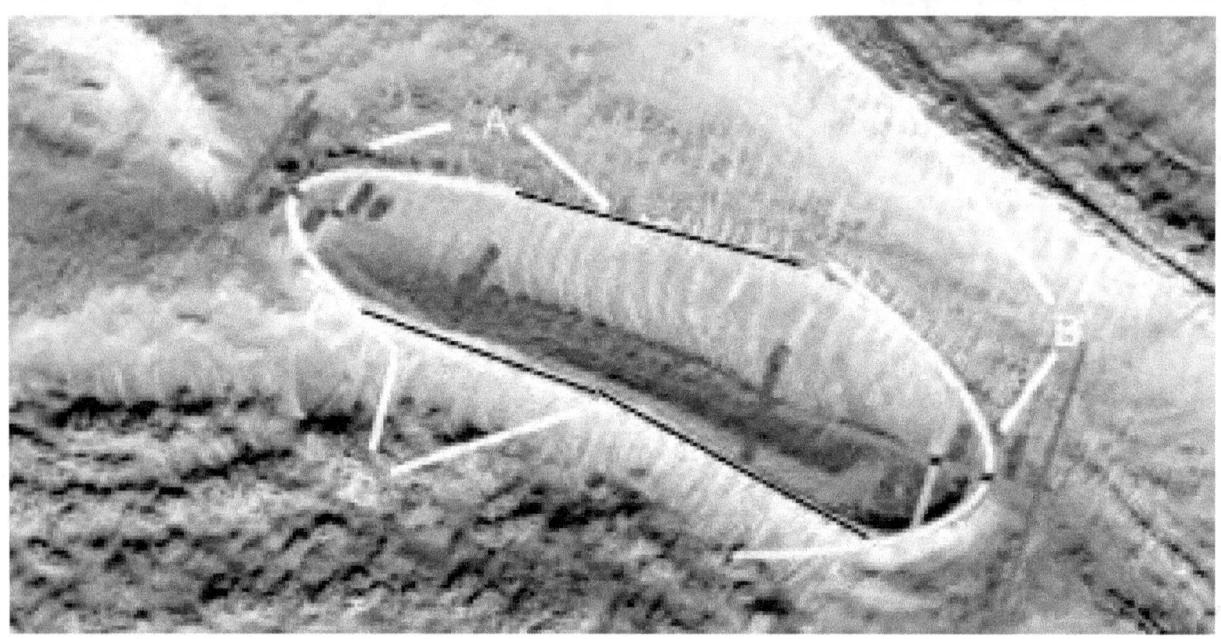

Prd965c

Hypothesis

These may have been canals or pit dams, they are highly geometric in shape. A shows a dam for water at 12 o'clock, another wall for a dam and channel at 3 to 5 o'clock. B shows a wall for a canal from 2 to 7 o'clock, it has a groove running along the top like a double wall.

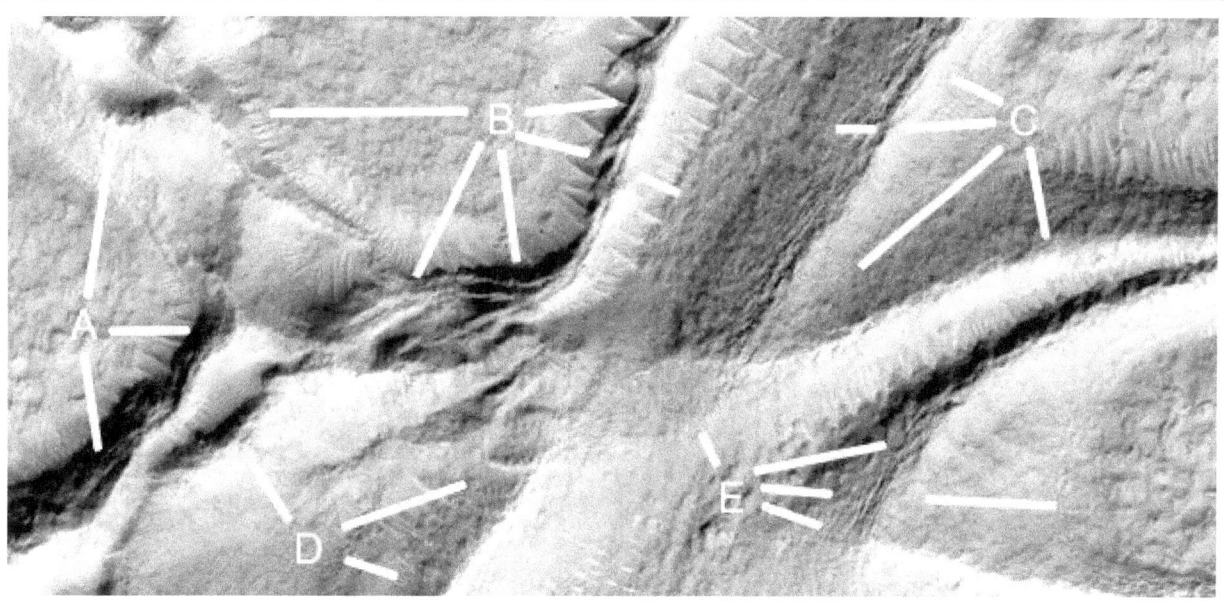

Prd965c2

Hypothesis

Part of a parabola is shown. The lines show how straight parts of the formation are.

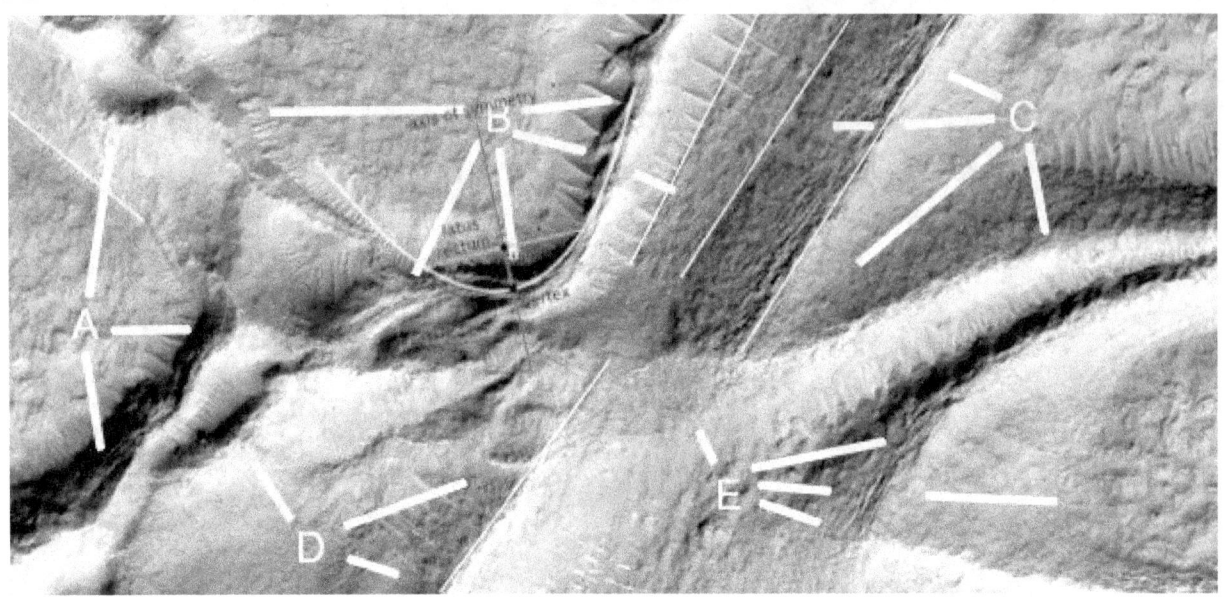

Prd965d

Hypothesis

A shows another double wall, the groove along it appears to be the same width and depth all along it. Between B and C the walls are quite straight, around B the inclines are fairly constant not randomly changing.

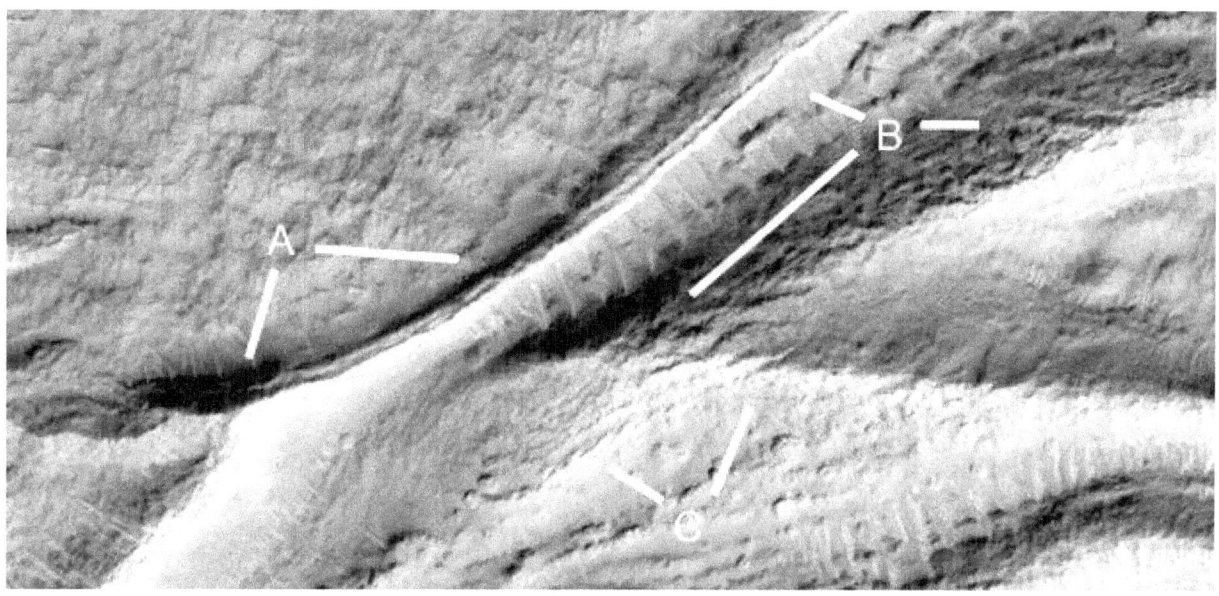

Prd965g

The eye like shape at A at 4 o'clock is unusual, B shows a pit dam with straight walls. C shows more straight walls as does D. E shows another pit dam.

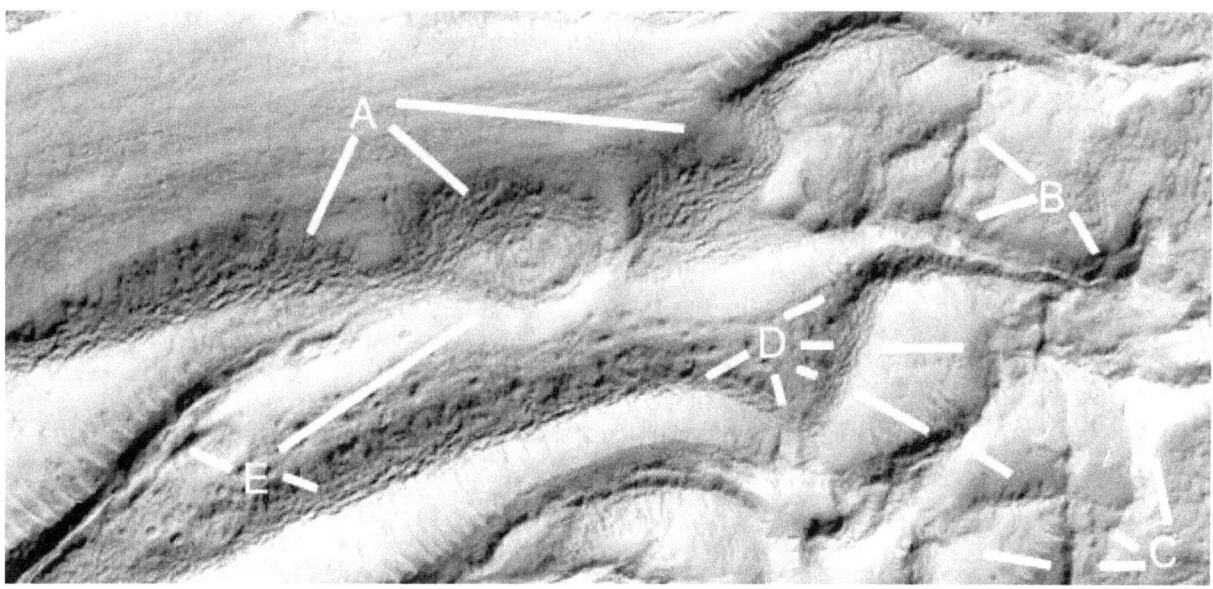

Prd965g2

Hypothesis

The lines show how straight the parts of the formation are. Also there is a right angled triangle close to the ratio 3:4:5.

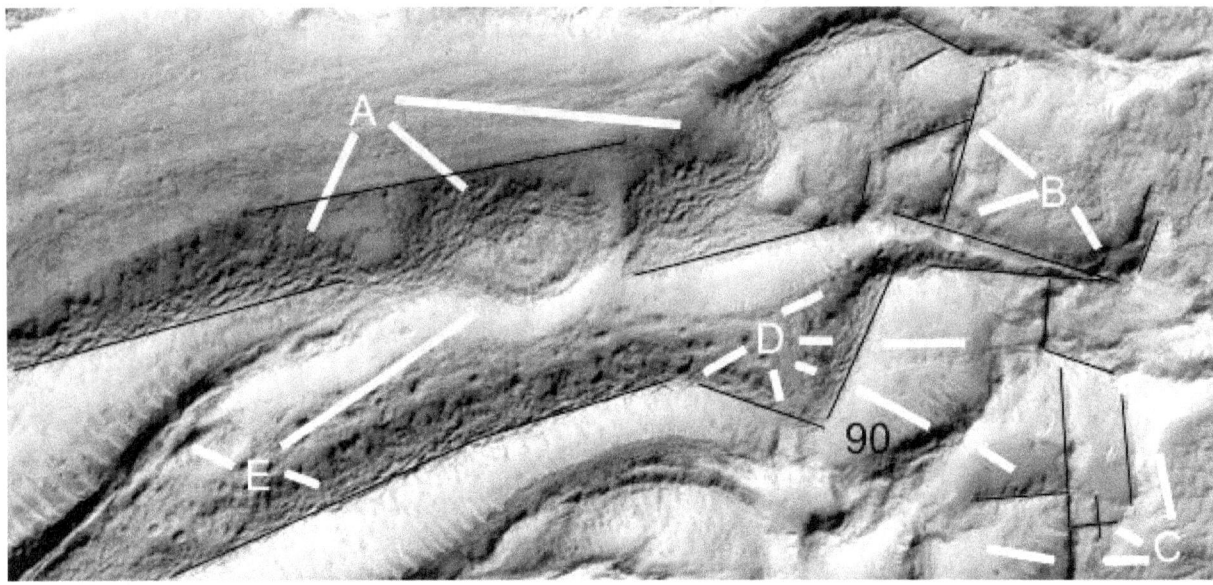

Prt978

Hypothesis

More pit dams are shown, water may have accumulated here from rainfall or from the water table.

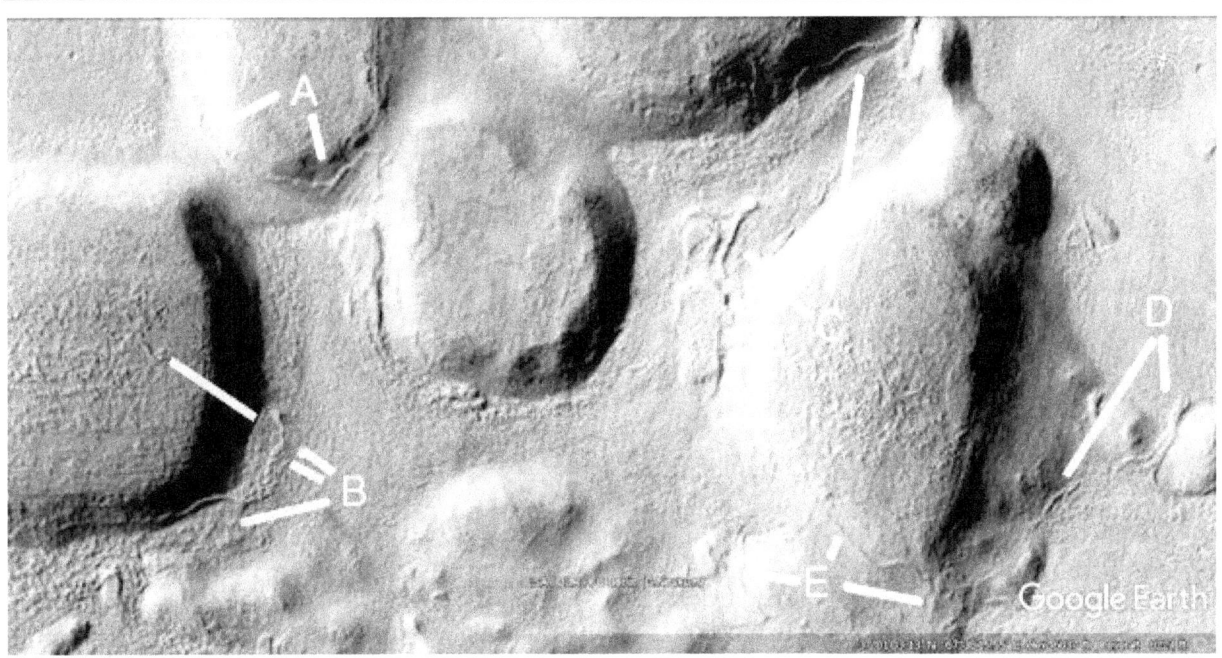

Prt978a

Hypothesis

Three parabolas are shown.

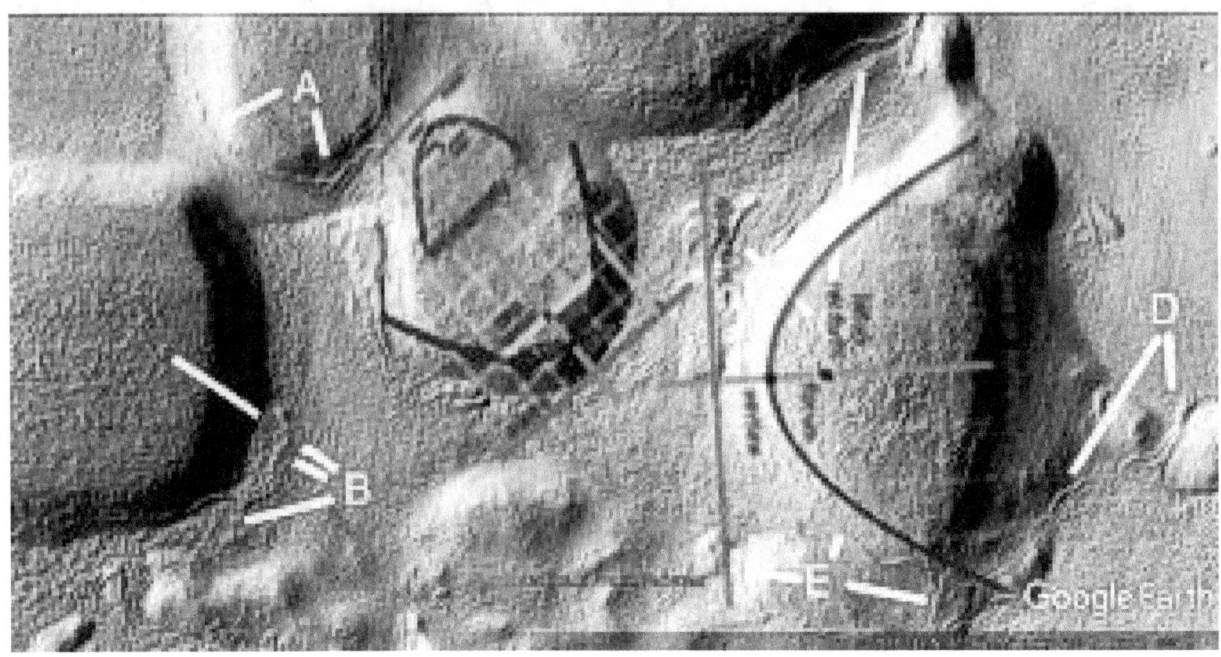

Prt986

Hypothesis

This appears to be a hollow hill, the roof has some straight segments perhaps as interior supports showing through.

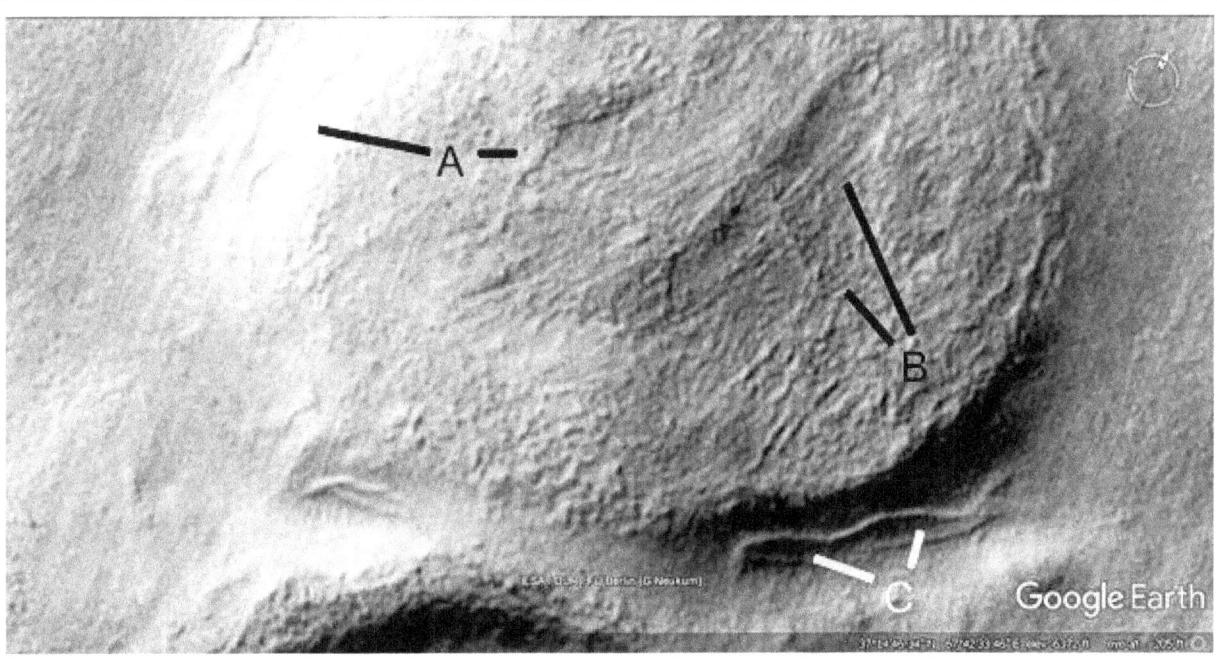

Prt986a

Hypothesis

A parabola forms one side of the hill.

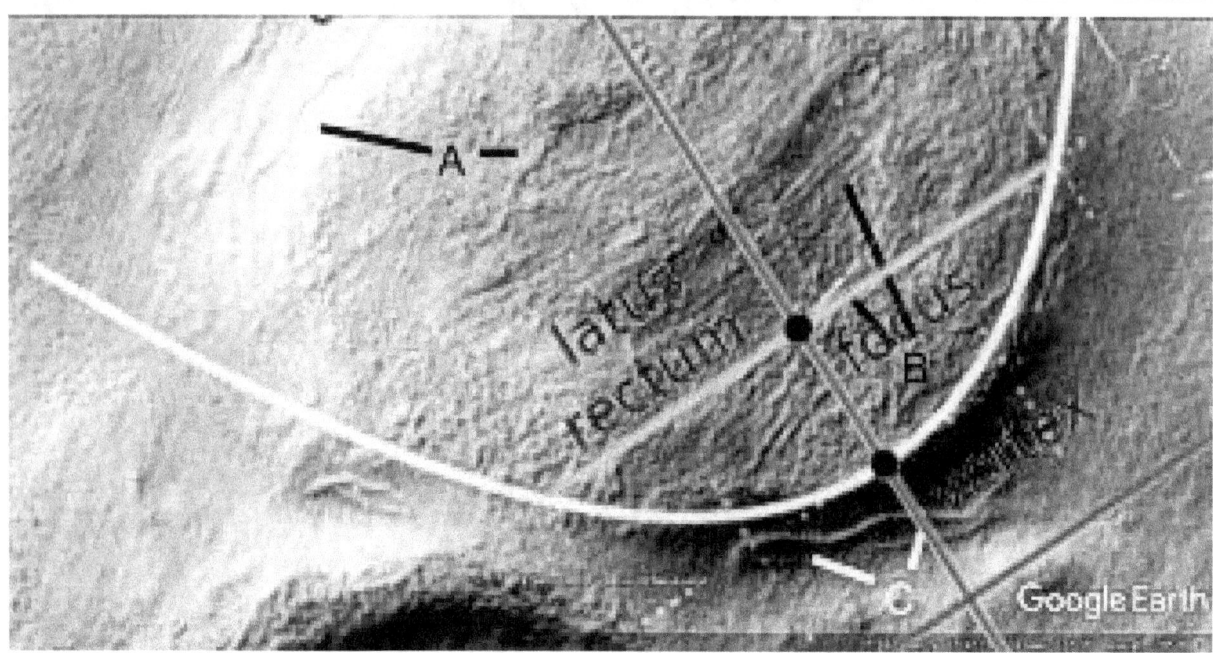

Prt987

Hypothesis

A shows a tube shape on the edge of the hollow hill, B may have been a collapsed hill.

Prt987a

Hypothesis

The hill's shape is bordered by two parabolas.

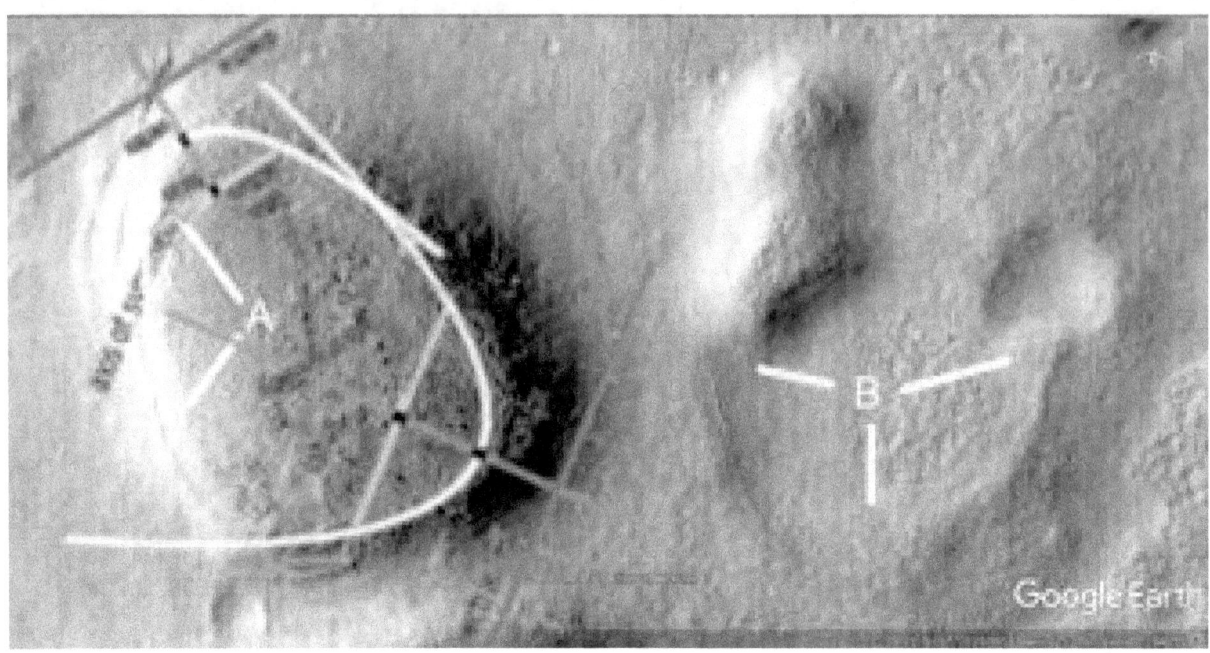

Prt989

Hypothesis

A shows more possible settling on the roof of the hollow hill. B shows a tube or a small walled dam, it goes into a small hill at 1 o'clock. At 11 o'clock there is a collapsed part of the hill or a dam, bordered by a wall at D. C shows another cavity or collapsed segment between 1 and 4 o'clock.

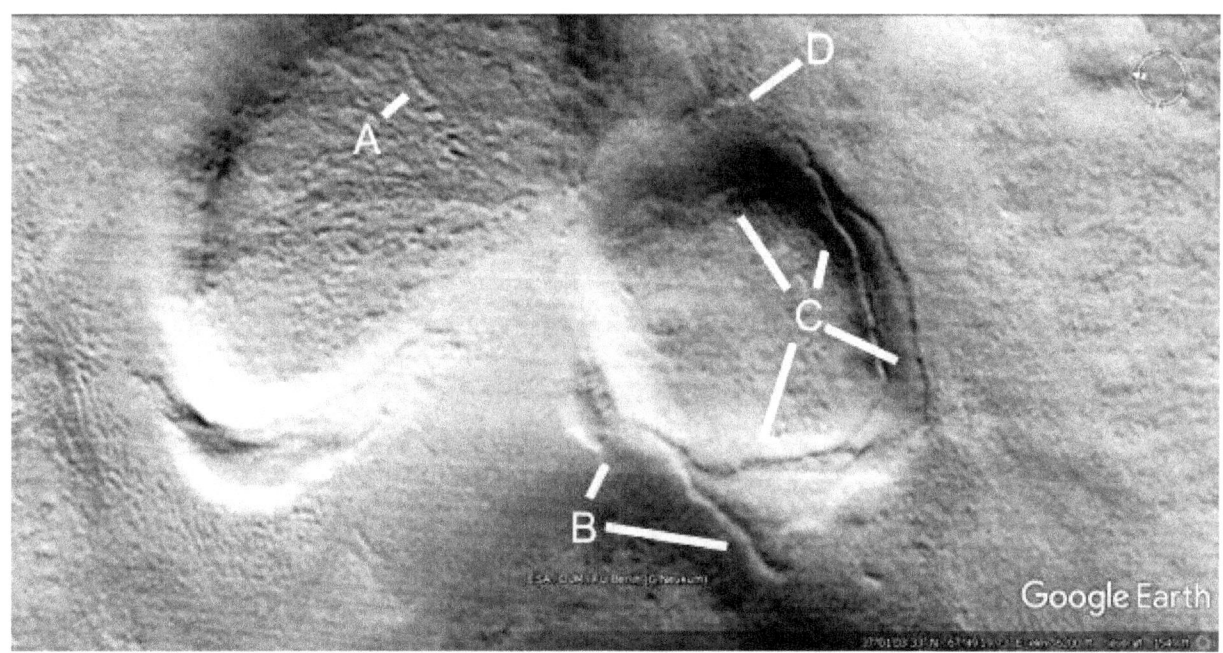

Prt989a

Hypothesis

A parabola forms part of the hill shape.

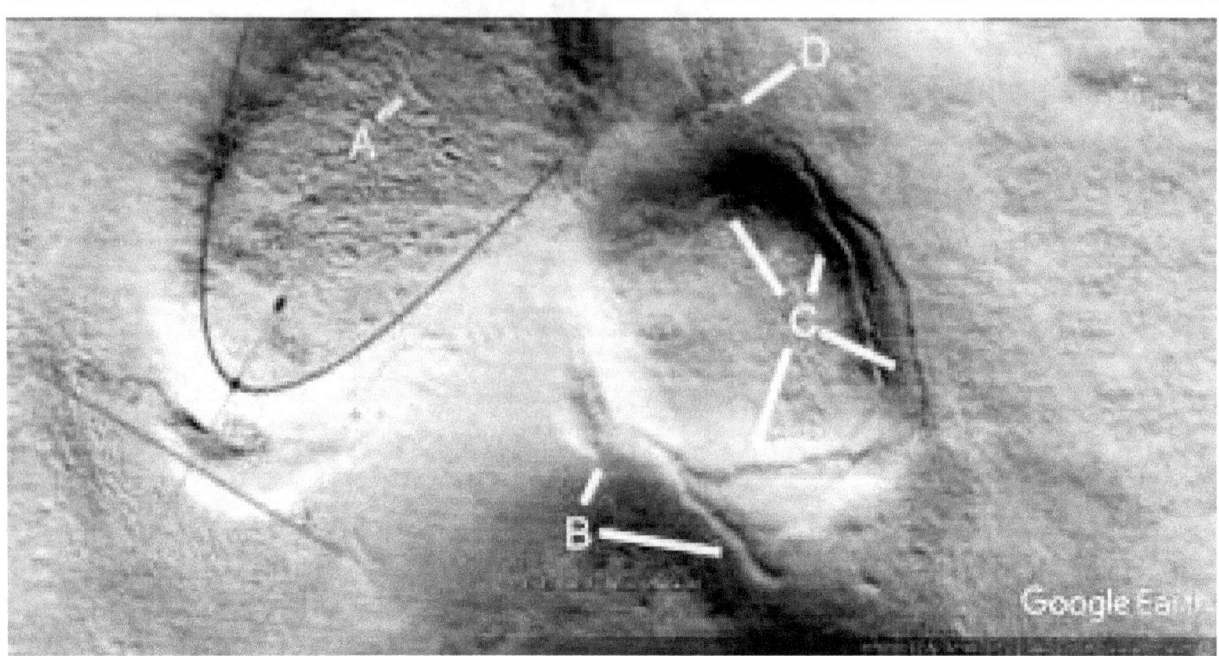

Prhh994

Hypothesis

A at 8 o' clock shows a wall on the edge of the hill, at 5 o'clock may be interior supports of a collapsed hill. B may be another collapsed hill, at 5 o'clock the sides are very straight. At 2 o'clock this may lead to another collapsed hill over to C at 5 o'clock. At 4 o'clock is another collapsed hill. D at 1 o'clock shows a collapsed hill with the outside walls intact. At 11 o'clock there are a trapezoidal shape perhaps interior supports.

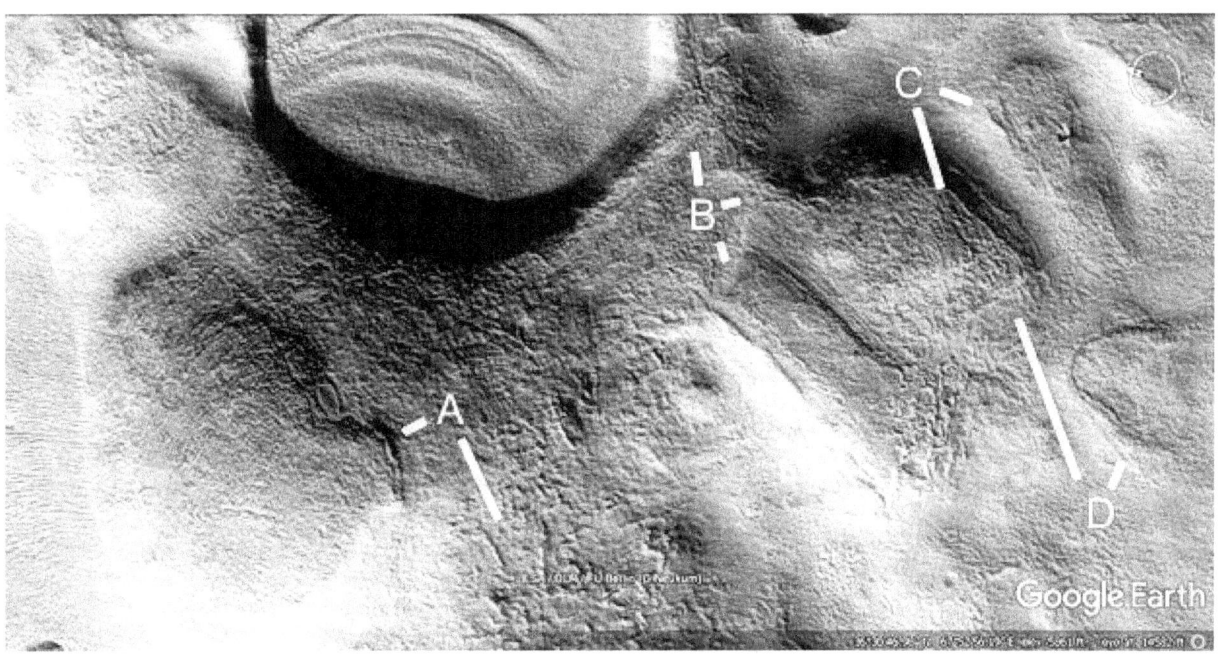

Prhh994a

Hypothesis

The lines show straight parts of the formations.

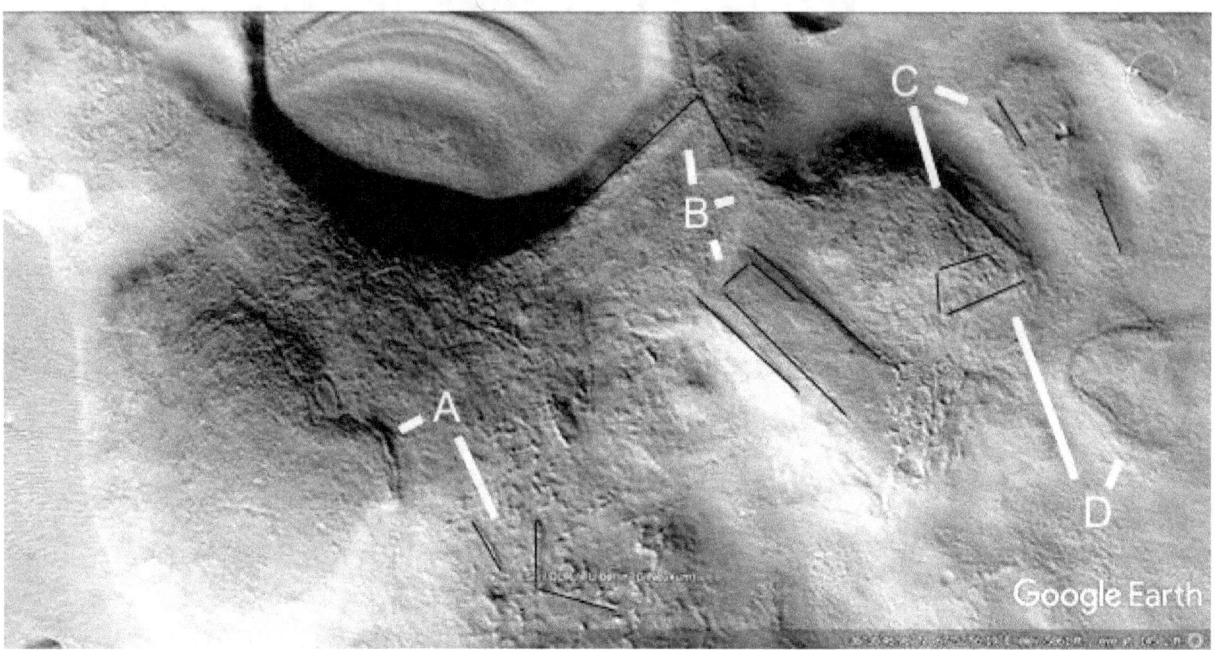

Prt998

Hypothesis

A may show the roof the hollow hill has settled or has been repaired. B shows a tube coming from the hill, C shows the skin may have peeled off the roof of the hill. D may show the edge of the peeled skin of the roof of another hollow hill.

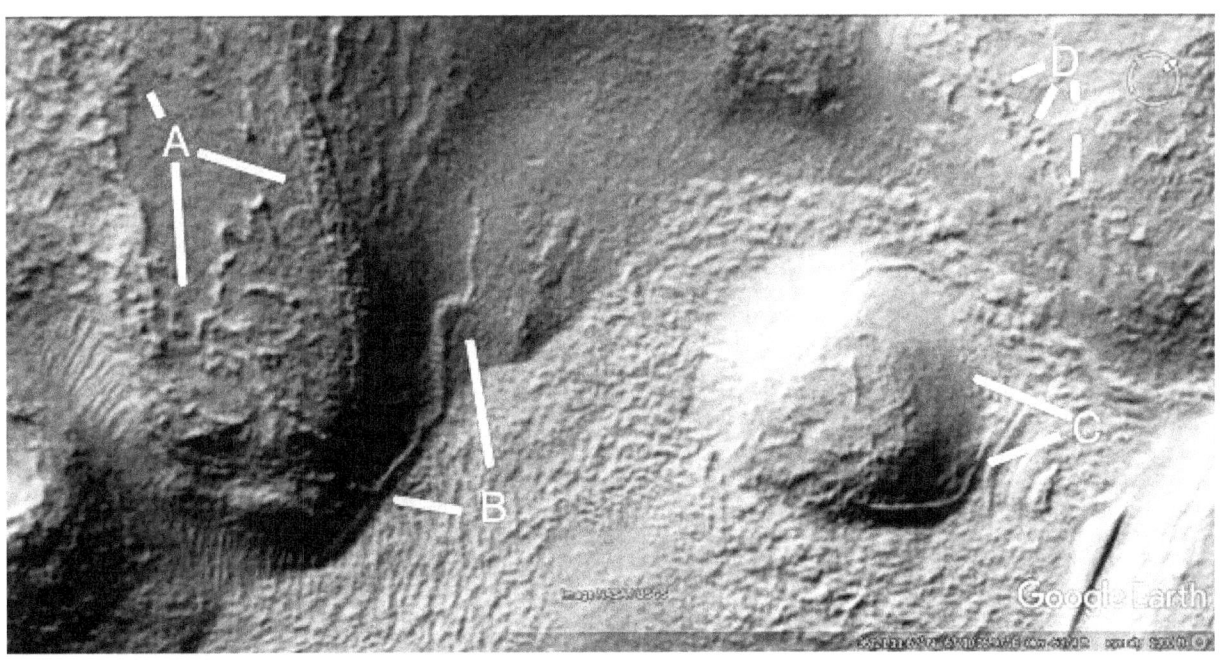

Prt998a

Hypothesis

Part of the hill is shaped as a parabola.

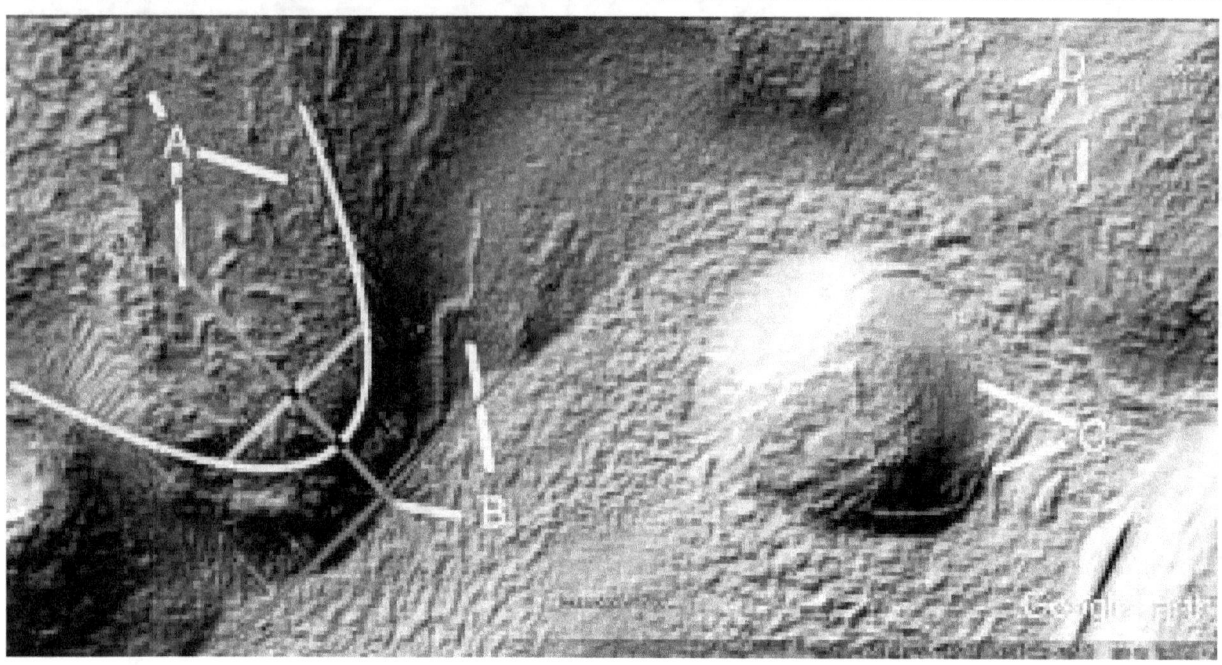

Prhh1000

Hypothesis

A shows a tube at 9 o'clock, another at 7 o'clock going to a crater. At 8 o'clock the roof of the hollow hill may be settling. B shows some straight walls, C shows more walls. D may be a collapsed tunnel into a hollow hill. E may show repairs to the roof.

Prhh1000a

Hypothesis

The walls form a parabola.

Prhh1001

Hypothesis

A at 9 and 10 o'clock may be cement flaking off the roof. At 4 o'clock tubes and a settled area connect to the crater. It is no longer round as if it has been repaired. B shows more areas where the roof is settling. C, D, and E show the boundaries of a large area where the roof has settled. D at 6 o'clock shows a right angled segment.

Prhh1006

Hypothesis

A shows a segment of the roof standing out at 6 o'clock, at 8 o'clock is another roof partially exposed. 4 o'clock shows a tube or straight collapsed part of the roof under it. B at 12 o'clock shows a wall on the edge of this roof, at 10 o'clock the darker area may be a repair or exposing materials under the roof skin. At 2 o'clock is a degraded tube going to a smaller hill. C shows a collapsed tube or parallel tubes at 9 o'clock, at 7 o'clock are walled segments around a hill

Prhh1014

Hypothesis

These lines on the roof may be where the interior supports hold it up.

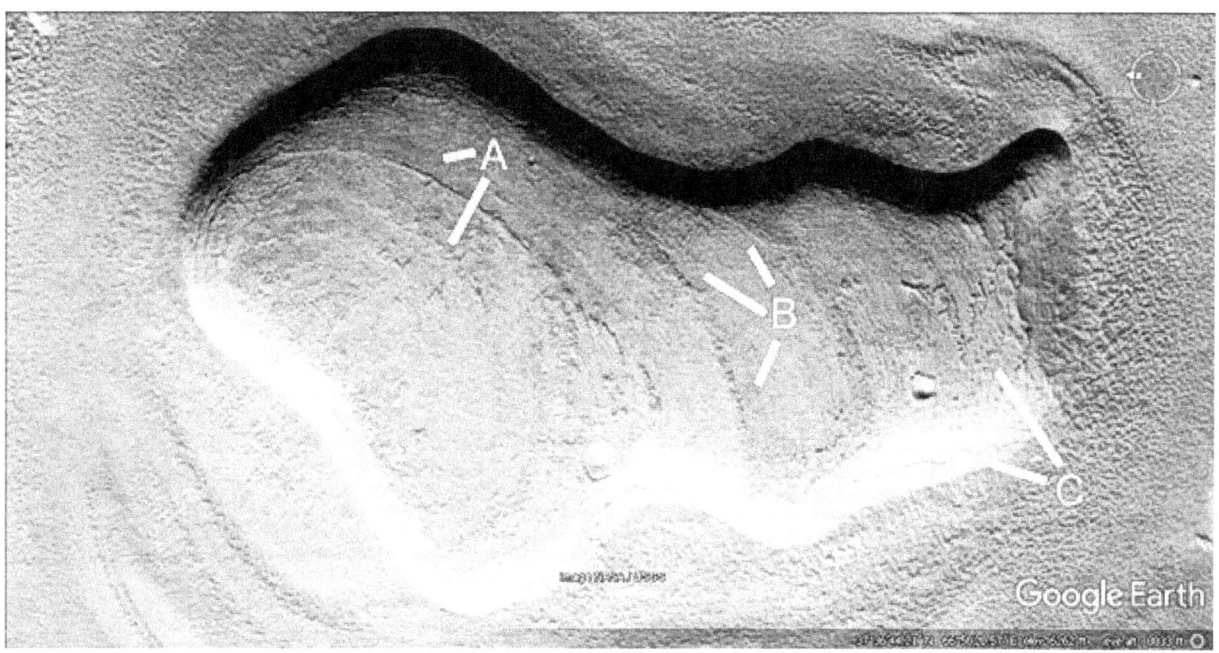

Prhh1014a

Hypothesis

The arcs on the roof line up with four parabolas as shown. The parabola would be the strongest curve for an interior support perhaps under these dark lines on the roof.

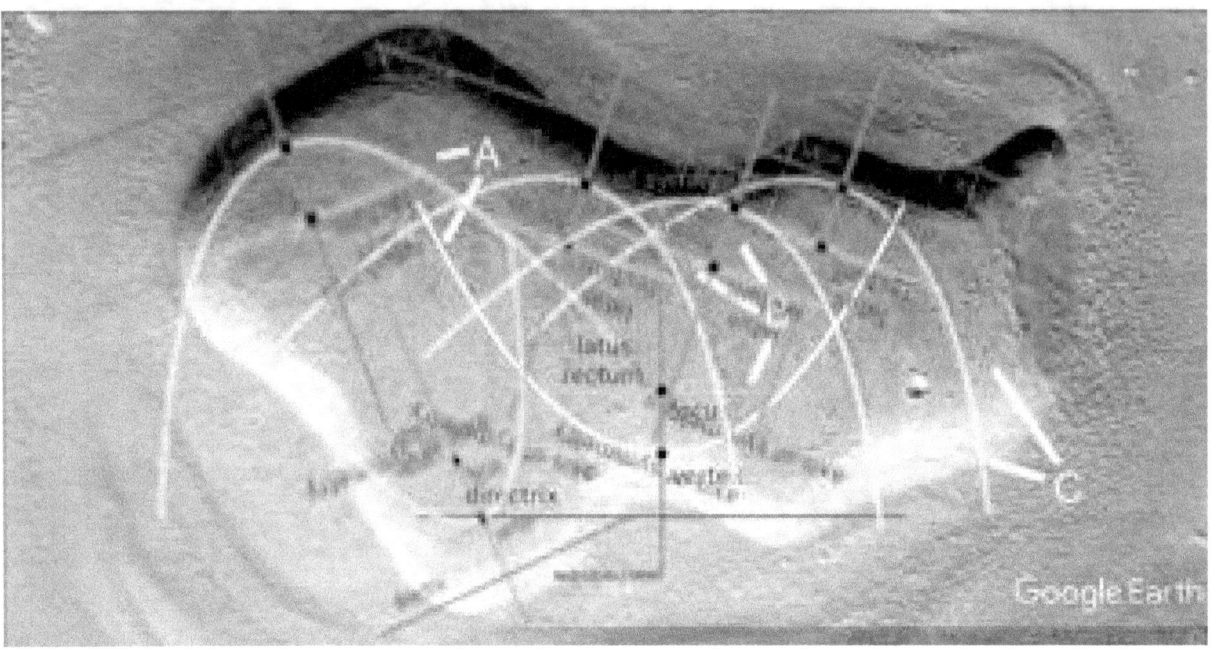

Prhh1015

Hypothesis

A and B appear to show settled roof segments on the hollow hill. The formation is also like a trapezoid with parallel straight sides.

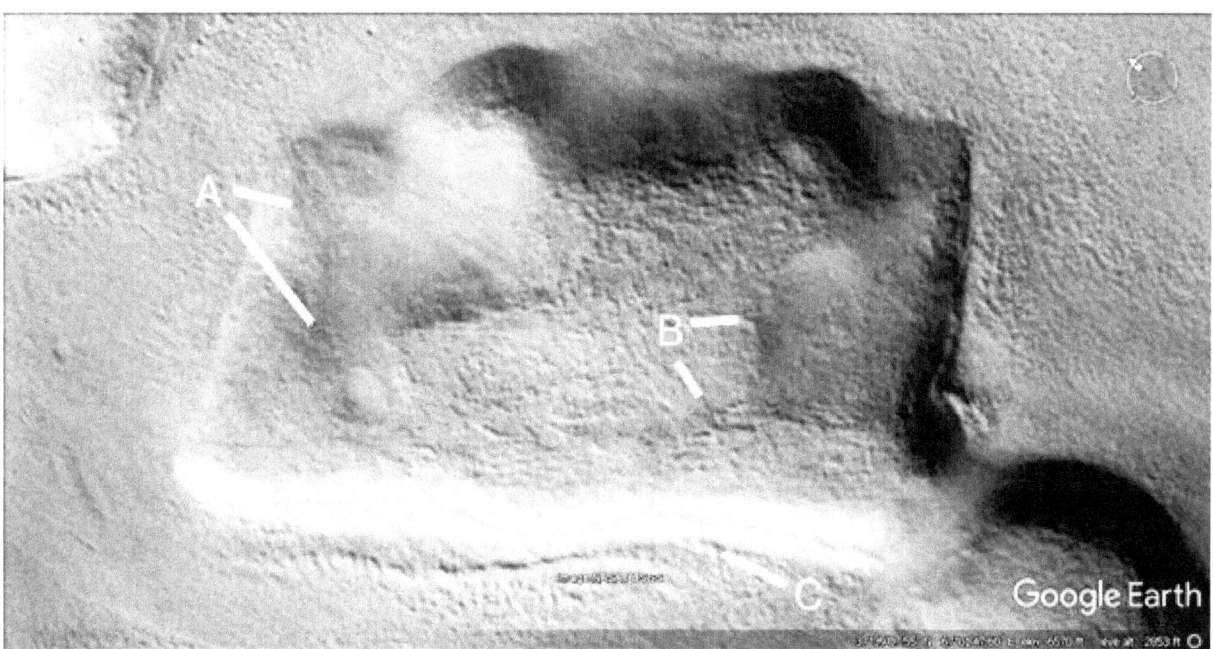

Prhh1018

Hypothesis

Many tubes come out of this formation, A at 8 and 9 o'clock shows a tube intersection. At 3 o'clock is another tube from the pit wall. B shows two more tubes, below the one at 4 o'clock are two small enclosures, also another two between there and C at 8 o'clock. These may all be dams including the large pits. C at 7 o'clock shows many faint tubes coming out of the pit wall. D at 9 o'clock shows the pit wall is doubled with a groove between them. At 5, 6, and 7 o'clock the pit wall is very even and rounded, at 3 o'clock is another tube coming out of the pit wall. E at 12 o'clock shows one of the pale formations inside the pit, these may have been hollow hills and have a similar albedo to parts of the pit walls. At 2 and 9 o'clock the pit wall gets thicker, this part has a roof like a tube but to the right and left it becomes a groove again. It's likely then most of these pit walls are hollow.

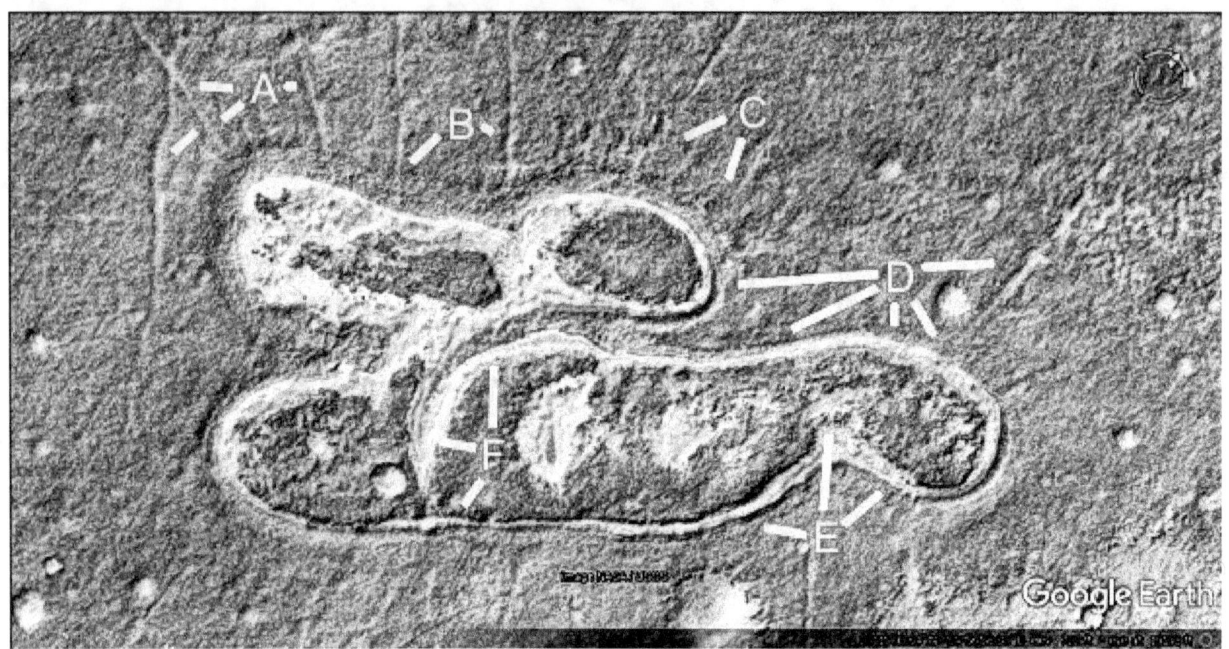

Prhh1018a

Hypothesis

The lines show how straight the tubes are. Also six parabolas are shown to fit onto the edges of the pit dams.

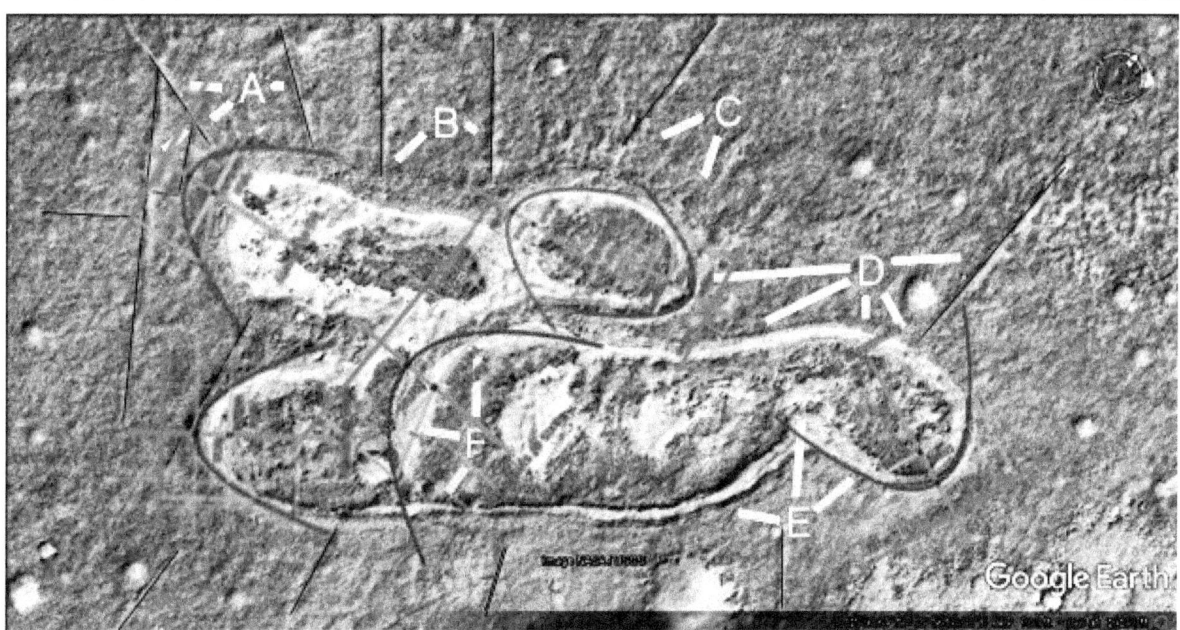

Prhh1019

Hypothesis

An unusual shape looking like a leaf, this is near the Ferns so it may be related to them. A and B show straight grooves parallel to each other, these connect to C which are also parallel to each other. The spacing between the grooves, the depth of the grooves and their overall shape remain consistent throughout the image. The groove at A appears to connect to a tube shape at F at 4 and 7 o'clock, which then ends near a tube shape at right angles to it. This closes the parallel lines at C. E shows other tubes on the formation, D shows there is a groove around the leaf shape with a consistent depth and albedo.

Prhh1019a

Hypothesis

This shows the leaf shape is formed by a double parabola, the tube at E at 2 o'clock divides the parabolic shape in half. F can be regarded as a chord or line which connects the two parabolas, this may have geometric significance. There are also shapes which may have defined the foci and a line between them on their respective axis of symmetries. If the parabolas were moved a small distance to the right they would still be consistent with the leaf shape and this may then connect the foci. The arrangement of the parallel lines and the double parabolas may mean this is some kind of geometric statement overall.

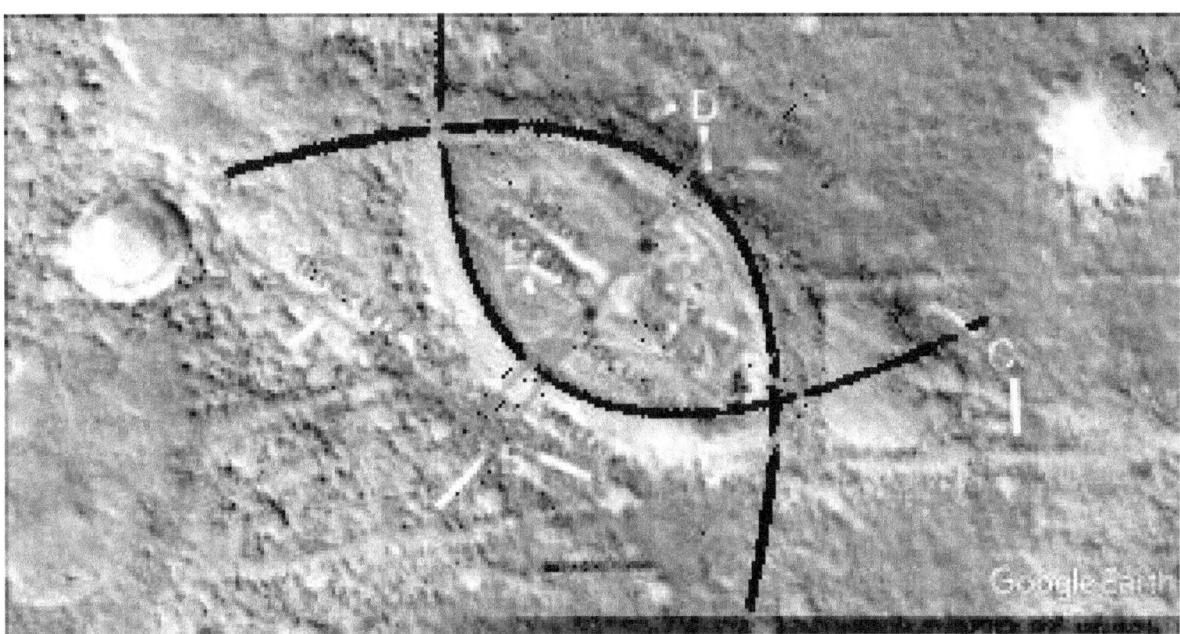

Prhh1021

Hypothesis

This shows an ellipse and approximate circle connected with grooves, A at 7 and 8 o'clock forms a tangent to the ellipse that intersects another tangent 4 and 7 o'clock. These tangents are close to the major and minor axis of the ellipse and again may be a geometric statement. A at 6 o'clock second leg shows another groove or road connecting to the circle. B shows the sides of this ellipse at 10 and 12 o'clock, at 2 and 4 o'clock there are many grooves which may have been tubes or roads.
C, D, E, and F show many more of these coming out of the circle, they do not seem to be a random chaos or series of cracks but come out of the circle perhaps as roads. G shows an inner wall of the circle at 7 and 11 o'clock, a similar wall for the formation at 2 o'clock. These may have been dams or for growing crops such as rice that required a shallow wall around them to hold in water. H and I show more of these walls and shallow pits. This is little evidence to show what these were for, but they do seem to have a special purpose which is likely to be food related.

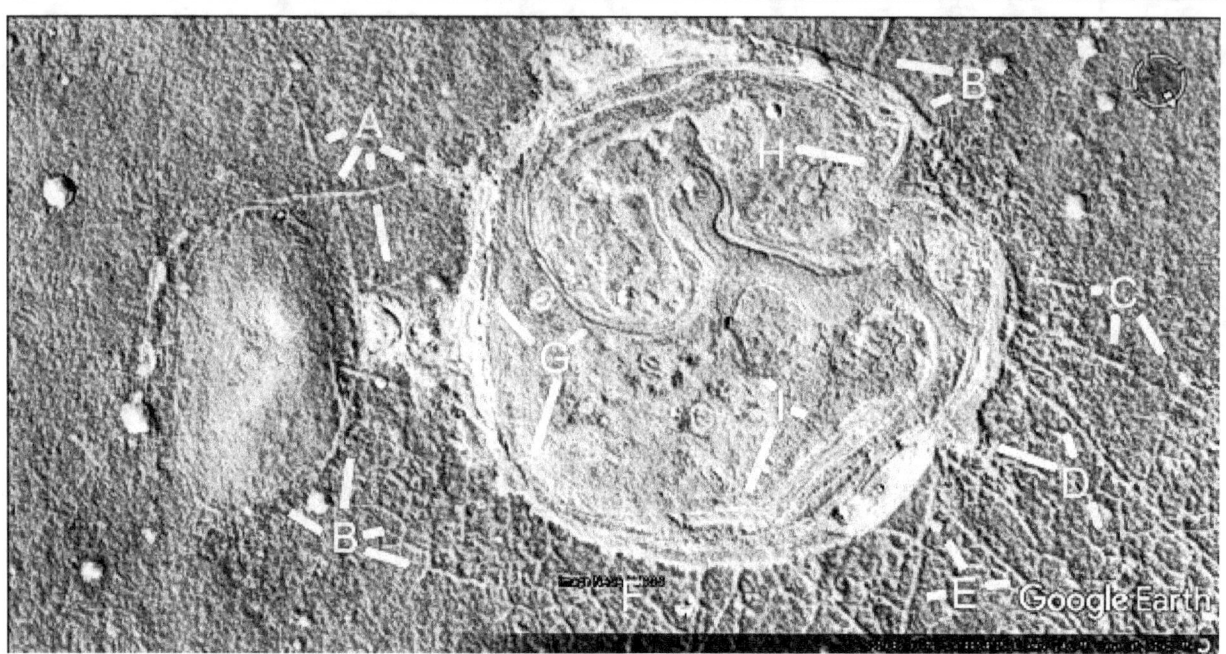

Prhh1021a

Hypothesis

This overlay shows how close the ellipse is to a perfect geometric shape. The circle on the right is also partially elliptical but as can be seen is also close to perfect. The curved shapes in it may give an optical illusion that its boundary was more irregular. The small deviations may have occurred from the edges subsiding or eroding, also the ground could have moved in this area deforming both shapes.

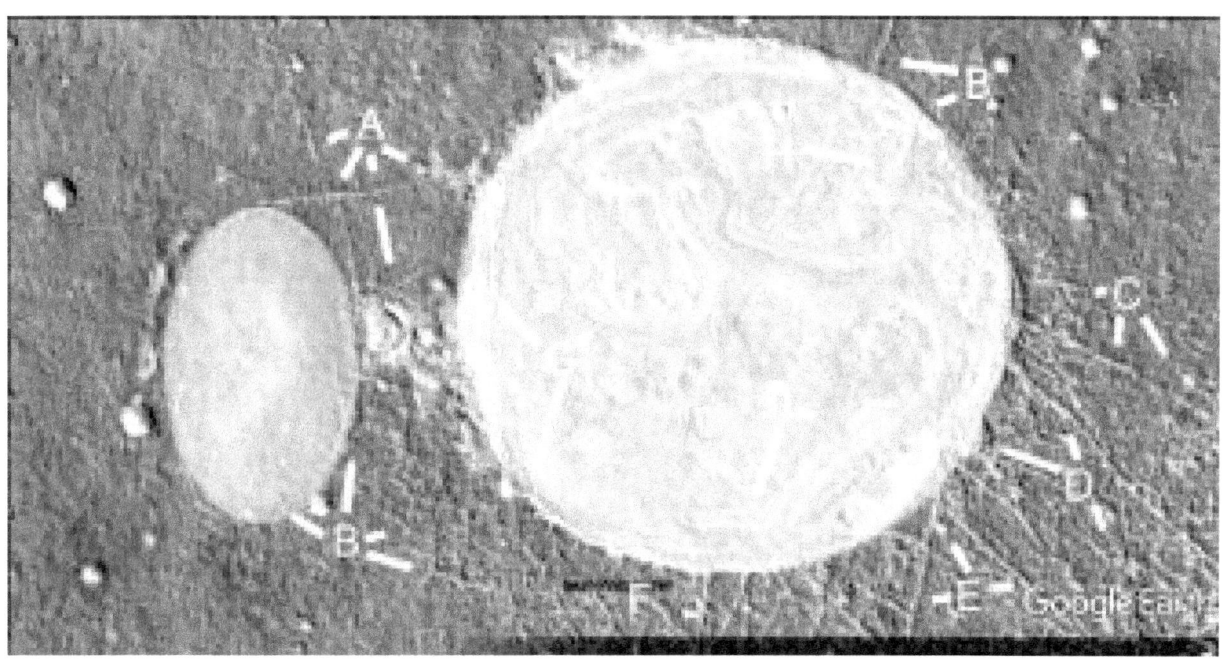

Prhh1030

Hypothesis

A shows the pit wall of a possible collapsed hollow hill, B shows other parts of this pit wall. At C the wall is more eroded. D shows eroded tubes near an interior support.

Prhh1030a

Hypothesis

This shows how part of the pit wall is a parabola.

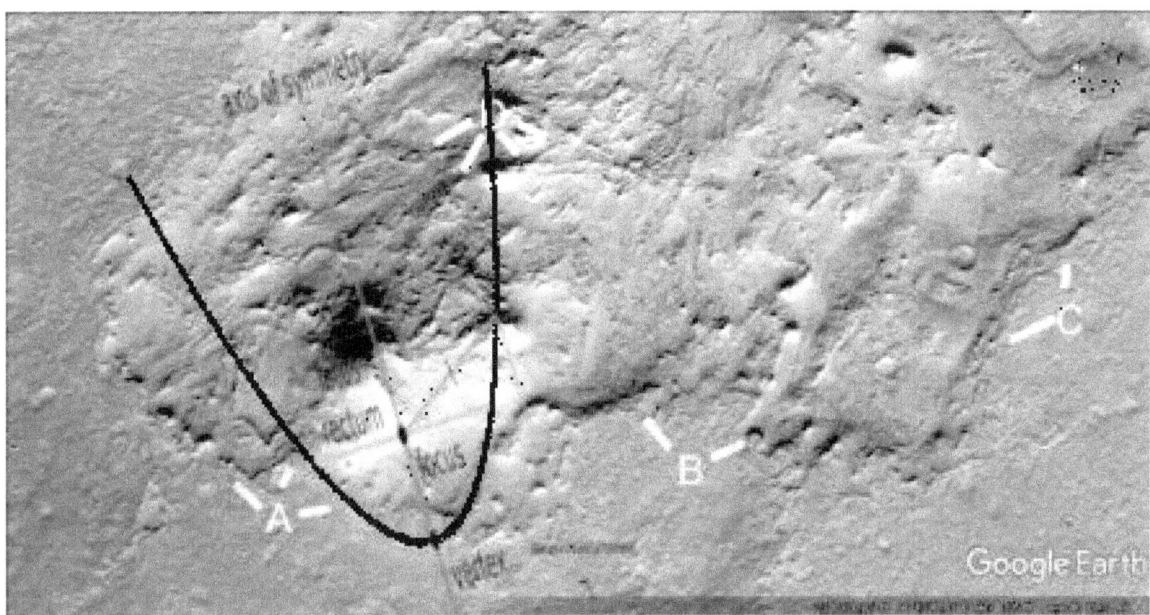

Prhh1031a

Hypothesis

A shows a pit wall at 7 o'clock with an angular walled segment, rounded inside it. From 2 to 5 o'clock this may be a large collapsed tube going into the crater, to the right of this above B it seems to merge into a more intact roofed segment. B shows a smooth slope to the wall in this segment, C shows a curved tube that goes around in a loop. To the left of C at 10 o'clock is a groove which may have a collapsed roof extending to the left into a hollow formation.

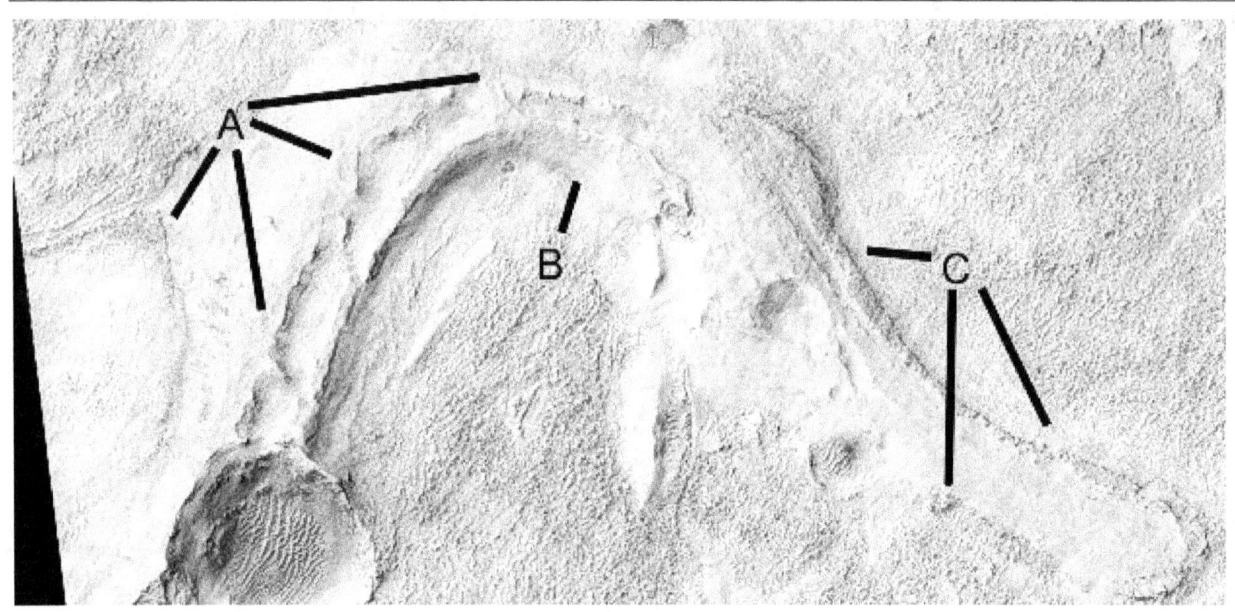

Prhh1031a2

Hypothesis

This shows how the left side is a parabola.

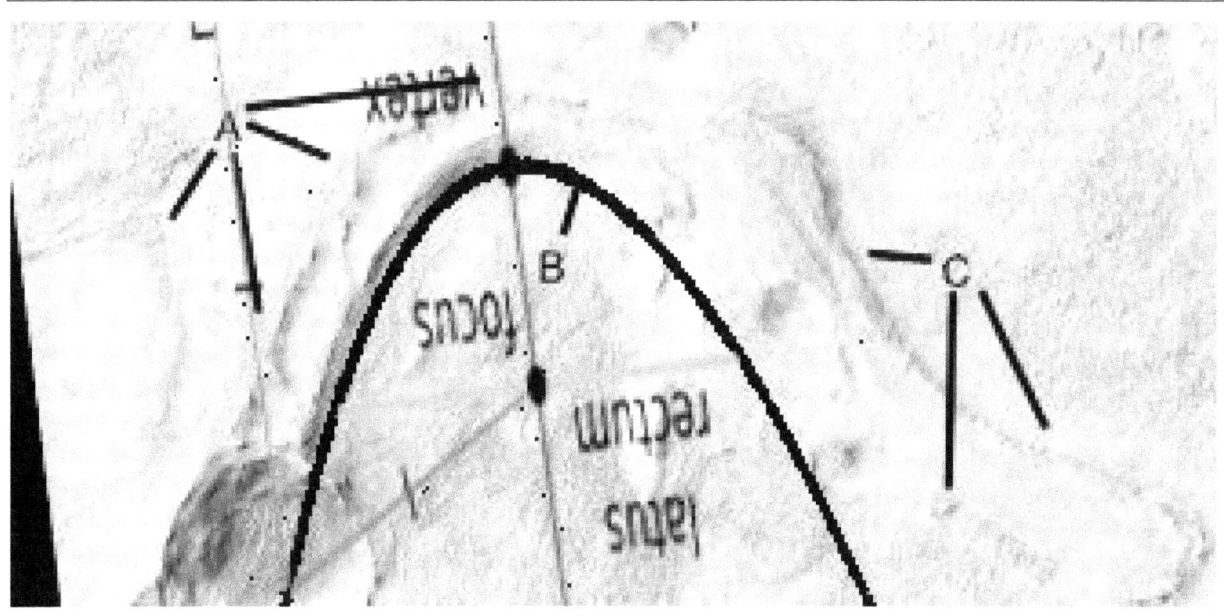

Prhh1031c

Hypothesis

A at 12 o'clock shows a long dark line like a road which connects to D at 10 o'clock, where it becomes a tube. At 2 o'clock D shows a tube connecting to it, A at 4 and 5 o'clock show other dark lines. B shows another dark line merging into the main one, these are unlikely to be dust devil tracks because they are so straight. There is also a right angled tube or wall from B at 10 o'clock to A at 11 o'clock.

Prhh1031d

Hypothesis

A shows a curved tube with a narrow roof, the cross section would be like a triangle. B shows a cavity at 10 o'clock like a collapsed hollow hill which may have extended down to C. At 1 o'clock B shows a part of the intact roof. C shows another tube or part of the hill wall. D shows parts of the former hill.

Prhh1031d2

Hypothesis

This shows how two of the curved tubes are in the shape of a parabola at right angles to each other.

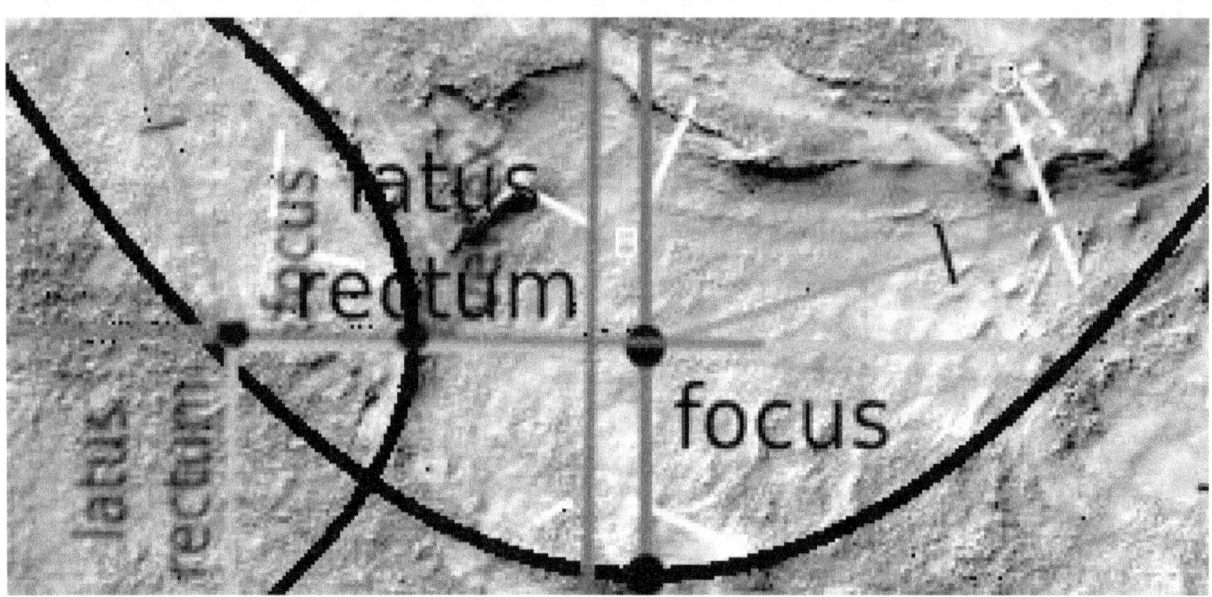

Prhh1031e

Hypothesis

A may be a collapsed hollow hill, the wall or a tube is shown at 4 and 7 o'clock. At 11 and 2 o'clock there is the darker material with a sharp boundary around the pit. B shows another probable pit wall, like at A it has regular grooves across it as if it is eroding between interior supports along it. C at 2 o'clock shows another wall, at 10 o'clock is another cavity. D shows another wall or tube at 5 and 10 o'clock, 1, 2, and 4 o'clock show more likely roads. E shows more roads, tubes, or pit walls.

Prhh1031e2

Hypothesis

These walls are also like two parabolas. G is closely aligned with the Latis Rectum.

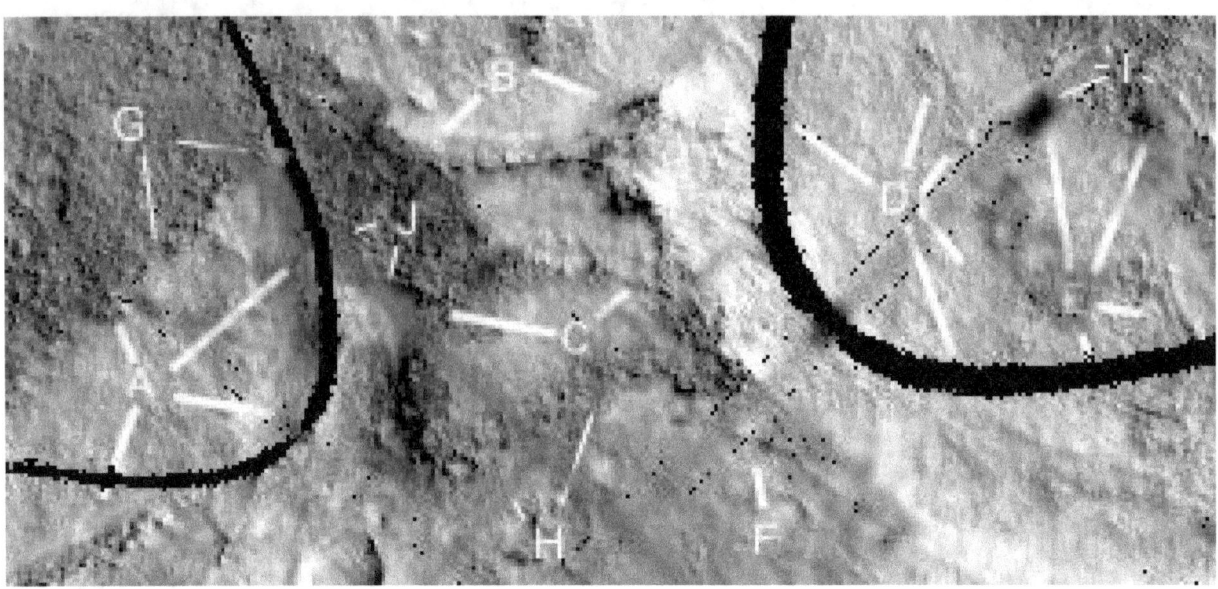

Prhh1031f

Hypothesis

A and B show curved walls or tubes. C shows a cavity like a collapsed hollow hill. D shows another degraded wall.

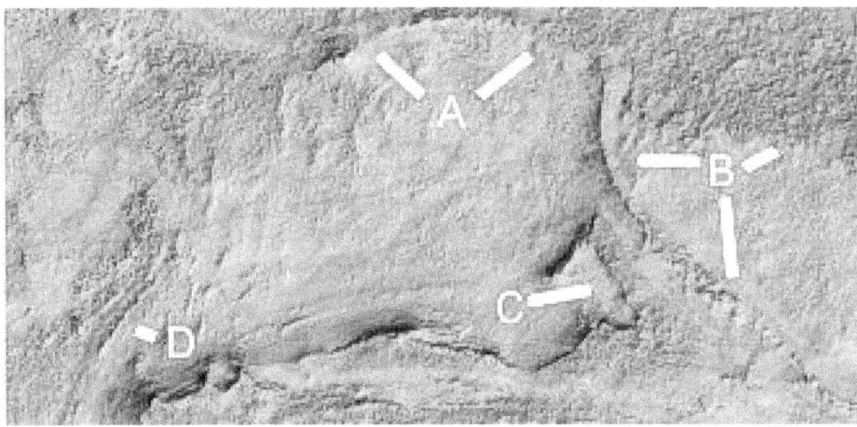

Prhh1031f2

Hypothesis

The two curved walls form parts of parabolas.

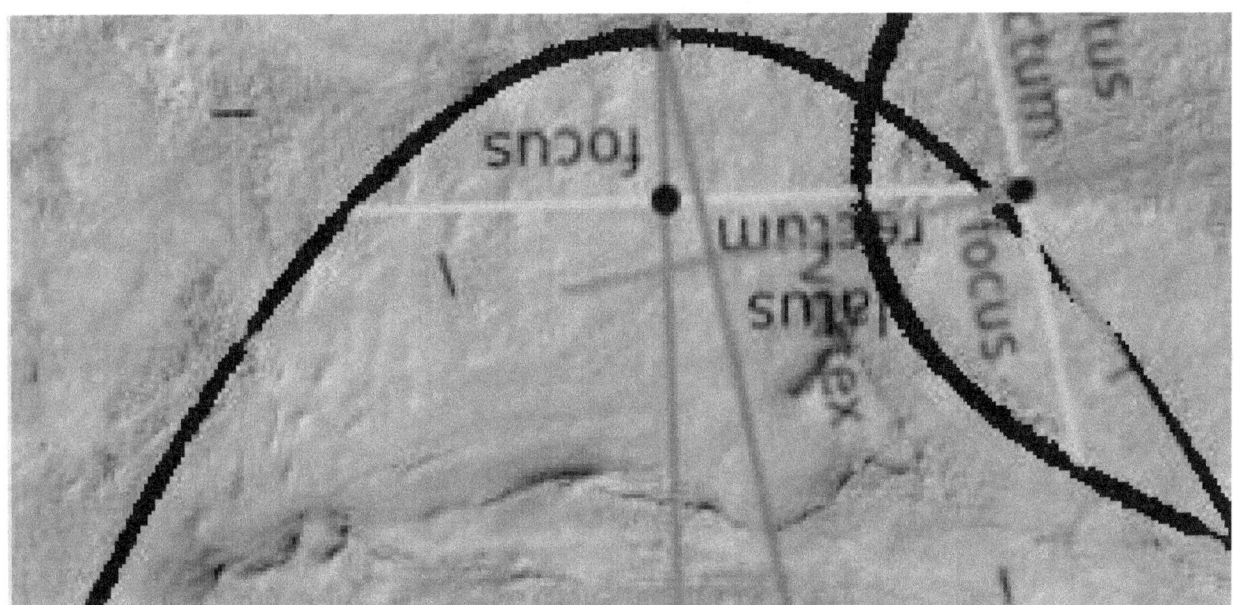

Prhh1031k

Hypothesis

A shows another pit wall, it also has regular grooves across it like the material between the supports has collapsed. It might then be like posts where the material between them has eroded away leaving a series of regular spacings. B is a right angled tube, C is a pit wall in better condition, D shows another wall connecting to C and over to A.

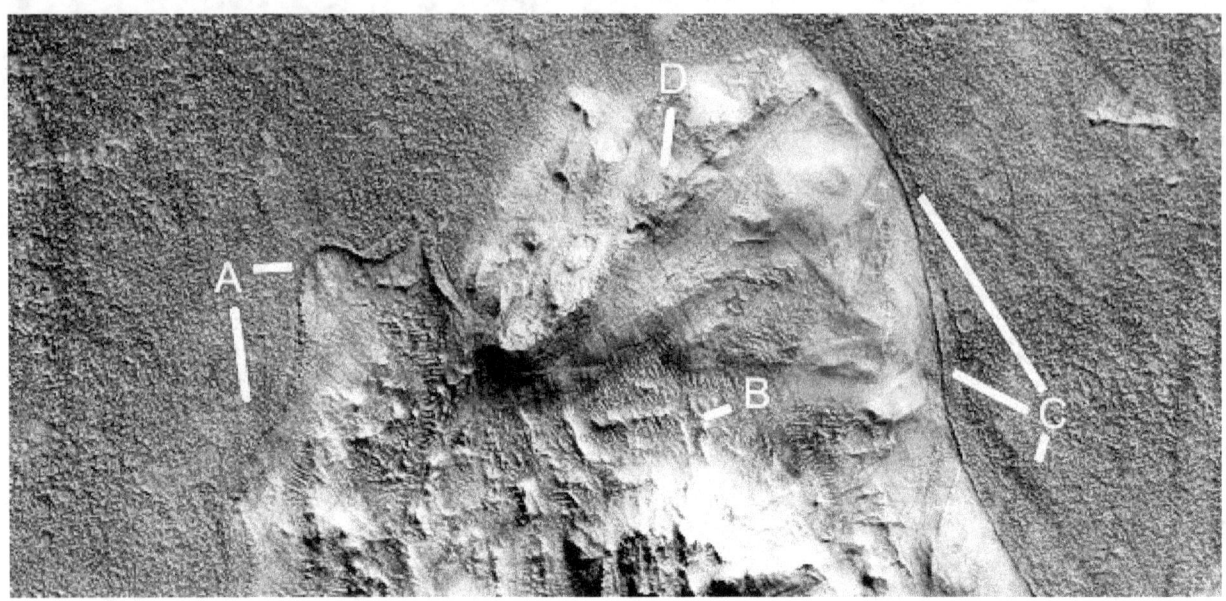

Prhh1031k1

Hypothesis

This shows how part of the formation is a parabola.

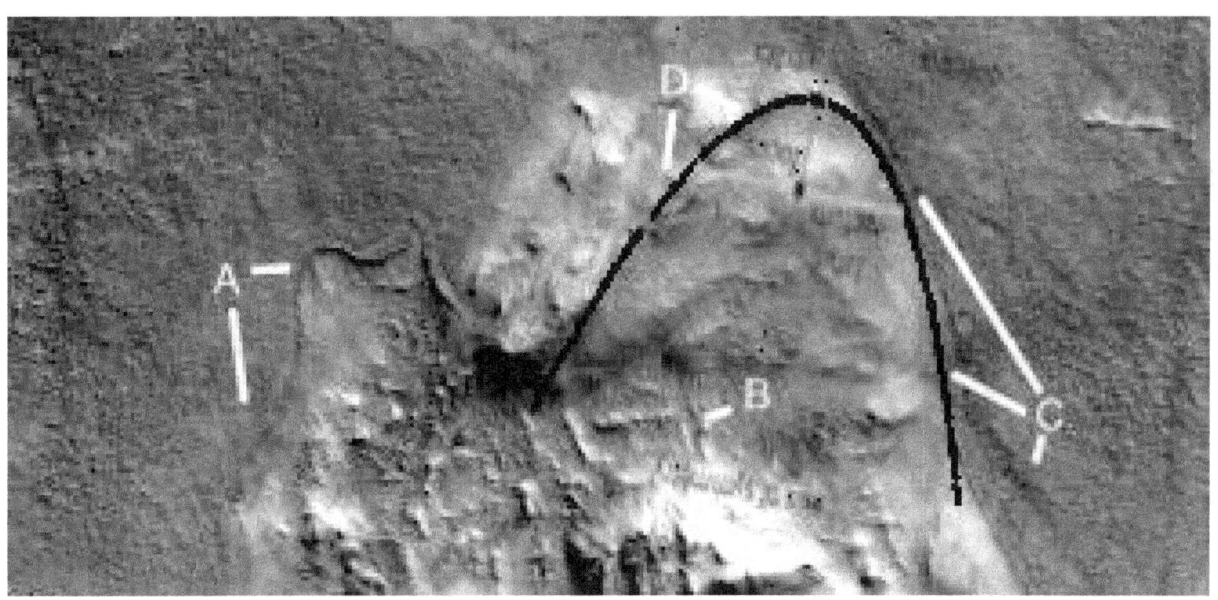

Prhh1031k2

Hypothesis

The outer pit wall also fits well into a parabola.

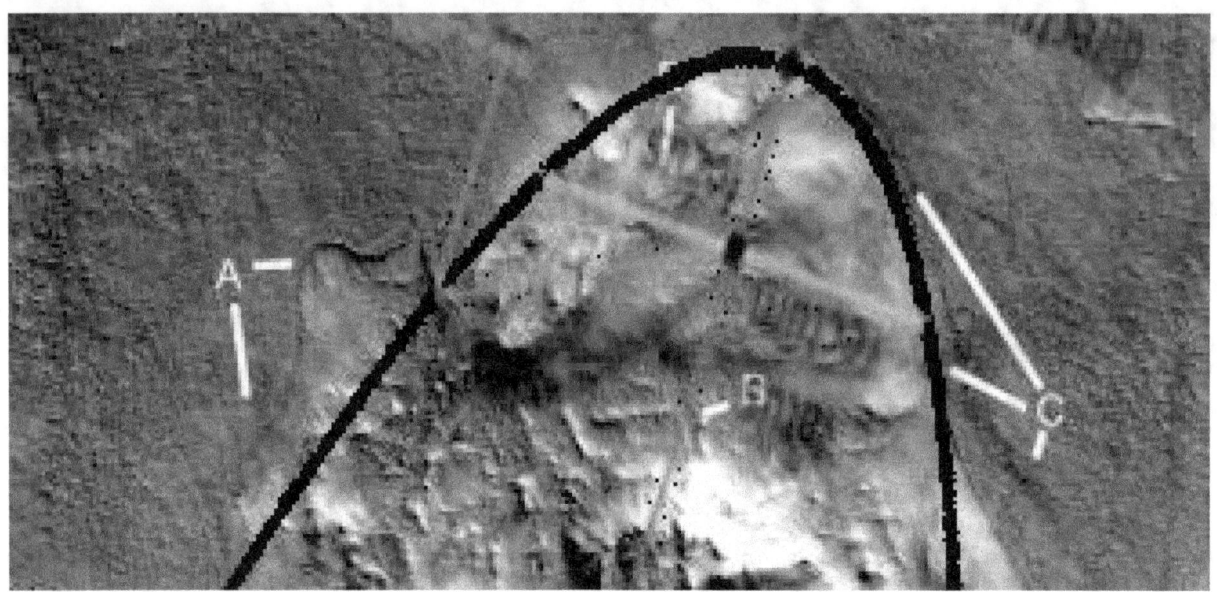

Prhh1031l

Hypothesis

A shows a probable road that comes to a T junction at D, B shows one of these roads going through a pit up the image. A from 9 to 5 o'clock closely bisects the large pale pit which has some symmetry, this may have lost some of its shape due to erosion. C shows a dark line like a curved road.

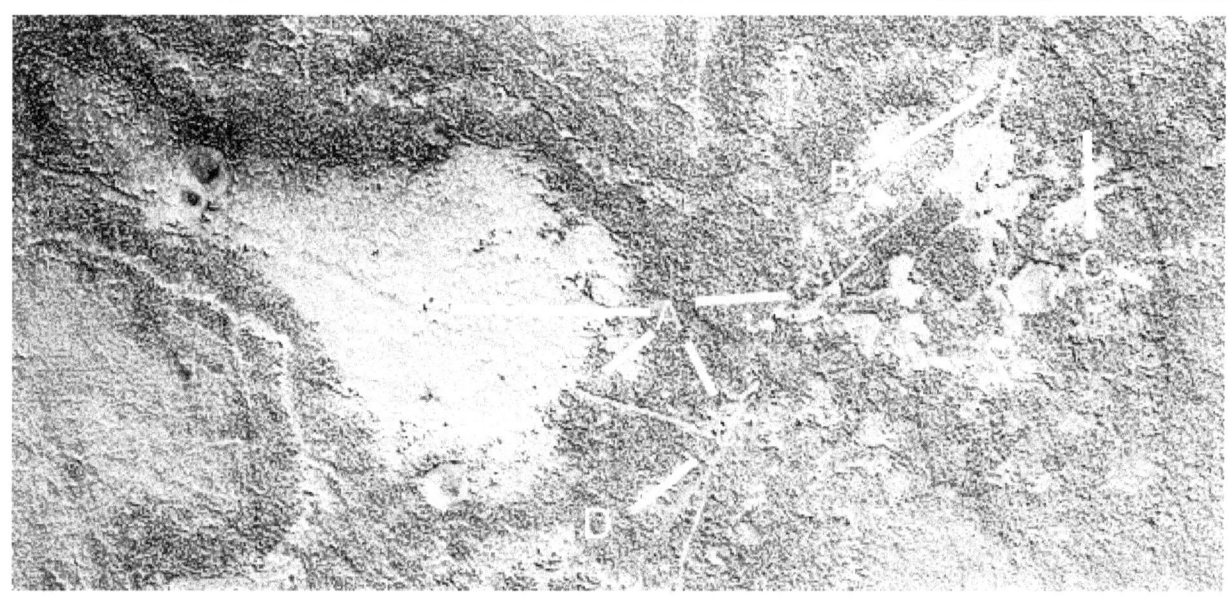

Prhh1031I2

Prhh1031l3

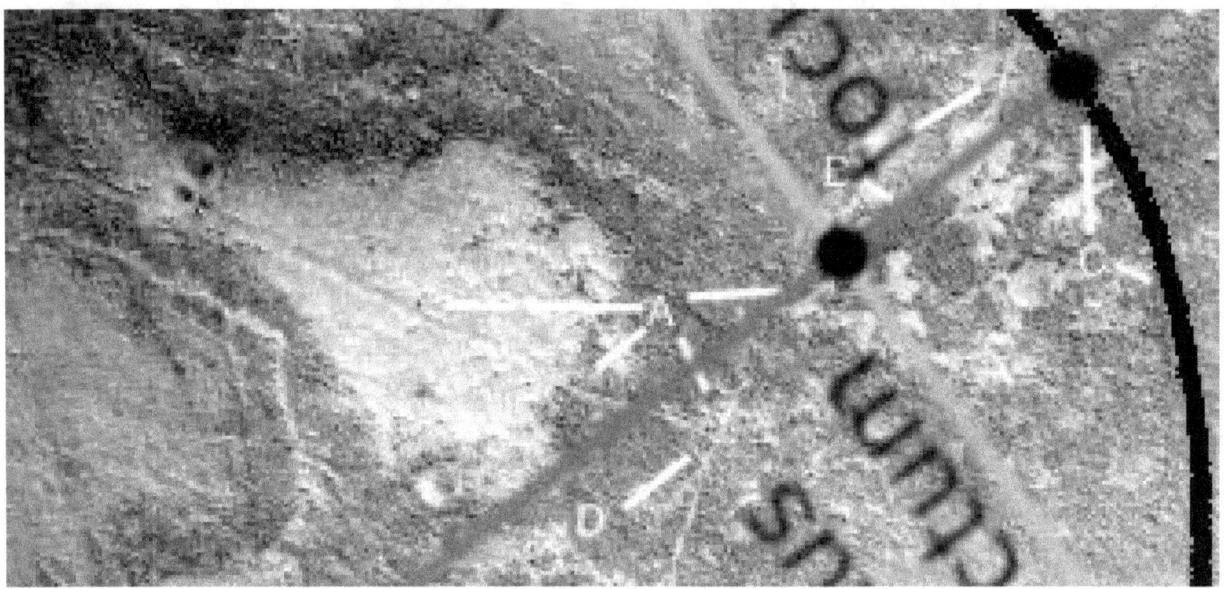

Prhh1032

Hypothesis

A shows some walls in close to right angles between them. B shows more walls, at 2 o'clock the slope is very even as is the height of the wall. At 7 o'clock and beyond there are many walls intersecting at right angles. C shows a cavity at 1 o'clock and 2 o'clock perhaps from a collapse, at 4 o'clock there is a curved wall or tube. D shows another possible tube, E shows a tube or wall intersection at 12 o'clock, more walls at 2 o'clock and a pit wall at 9 o'clock.

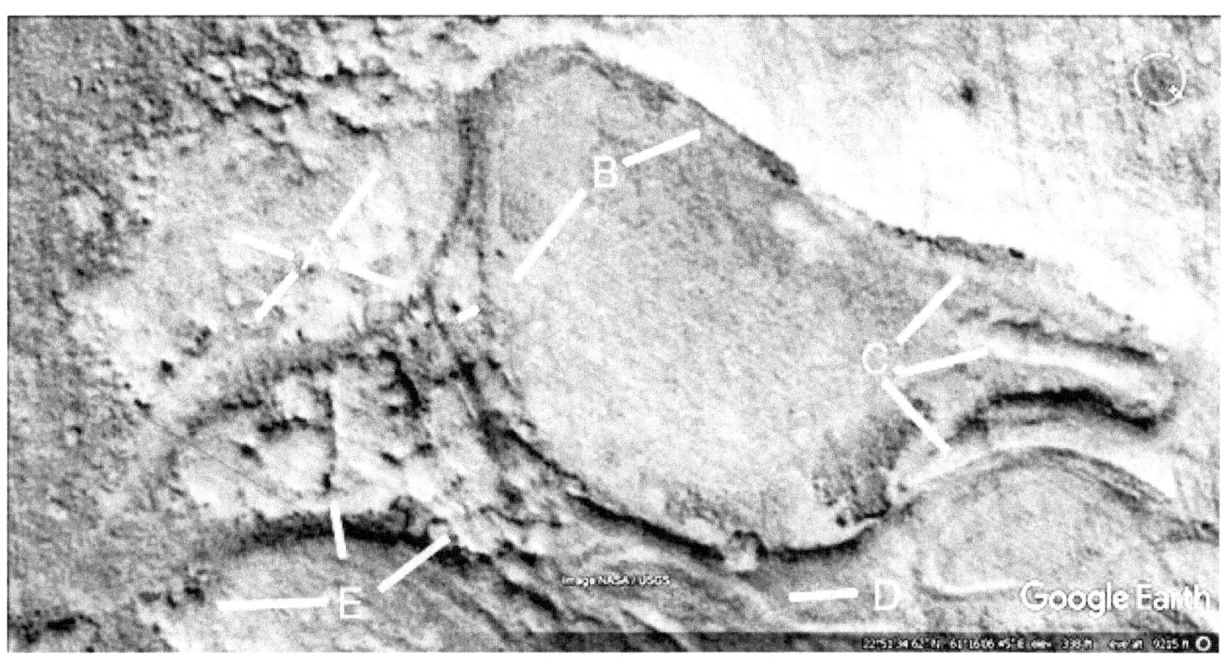

Prhh1032a

Hypothesis

Five of these curved walls form parabolas as shown.

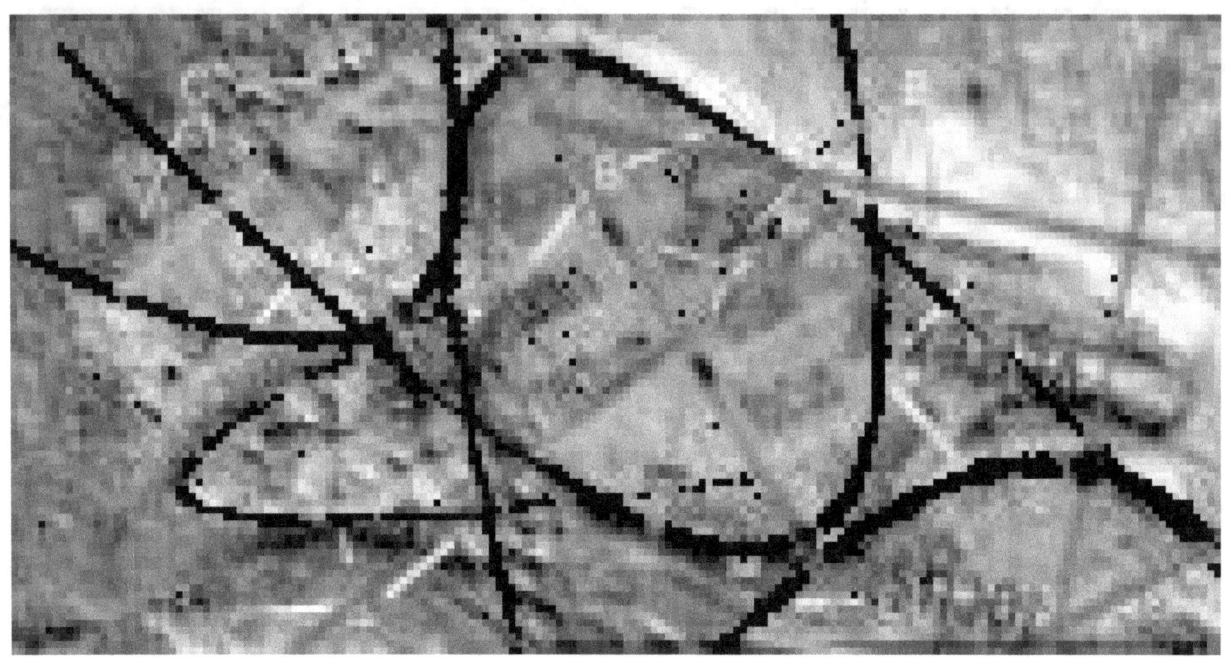

Prd1033

Hypothesis

This shows another series of pits or dams, A looks to be a collapsed hollow hill with a wall off at 2 o'clock. B shows another wall segment with a smooth slope as does C. D shows how the interior of this pit is much smoother than outside, as if the external erosion forces like water have been kept out. It implies then this has remained largely sealed with some material slowly coming in from the entrances at the top of the pit such as at E. F shows another pit which seems more concave, G shows walls or tubes connecting to it.

Prhh1033a1

Hypothesis

Part of the pit fits to a parabola as shown.

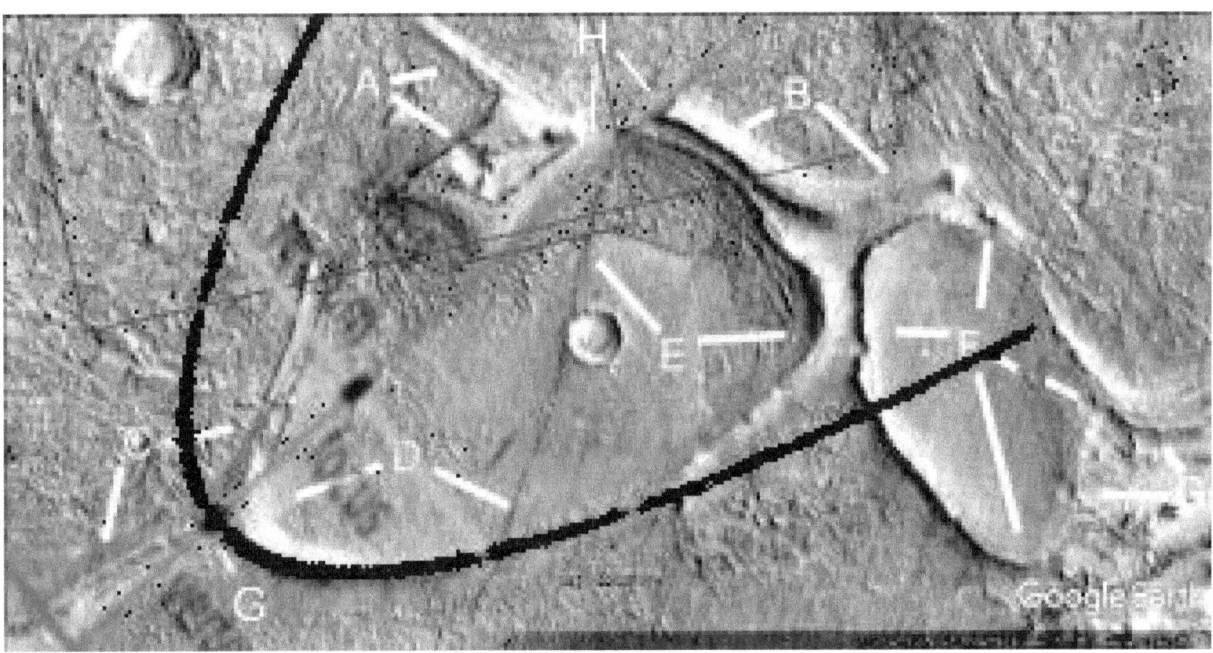

Prhh1033a2

Hypothesis

The other pit also closely fits to two parabolas with parallel latis rectums.

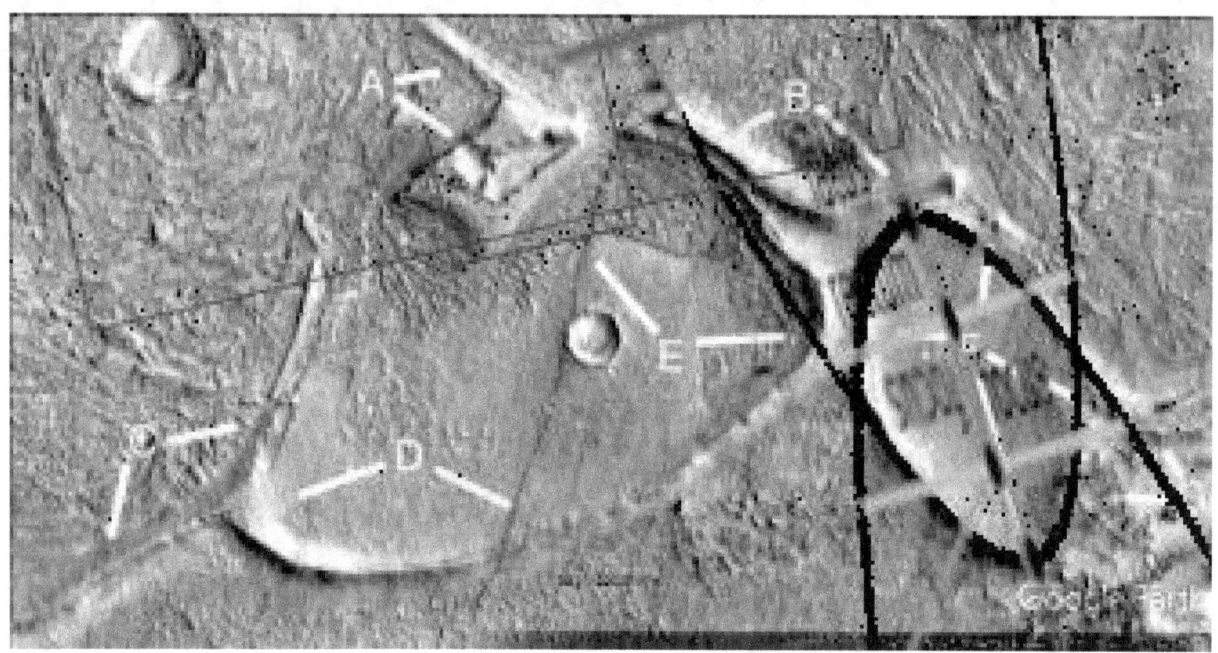

Prd1050e

Hypothesis

This is probably another tube, A shows how this goes into the higher segment at 10 o'clock as if it becomes a tunnel, then it is more eroded at 2 o'clock. The markings on the ground are approximately vertical in the image as if going through the tube, this may indicate it was built after these marks were formed. The tube would tend to block the wind or water flow otherwise. It goes into B at 7 o'clock, at 5 o'clock there is a pit which may have been a hollow hill. The polygons have formed in many areas but not in the tube, this may indicate it is cement and resisted the same geological forces like freeze thaw or drying out.

Prt1051a

Hypothesis

A shows another tube or wall with a gap in it at 10 o'clock, this may have been an entrance. At 12 and 1 o'clock there is a small walled enclosure. B shows another walled enclosure, extending up to B at 12 o'clock connecting to the smaller walled enclosure. At 6 o'clock the walls connect to another short wall which then connects to the longer wall along D. C shows another walled enclosure, a separate enclosure also is at 10 o'clock around the second leg. E also shows an enclosure, the wall or tube is wavy at 10 o'clock and at 4 o'clock the walls cross each other. F at 6 o'clock shows a wall connecting to a pit or crater, other walls at 9 and 12 o'clock.

Prt1051c2

Hypothesis

The curved wall on the left is close to a parabola with some deviations, the curved wall on the right is nearly a perfect parabola.

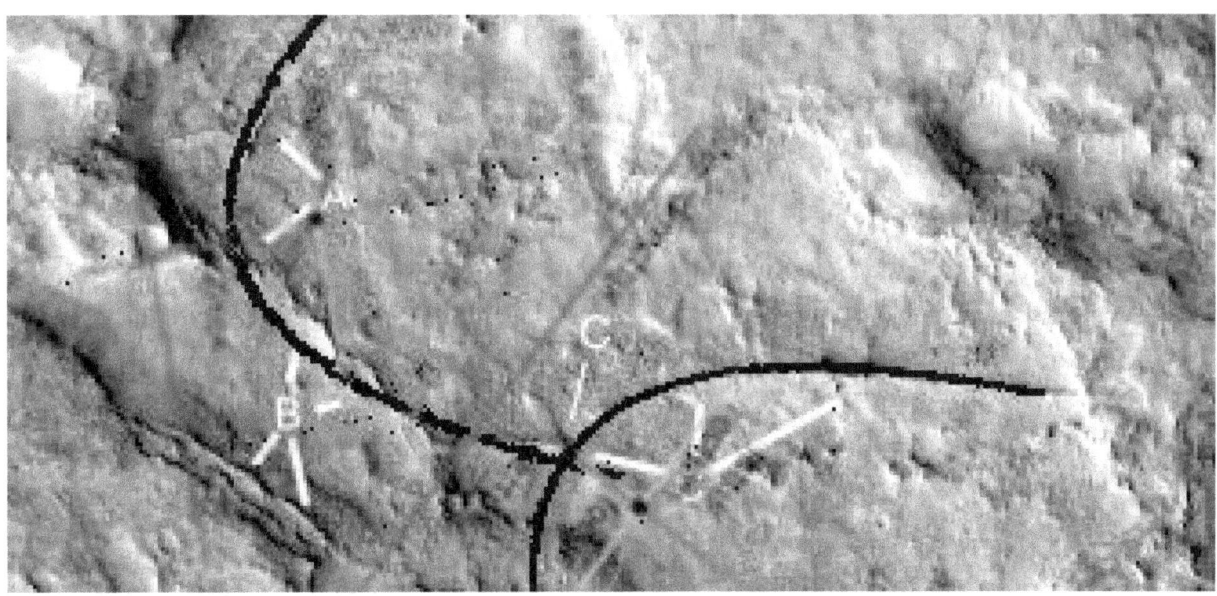

Prt1053

Hypothesis

A at 8 and 11 o'clock appear to show a tube running along the side of the slope, or perhaps a layer. At 4 o'clock the tubes connect together. This dark line in the cliff connects to a tube at B at 8 o'clock, this goes to the edge of a hill at 4 o'clock. C at 7 o'clock shows where tubes intersect at right angles, a tube then goes up through a crater to the hill at B. D at 8 o'clock shows the straight edge of the hill, and how the tube extends to the right through 11 o'clock. E shows a rectangular enclosure with walls or tubes extending from it.

Prt1054

Hypothesis

From A at 10 to 2 o'clock over to B at 7 to 4 o'clock is a road or collapsed tube. It connects to a hyperbolically shaped tube or wall from C past B at 12 o'clock. Then it connects to another wall at 11 o'clock. A at 5 to 7 o'clock and D show other walls.

Prt1054a

Hypothesis

The lines show how straight the walls and road are.

Prt1055

Hypothesis

This shows a nearly perfect hyperbola forming a tangent to the large crater, and to a smaller crater on the left.

Prt1055a

Hypothesis

This shows a hyperbola overlaid onto the formation, it shows it is nearly a perfect hyperbola. It deviates a small amount to the left at A as if affected by the gravity of passing near a planet or moon. B at the top of the image shows two other walls, C shows a road like shape connecting to the crater. B in the crater shows concentric circles which might indicate orbits around the sun, or the surface of a planet with the outer circle being the atmosphere. D is a line or chord drawn as a tangent to the smaller crater, it is at right angles to the vertical transverse axis, the dark line which nearly bisects the large crater. With the inaccuracies inherent from the age of this formation, also in fitting the hyperbola, this may have been intended to go through the center of the crater.

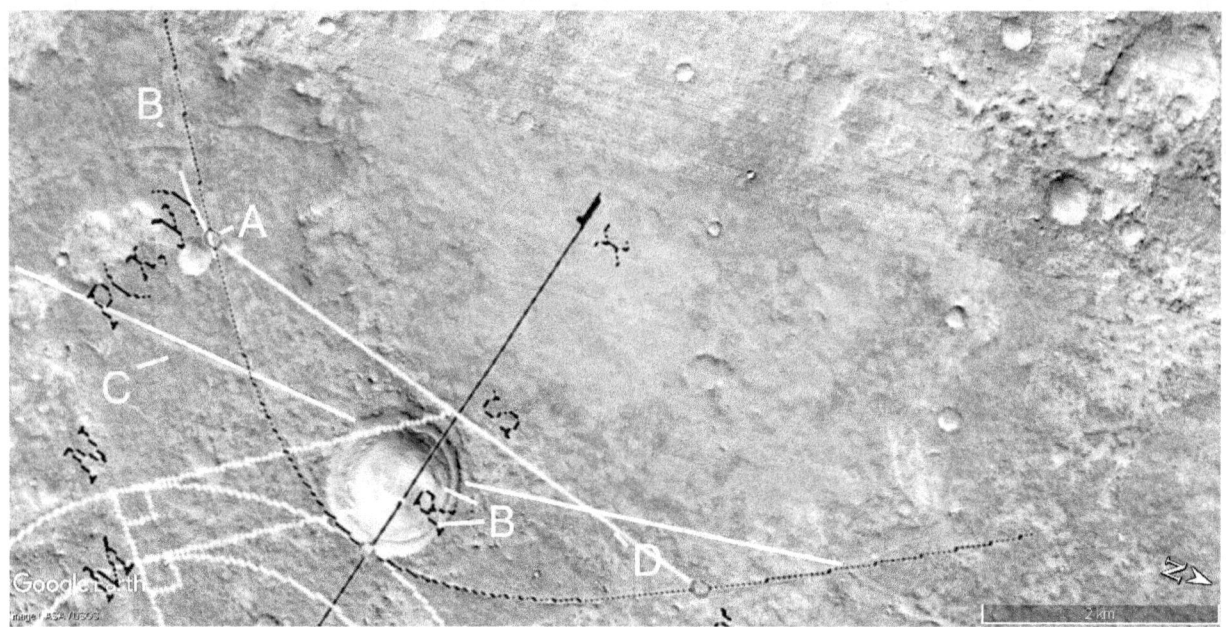

Prt1055b

Hypothesis

This shows another fit of a hyperbola onto the wall and crater, here the focus is on one side and the hyperbola is a tangent to the other side. It also bisects the crater more closely.

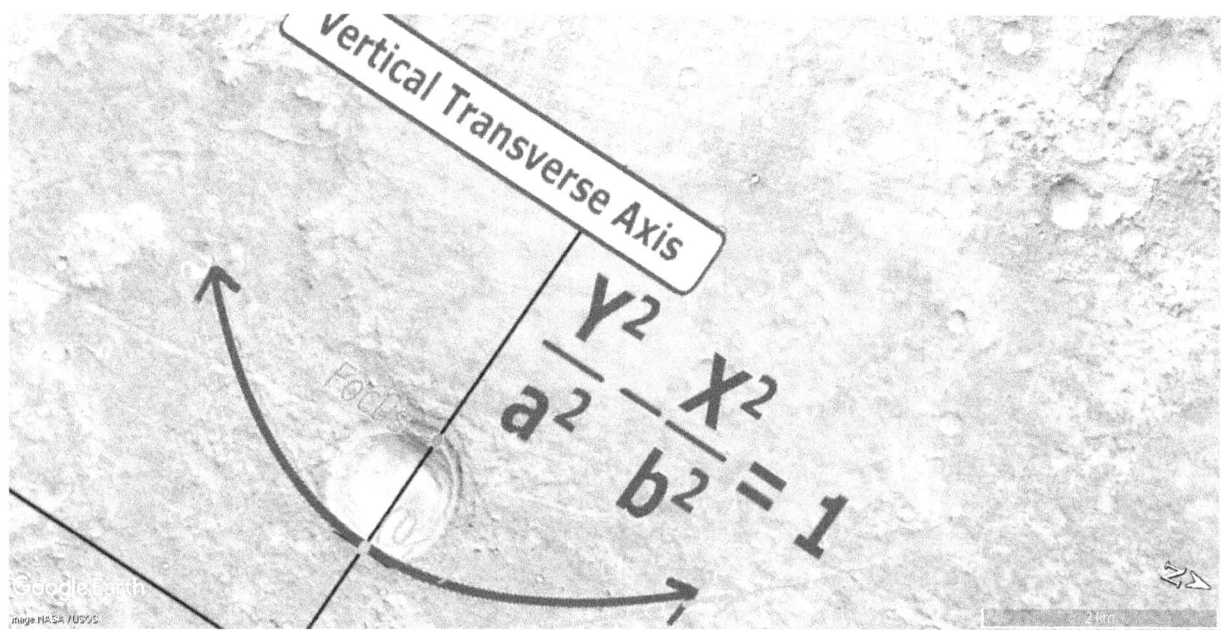

Prt1055c

Hypothesis

This is a closeup of how the hyperbolic wall grazes the edge of the crater. If this was natural it would have fallen into the crater or the impact would have broken the wall formation. A shows a groove around the crater but outside its slope, perhaps representing an atmosphere. The darker area in the crater might also represent the asteroid belt. C may be caused by the surface eroding.

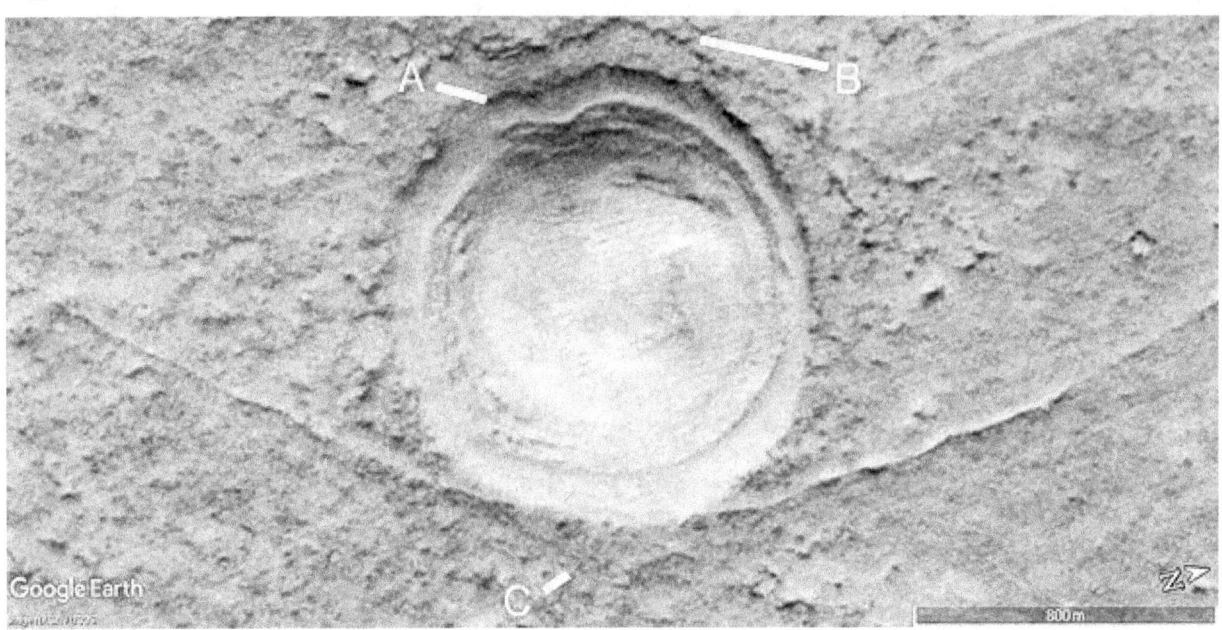

Prt1055d

Hypothesis

This shows in geometry how a circle and hyperbola can be joined in a mathematical pattern. This might then represent a mathematical statement and not a hyperbolic orbit.

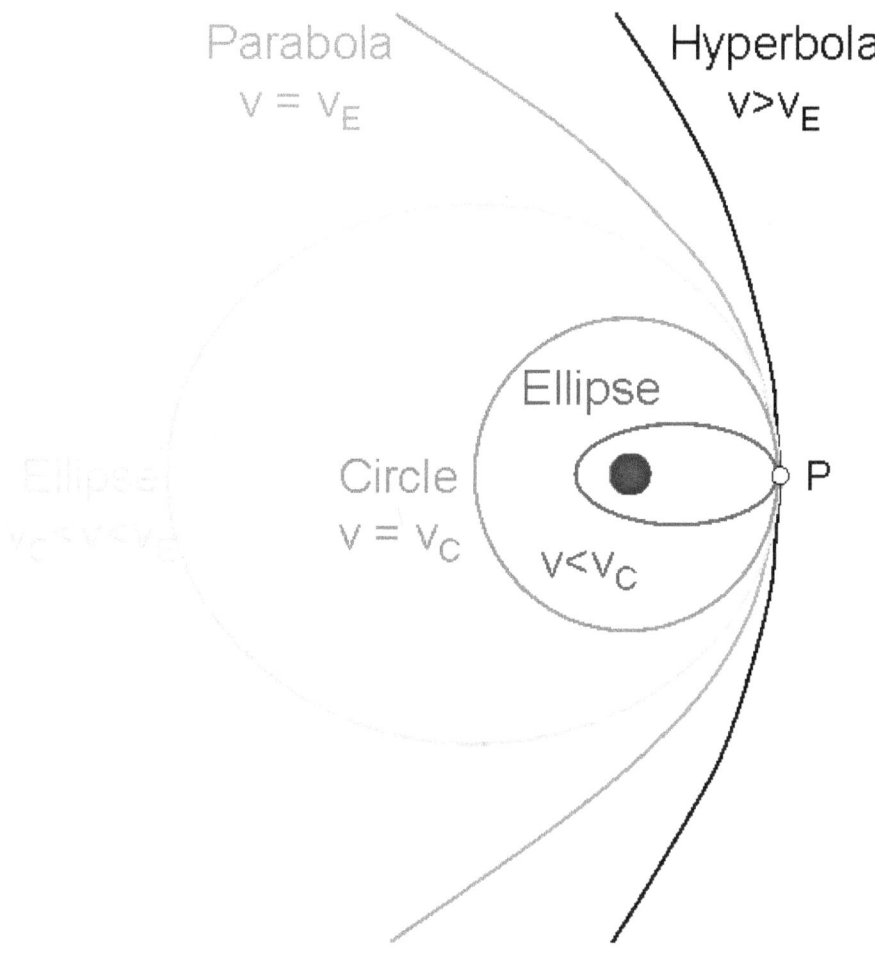

Prt1055e

Hypothesis

This shows some other formations analyzed in these books. The hyperbola is a dark line at prt1055, the white horizontal line is the old equator. This has the Cydonia Face, Nefertiti, and the Crowned Face on it. This also has the ferns formation on the old equator, they look like large plants with Fibonacci branches. These should not occur by chance unlike how rivers can randomly form a root pattern.

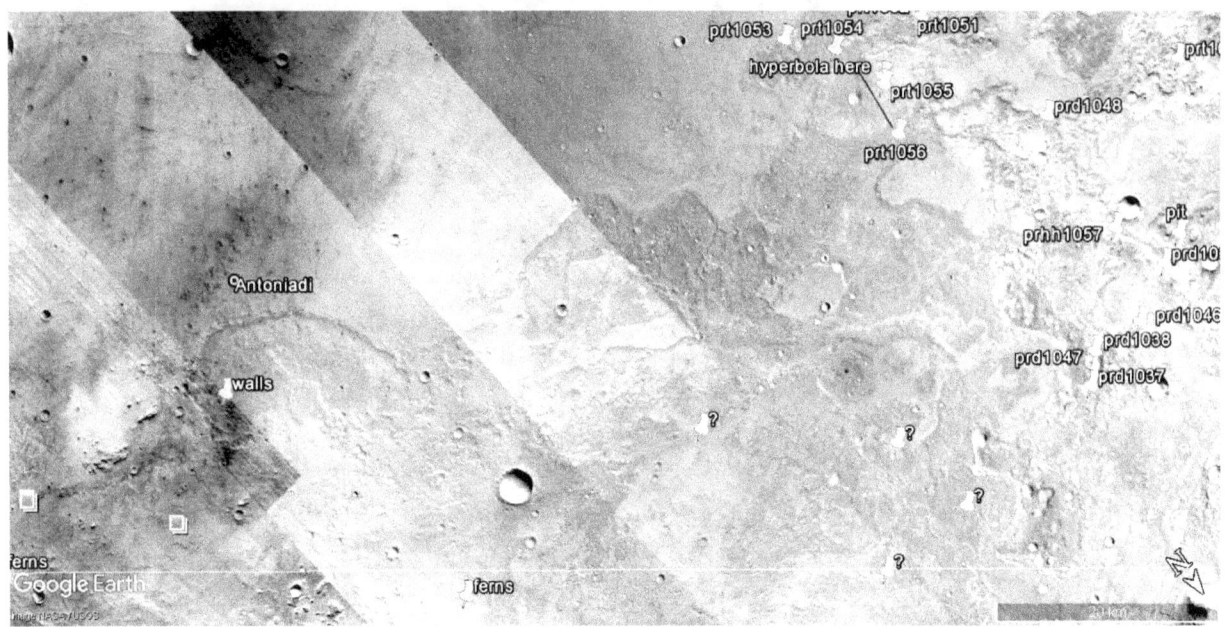

Prt1056

Hypothesis

This is a closeup of the road going across the hyperbola from A to B. At A the wall goes over another wall as if making a knot. At B at 5 o'clock it crosses another road or wall.

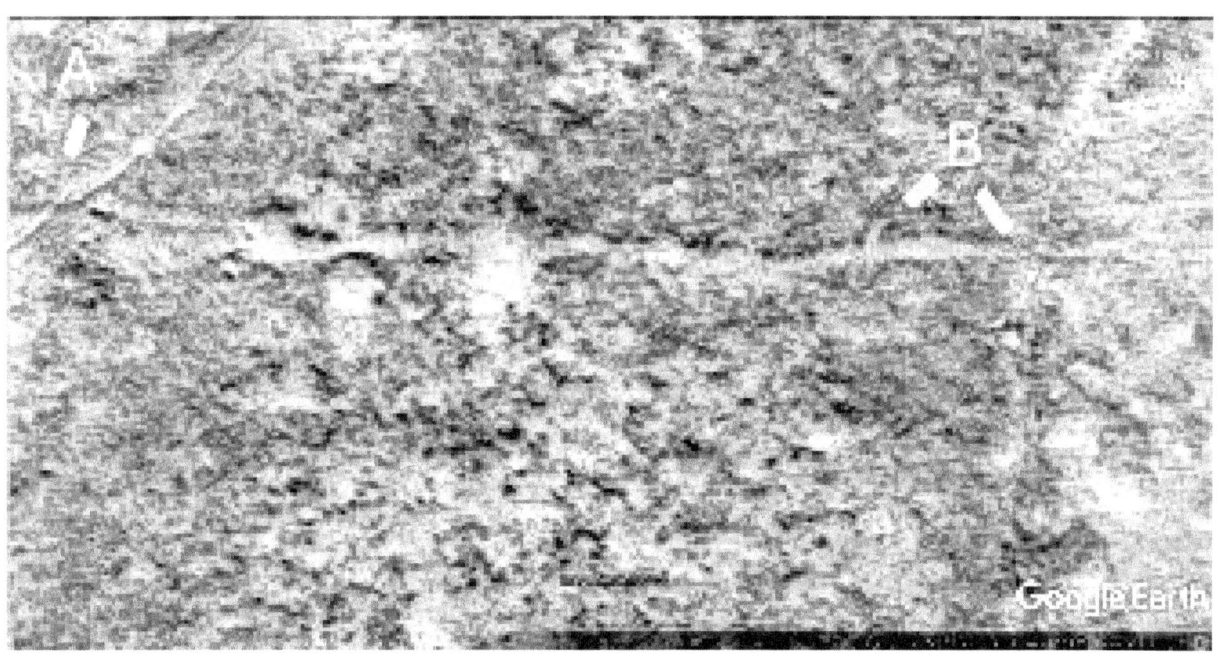

Prd1059

Hypothesis

A, B, and C appear to be walls. D may be a pit dam to collect water. The surface is smoother than the surrounding terrain, it may be cement.

Prt1059a

Hypothesis

Four parabolas fit the shapes of these walls.

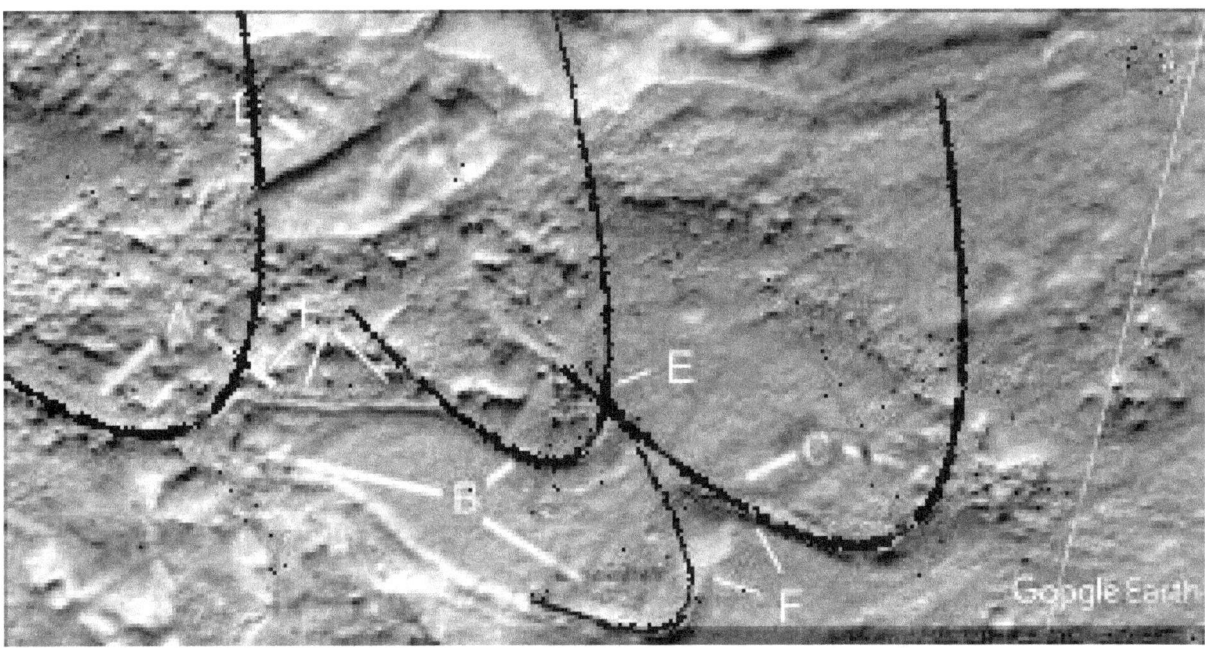

Prd1062

Hypothesis

A and B show walls enclosing a smooth interior, perhaps cement. This may have been a pit dam or large hollow hill. There may also have been a hollow hill between 7 and 10 o'clock at B with a long entrance coming out of it at 8 o'clock. C may be another pit dam at 10 o'clock, at 4 o'clock there is a wall or tube. D may be another parabolic wall, the wall at E at 2 o'clock is fairly straight. There may have been other walls and canals for water at E, F, G, H, I, and J.

Prd1062a

Hypothesis

The lines indicate how straight parts of the formations are. Also four parabolas are shown.

Prd1065

Hypothesis

A, B, C, D, and E show pit dams, the walls should have been able to keep water inside them. F shows one of the walls, G and H show other external walls.

Prd1065a

Hypothesis

The lines show how straight the walls are.

Prt1066

Hypothesis

These formations are right next to the ferns as can be seen on the right at C. A shows walls which could have kept in water or perhaps animals, enclosed a farm, etc. B and D show more walls.

Prhh1068c

Hypothesis

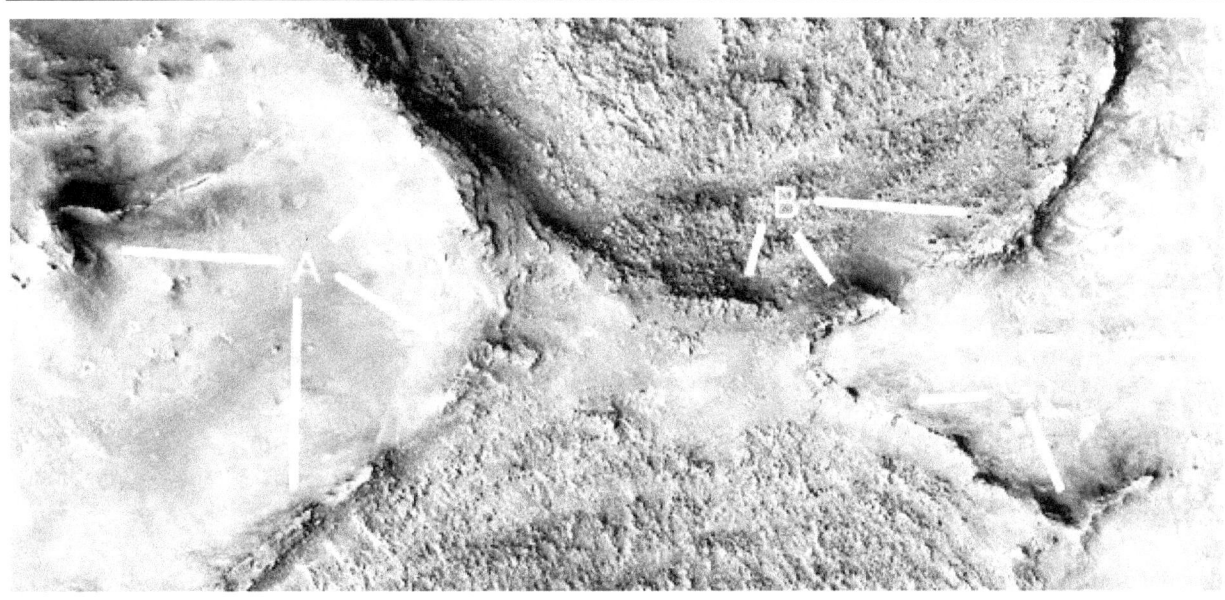

Prhh1068c2

Hypothesis

Two parabolas are shown here.

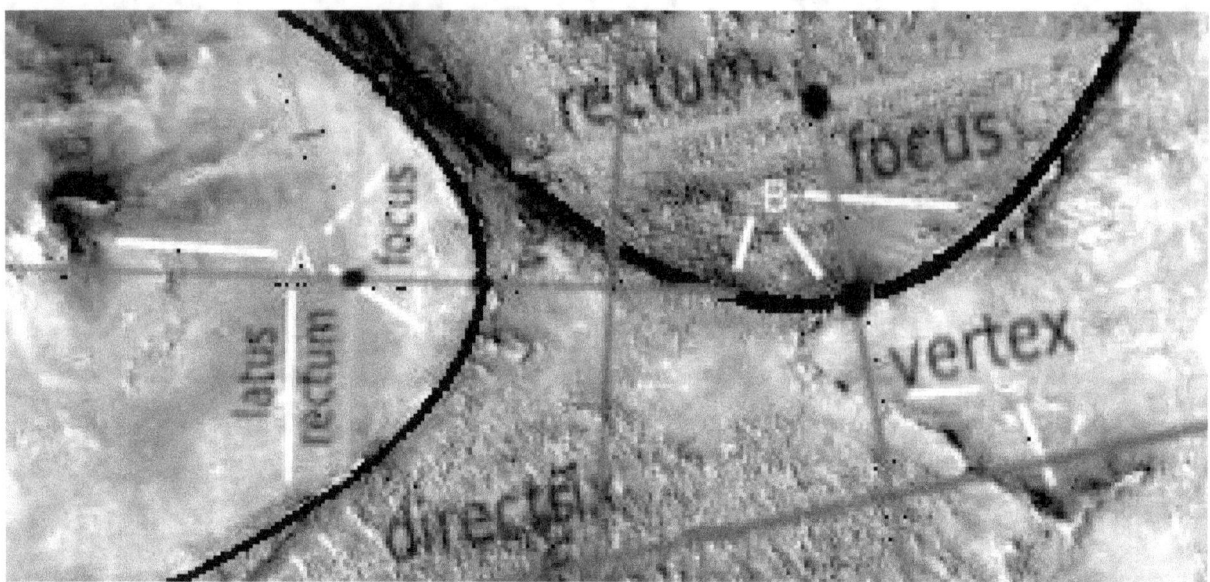

Global Hypothesis

Many people have seen, or heard of, the discovery of faces on Mars. Often they are sceptical about this. One common objection is the faces look too much like us to be an alien race, so researchers are recognizing faces in the terrain that aren't there. This has also been an objection to possible discoveries of bones, statues, even small animals. The mainstream view is that these are the products of people's imaginations, often this is a fair comment. Historically though, people have believed in a Martian civilization, whether still existing or extinct. This was explored in many science fiction books from Edgar Rice Burroughs and Arthur C. Clarke to Robert Heinlein. Many expected Mars to be habitable, or even inhabited, when the Mariner 6 and 7 spacecraft went to Mars in 1969. What was found instead was a near airless world devoid of water. The conventional wisdom was turned on its head, that Mars had never been inhabited and probably never had any life at all.

From this time forward the mainstream scientific opinion was that Mars had always been devoid of life much like our own Moon, so anything that looked artificial was just people seeing things. This is called Pareidolia, seeing illusory faces and animals often in clouds and random patterns. The problem in overcoming these legitimate objections was that spacecraft imagery was low resolution, it could only map the surface of Mars very slowly. So if signs of an extinct Martian civilization did get imaged then they would likely be ambiguous in this low resolution, and be dismissed as fringe science and illusions. But these anomalies have kept turning up as the spacecraft imagery became higher in resolution, more able to see signs of this civilization if they existed. Mars is now largely mapped to a fairly high resolution, called the HiRise and CTX images, so many unusual formations have been found. The situation has also continued to be toxic for mainstream science, some use their imaginations too much and see things that really are not there. This tends to scare away mainstream researchers, they are rightfully concerned that too much speculation can damage their careers. But other formations are not so easily dismissed.

Another complication is that this hypothetical Martian civilization would have died out perhaps billions of years ago. This is because Mars had a warm climate and oceans long ago according to NASA, but being further from the sun it cooled with the atmosphere and oceans freezing at the poles. With billions of years of erosion many possibly artificial formations look more natural over time. The evidence has then been ambiguous and highly eroded, but with thousands of possible artefacts being found.

One problem for mainstream science was in understanding what was actually being claimed by researchers. Mixing more plausible artefacts with illusions also makes the claims less logical. For example finding skulls and boats runs into the objection of bone and wood quickly eroding under the surface conditions. They might also give the impression that boats may have been used in an area that had no oceans or rivers. Separating the more plausible artefacts then improves the quality of these hypotheses. This may help to answer the questions of who constructed them, where they lived, how they created these formations and why. If hypothetical aliens came to Mars, then why would they build faces and not another kind of formation. Some might have preferred finding large geometric shapes or perhaps a representation of an equation. These have been found as well. But the problem then was not just what was found made little sense, but that it did not fit into the preconceptions of mainstream science of what they should find.

It became necessary to try to connect these ambiguous formations together into a global hypothesis. In that case mainstream scientists and others could see all the evidence and how it connected together. As will be shown, the evidence looks like a civilization but one profoundly alien in some ways. It likely covered most of Mars, life tends to extend to wherever it can survive. So, to understand this global hypothesis, images from all over the globe of this evidence need to be viewed and seen holistically. Sentient creatures should have learned to tame the climate and can live in wider temperature ranges, also where water is plentiful or scarce. We should expect a hypothetical Martian civilization to do the same. In different areas the evidence should point to different adaptations.

Methodology

The main methods used with these hypotheses are falsification, the law of large numbers, and the reduction to the absurd. Falsification means that the null hypothesis, that these formations are random geology, cannot be true. This is because geology perhaps could not create structures like this. The other method is the law of large numbers. That there are too many of these structures to be from the occasional coincidence. For example the parabola appears to have been used extensively in these formations, it has been used on Earth in many dams because of its load bearing properties. It is also used in parabolic domes. In these Martian formations there are 945 parabolas which are shown and outlined. These outlines are from geometric parabolic shapes, in some cases they might be widened or narrowed. This does not affect their load bearing properties, they are still described by a simple mathematical formula $y=ax^2$ where a is a variable. This is a large number, there are formations like dams in many craters and most of them are parabolas as will be shown. It would seem highly unlikely that they eroded into parabolic shapes as these dams are formed in many different ways. Parabolas are not known to be associated naturally with formations like these. In some cases a reduction to the absurd might be applicable. This might be hard to define scientifically but it may be apparent to some readers that a natural explanation is absurd. This should be used with some caution as some patterns can form by random chance or be illusions. However the human eye is good at seeing real patterns and is not so easily fooled.

A basic global hypothesis

The next section goes through a number of different types of hypothetical artefacts. These should be looked at as a whole, how each connects to the others. They can be regarded as components of a viable civilization such as buildings, water supplies, farms, roads, artistic works, etc. The significance of a hypothetical road then is also what possible buildings it connects to. A farm is significant in the context of possible buildings near it. Possibly artificial canals and lakes are significant in terms of their proximity to ancient oceans, also to dams in craters collecting groundwater.

Faces

The Queen Face

One of the most controversial problems with the evidence accumulated has been the discovery of Martian Faces. That they appear to look like us raises the suspicion of Pareidolia, like seeing faces in clouds. However Mars and Earth would have had their ecosystems connected by panspermia, this is where life can be transferred from one planet to another by meteors. We may then have had a similar genetic background, and so plants and animals may have evolved to look similar on both planets. Panspermia is a just a hypothesis, but we don't know whether DNA from Mars might have caused us to evolve later looking similar to Martian life. The Queen Face was discovered by the author recently, it is close to the Cydonia Face which was the first Martian Face discovered in 1976. There are about 30 Martian faces of varying degrees of plausibility. Some might see these reducing to the absurd, that the idea these could all form naturally as absurd in a way that is hard to define. Others might see the number of faces as statistically significant, a product of the law of large numbers. Still other might be unconvinced or believe they are random or illusory. Some find them quite shocking with the impression of artificiality they give.
This shows two versions of the Queen Face from different CTX images. It appears to have hat like a crown, like most of the other Martian faces.

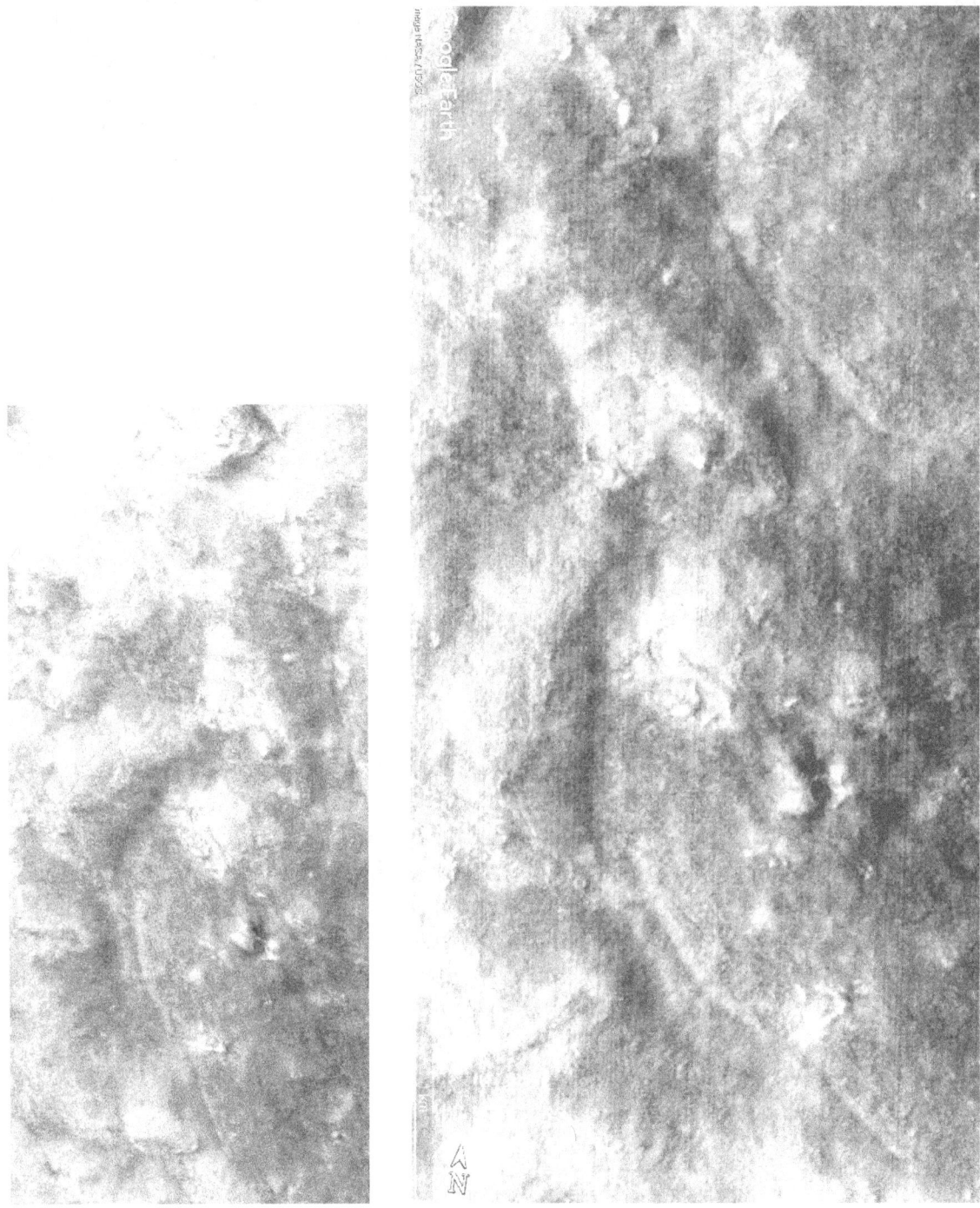

The High Face

Most of the Martian faces are found in a small valley in Libya Montes, near the better known Crowned or King Face. This is often referred to as the King's Valley, a similar name to the Valley of the Kings in Egypt. The High Face is named because it is high on a cliff overlooking the valley. The faces are discussed in two papers in Martian Hypotheses Volume 11. A statistical argument can be made, as to why so many faces would be found next to each other or to be on a great circle bisecting Mars.

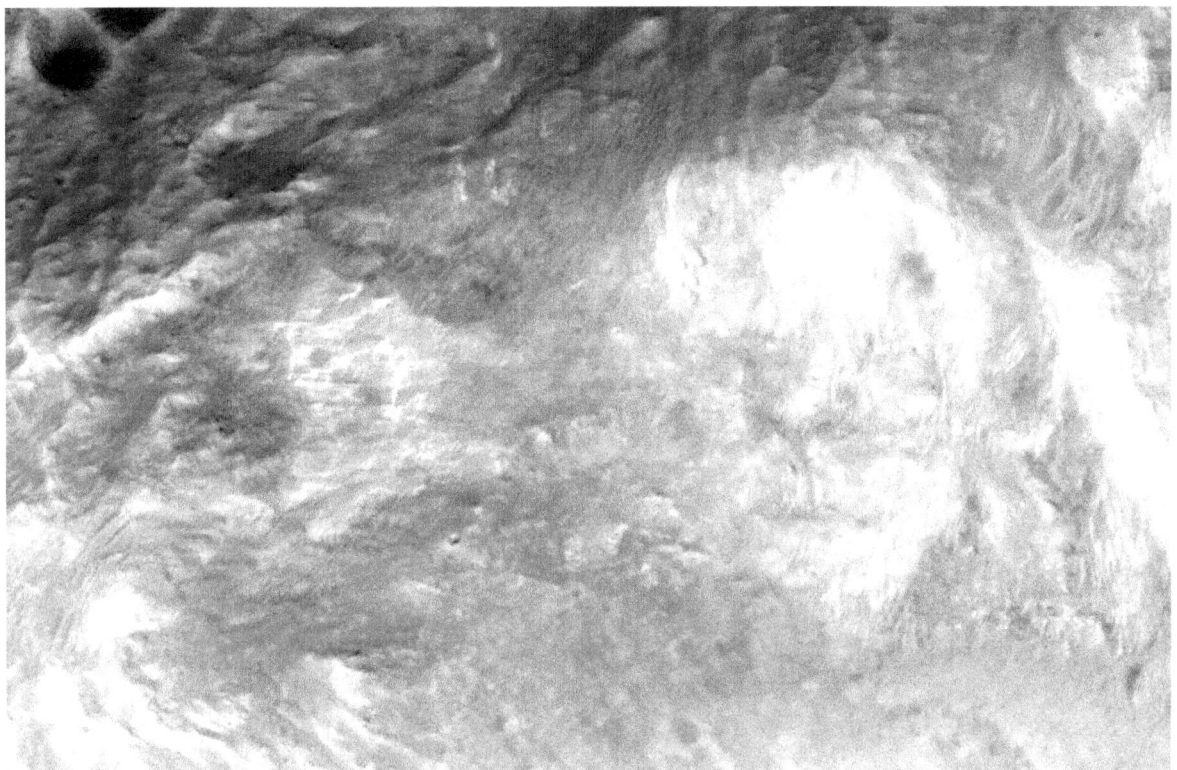

The Meridiani Face

This face was discovered in a Viking image by a Martian researcher Terry James. It is also discussed in Volume 11.

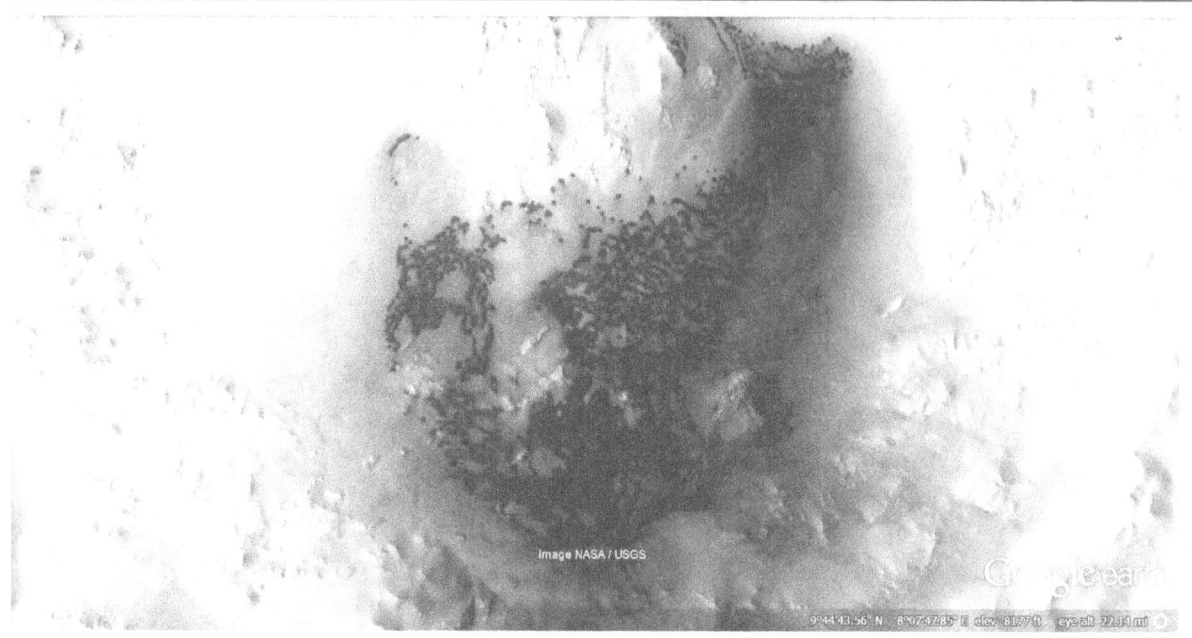

Nefertiti

This face was discovered by JP Levasseur, it is discussed in Volume 11. The two inserts are from higher resolution images that were recently taken by the HiRise orbiter, they were added by the author. It missed the whole face but shows some of the hat and face. It represents a successful prediction, that higher resolution imagery would make these formations more face like rather than appearing more natural.

The King Face

The King Face was discovered by the author in June 2000. It has been called the Crowned Face, however with the discovery of the feminine looking Queen Face the name King Face may be more appropriate. Whether they had sexes or if we could tell the difference is another hypothesis.

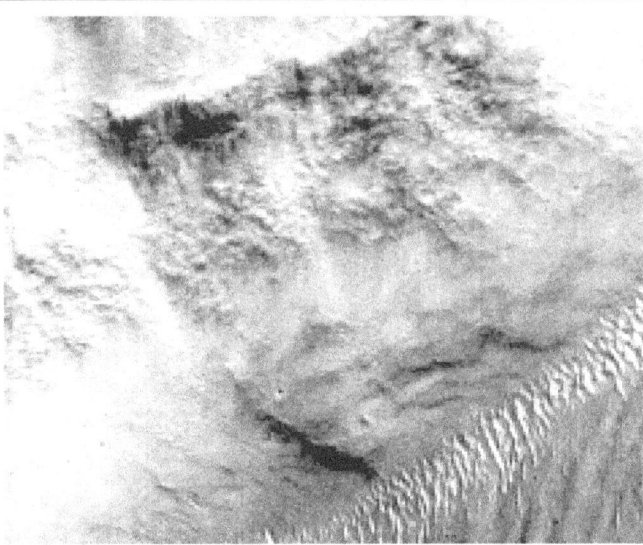

Dams

In many craters there are formations that look like dams, these seem to follow an old Martian equator implying that water may have been liquid in an equatorial zone. This old equator hypothesis is discussed more in Volumes 11 and 12. Most of these dams are parabolic in shape, the hypothesis is that parabolas are well suited for load bearing in dams. From here the analysis from the book is included with each example image.

Cymd259c

Hypothesis

These dams are in the same crater, A which appears parabolic and B have smooth walls with a few cracks as shown. B at 4 o'clock has a sharp edge to the dam wall in good condition. C at 4 and 6 o'clock show a secondary dam perhaps to catch the overflow, the second line at 6 o'clock shows the base of this wall. D shows another section, perhaps parabolic, with a cracked wall at 5 o'clock. C at 10 o'clock shows a probable parabolic arch. There appear to be faint vertical ridges on the upper part of the dam walls as seen in other dams, these may be for strengthening the wall such as there being pillars inside.

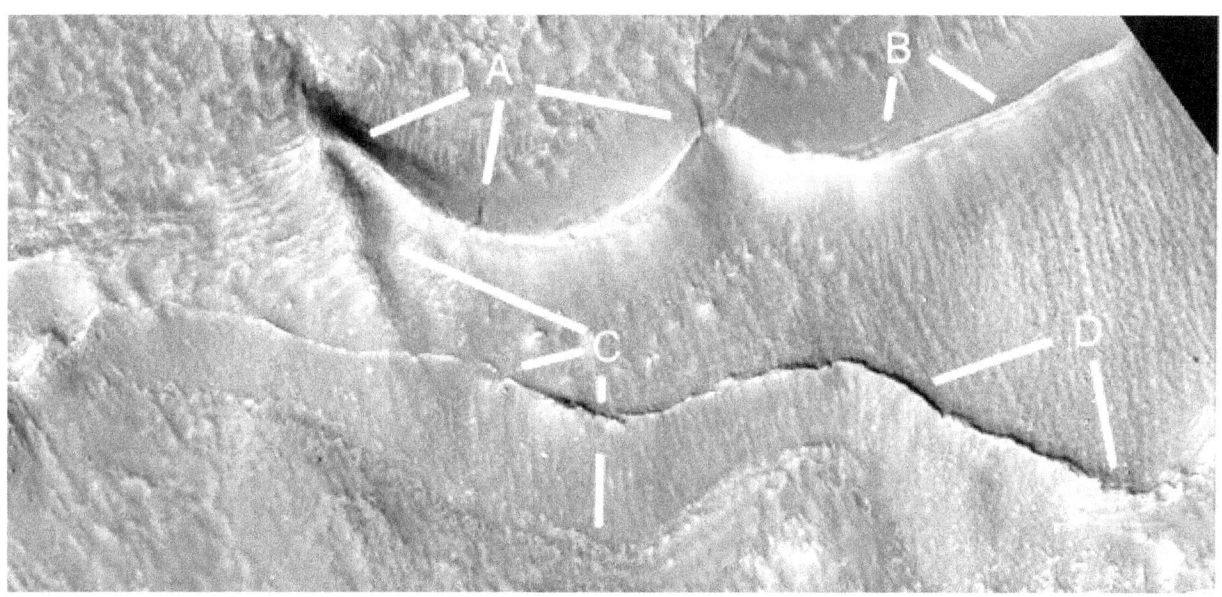

Cymd259c2

Hypothesis

A parabola is shown.

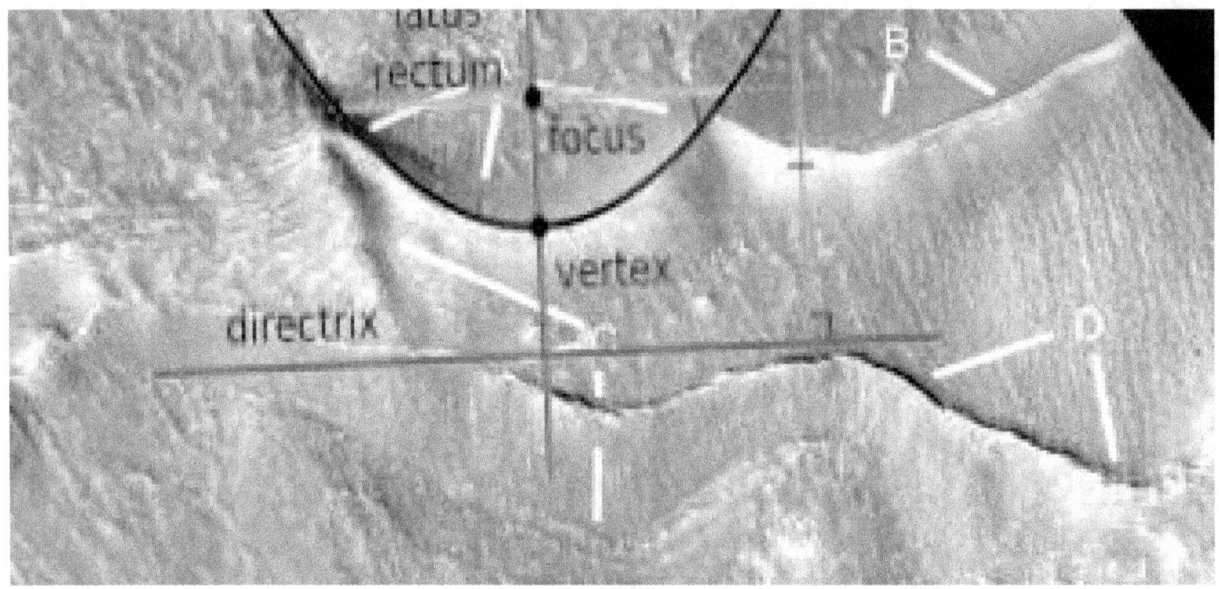

Cymd280a

Hypothesis

A shows how the skin on the dam wall is peeled off, at 3 o'clock is has many pits like on the skin of hollow hills. At 4 o'clock this rough interior is exposed but just below it the skin is smooth. At 6 o'clock is another edge of the smooth skin. B shows at 8 o'clock. How it is peeling off, at 5 o'clock it is more stable. At 10 o'clock there are many pits as it degrades, at 2 o'clock it shows the lip of the dam has broken off. C shows a smooth area that goes up to the broken lip of the dam wall like an external layer, perhaps a patch.

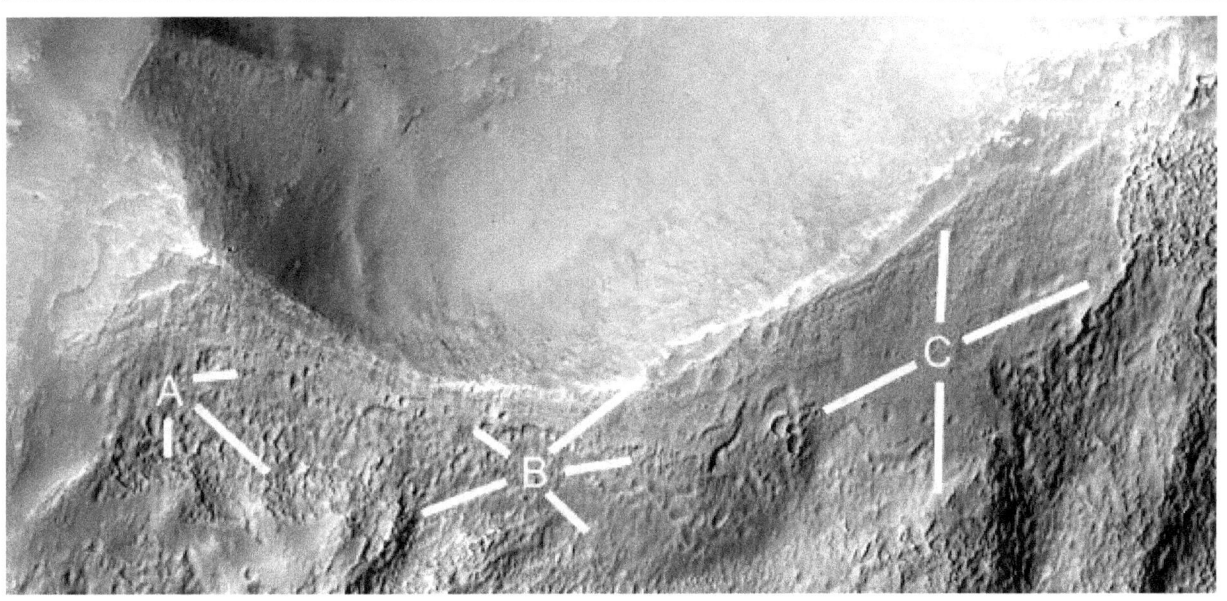

Cymd280a2

Hypothesis

A parabola is shown.

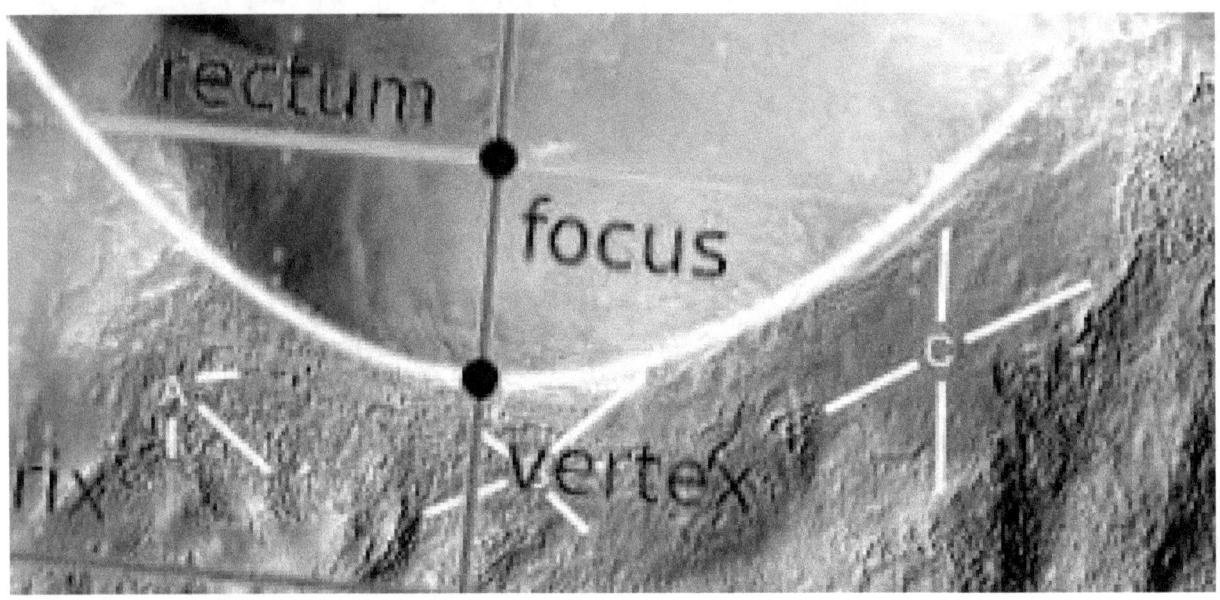

Cymd280i

Hypothesis

Engineers might examine how this wall is fracturing at A to D, Also D at 2 o'clock shows the thicker base holding the dam wall in place. Above C the dam floor is smooth like cement, higher up and outside the dam the terrain is much rougher.

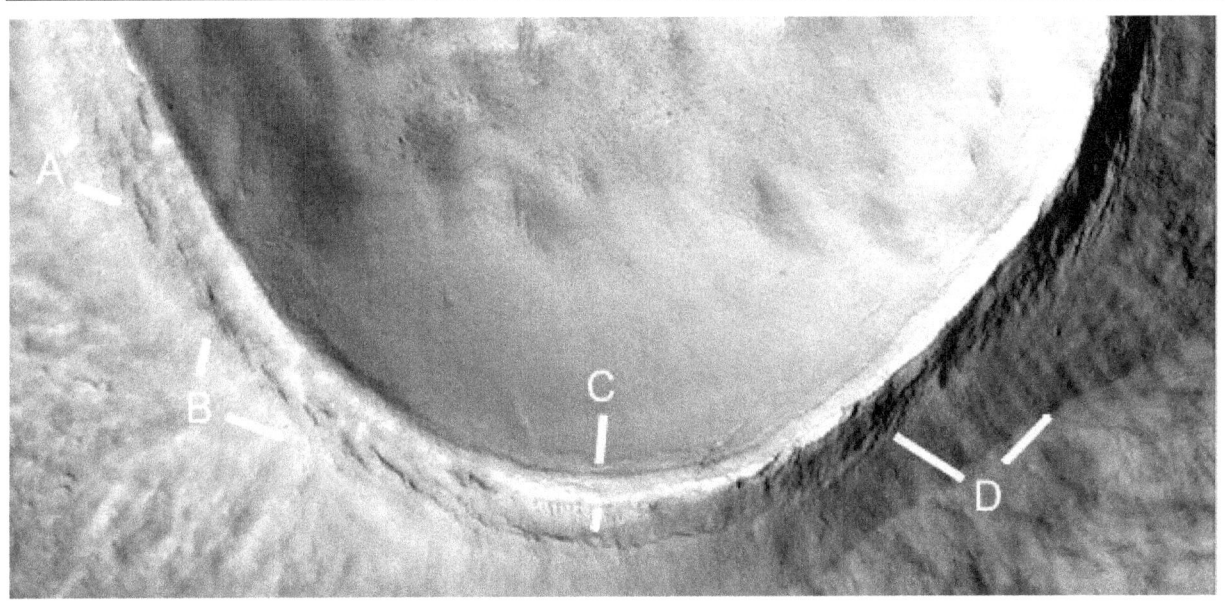

Cymd280i2

Hypothesis

A parabola is shown.

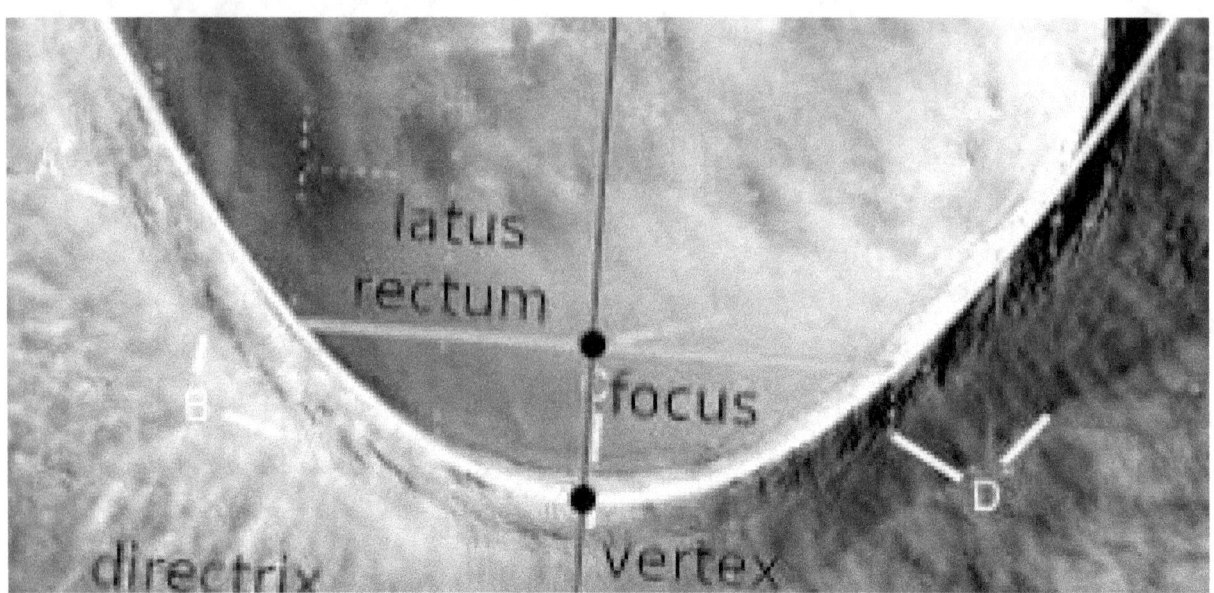

Cymd408a

Hypothesis

An unusual shape pointing up the crater wall, A is one dam, B may show some creep or cold flow in the dam, this where over time rock might slowly flow like a viscous liquid. C shows a smooth dam floor like cement, different to the terrain outside the dams. D at 7 o'clock also shows the smooth dam floor compared to the ground above it. At 2 o'clock the wall is eroded or breaking.

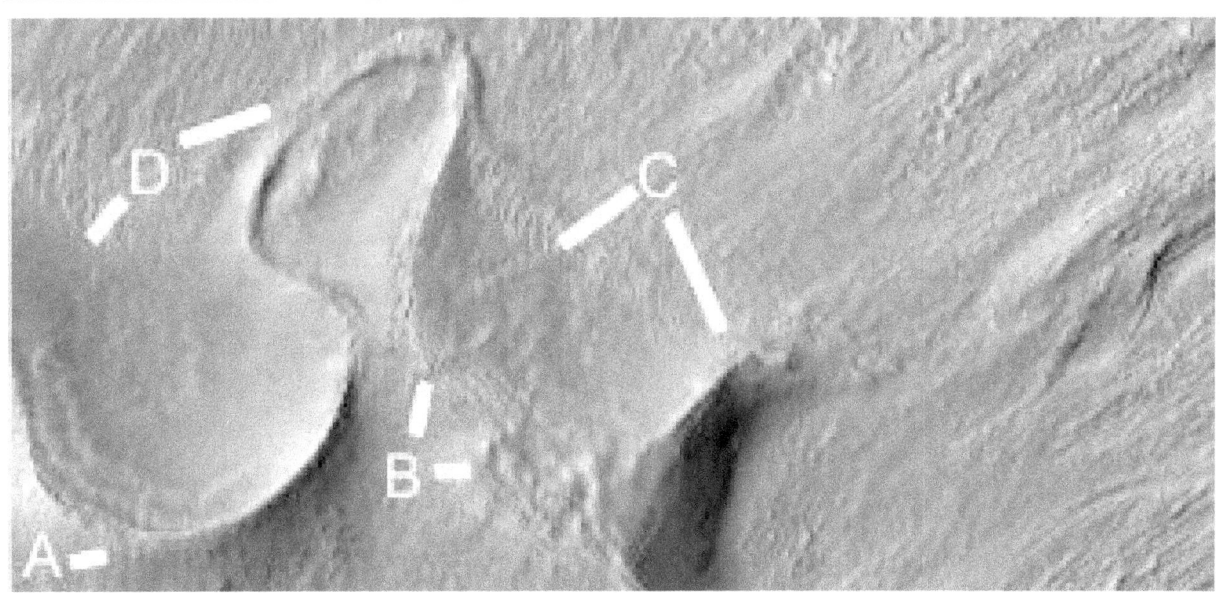

Cymd408a2

Hypothesis

This shows 4 parabolas making up the formation. These would have used the load bearing properties of the parabola to resist erosion. The straight dam at B may have broken because it did not use a parabola.

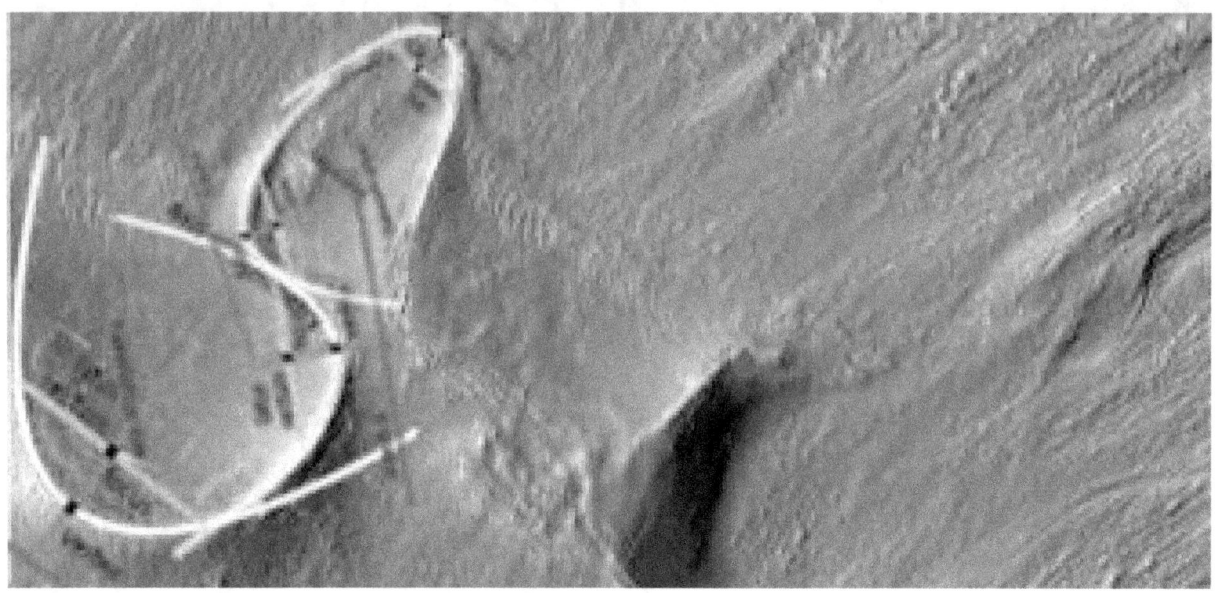

Argd1444a

Hypothesis

Eighteen parabolic dams are shown. A few others are too eroded to determine their shape. It would seem impossible for eighteen mud slumps to happen to form perfect parabolas, above them the materials look highly random by contrast.

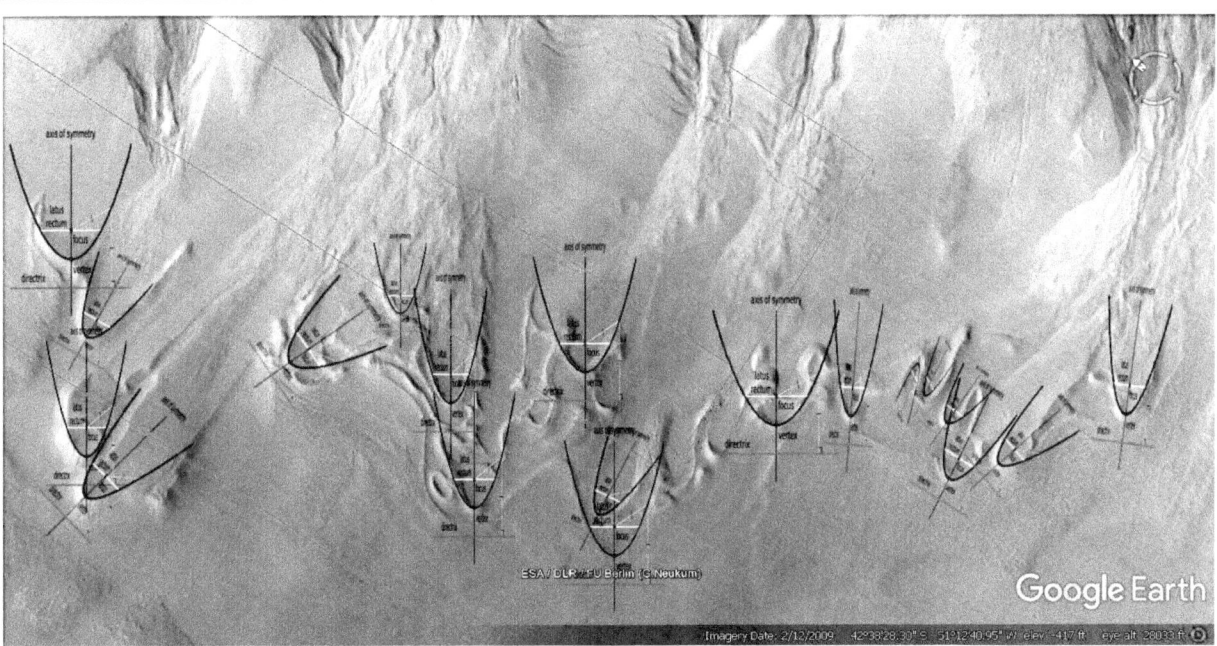

Canals

Some areas near hypothetical Martian buildings and dams have these canal like formations. The hypothesis is that water was important in this civilization, they used dams in craters to collect water often associated with water channels and perhaps pipes. In other areas canals may have brought water from the lakes and oceans, perhaps irrigating farming and residential areas or even for transport using boats. This is what we use canals for on Earth.

Prca480

Hypothesis

More of these tube shapes, A shows dark spots along it like it is breaking up. B at 9 o'clock is like a hollow hill as seen in many other areas, the dark patch on top may be the roof. B at 5 o'clock shows more collapsed areas. C at 7 o'clock shows the bank is well defined, at 4 and 8 o'clock the tube shape changes from dark to pale. At 10 and 4 o'clock the bank is also well defined.

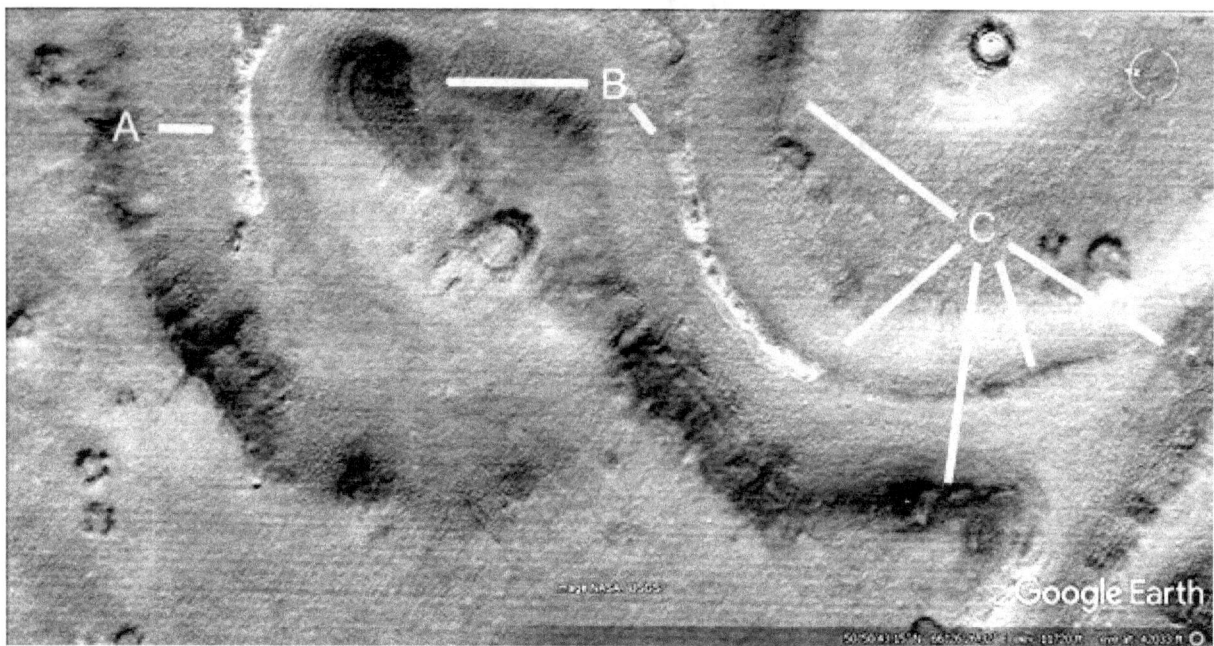

Prca480a

Hypothesis

This part of the tube shape is a near perfect parabola as shown, unlikely to occur by chance. The tube shape is also about the same height and width wherever seen, it does not vary much randomly like a natural formation from weather erosion. Also parabolas are shown in canals as well as dams, a natural hypothesis would need to explain how geological processes formed parabolas in each. They also appear in hypothetical buildings and as walls around possible farms.

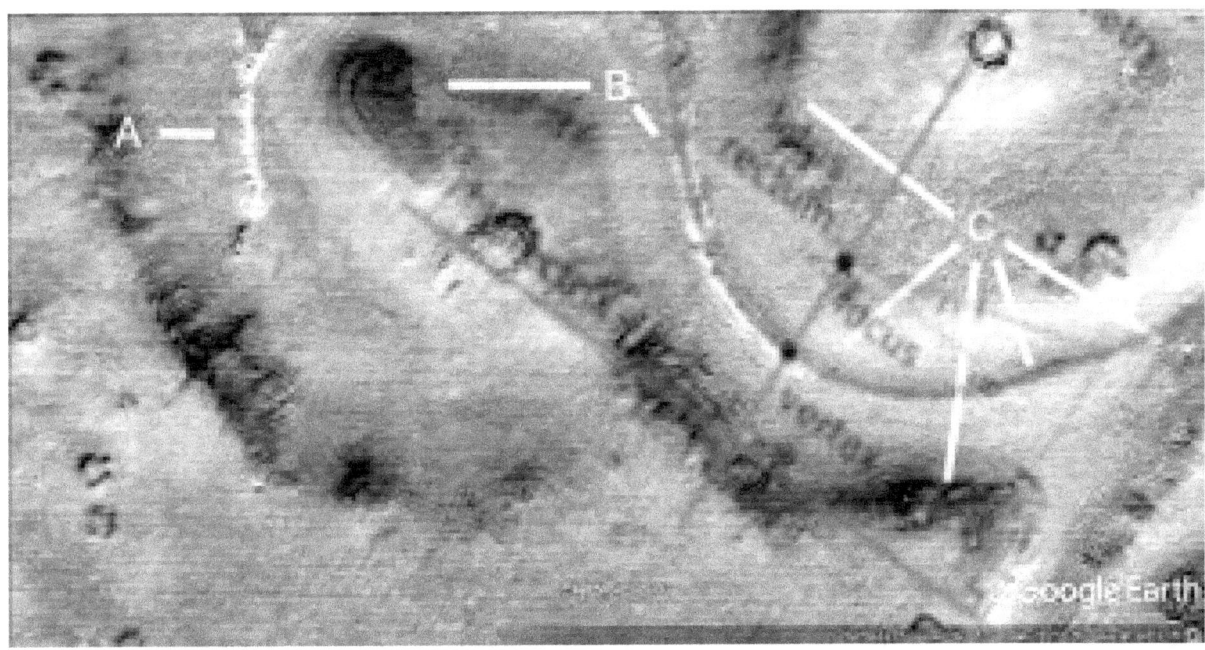

Ect1619

Hypothesis

A shows a much thicker wall with a line running along it as a peak, from 4 o'clock to B at 5 o'clock, up to E. This may have been a habitat connected by hollow walls. At 2 and 6 o'clock A shows a clean edge like cement to the dam floor. B at 9 o'clock shows a double wall like a collapsed tube. At 3 o'clock B shows a small hill or dark area. C may be a collapsed hollow hill, the ridge shown may have been an interior support and part of the larger hollow wall. D shows a darker line perhaps a collapsed wall, also a narrow wall like those in Hellas at 1 o'clock second leg.

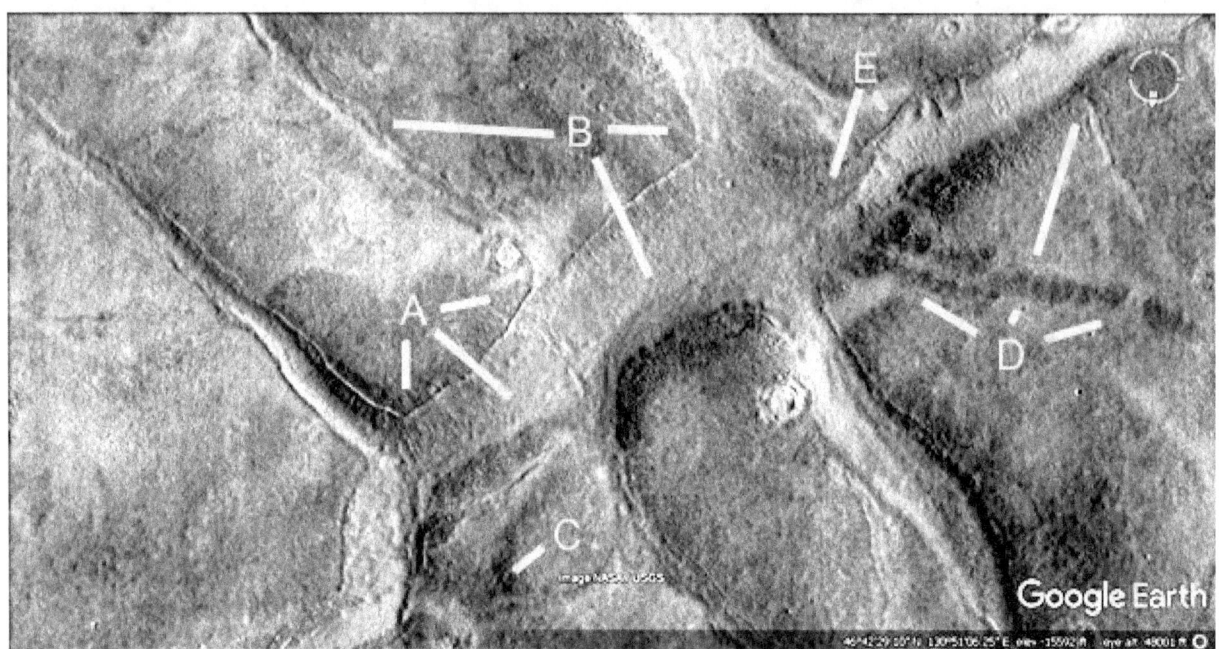

Ect1619a

Hypothesis

Four parabolas are shown.

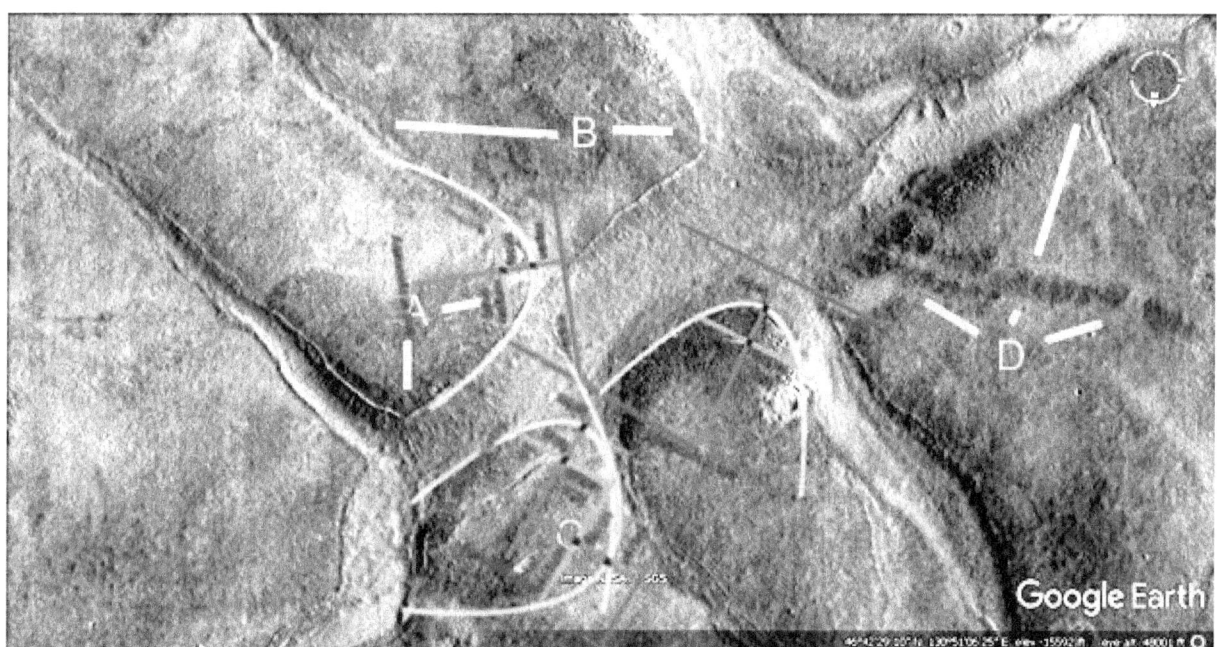

Ect1643

Hypothesis

A shows more ridges like grout, these connect into the canal wall at B but do not extend into the canal embankment. C shows regular spacing like tiles at 11 o'clock, squarish tiles at 3 o'clock, and a collapsed tile segment at 6 o'clock. D shows a gap growing between the bank and the wall, also with regular tile spacings. At 6 o'clock second leg there is a ridge like grout. E shows more grout connecting to the canal wall like a single segment. This cannot be cracks then because it must be the same material as the wall, probably cement. F shows more tiles.

Ect1643a

Hypothesis

A parabola is shown.

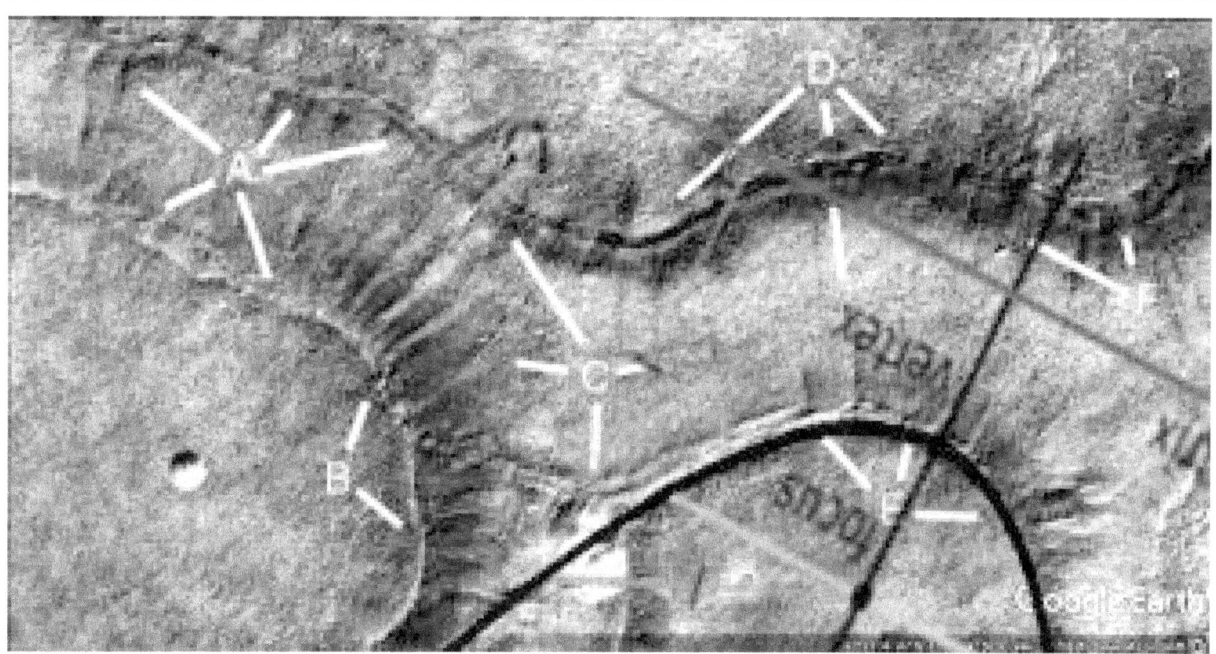

Water channels

Water channels can encompass the conduits feeding dams in crater, they can extend up to the hypothesis of large scale canals. They would have been important, to direct water into dams instead of being dissipated into the ground. Also there are overflow water channels which appear to direct water from an overflowing dam to another so as not to waste water.

Prd965c

Hypothesis

These may have been canals or pit dams, they are highly geometric in shape. A shows a dam for water at 12 o'clock, another wall for a dam and channel at 3 to 5 o'clock. B shows a wall for a canal from 2 to 7 o'clock, it has a groove running along the top like a double wall.

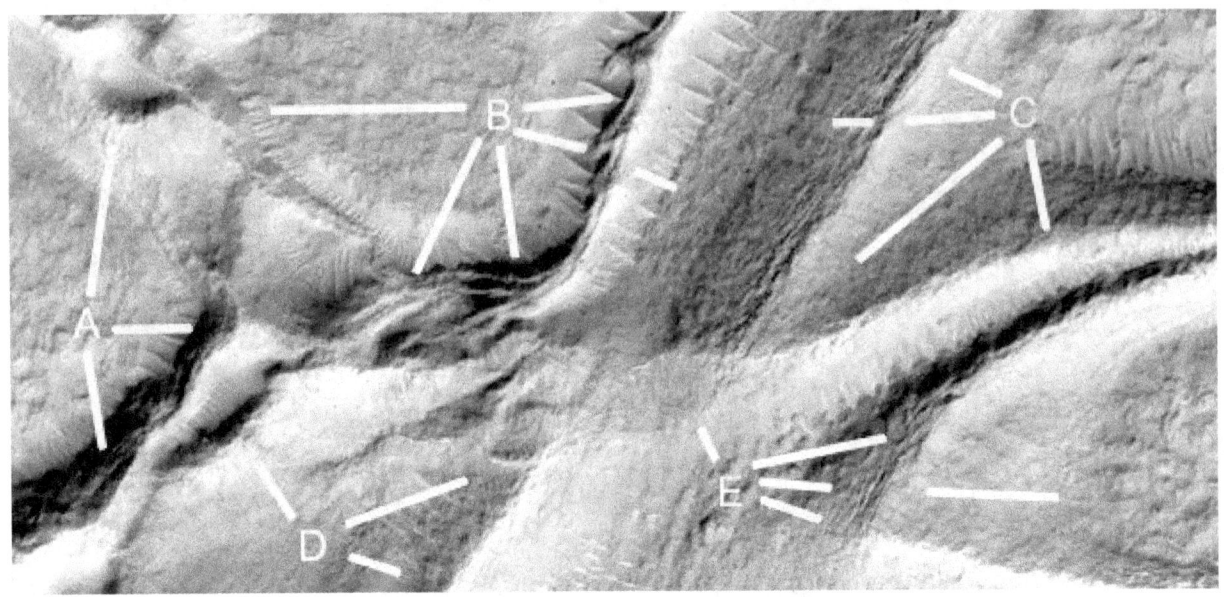

Prd965c2

Hypothesis

Part of a parabola is shown. The lines show how straight parts of the formation are.

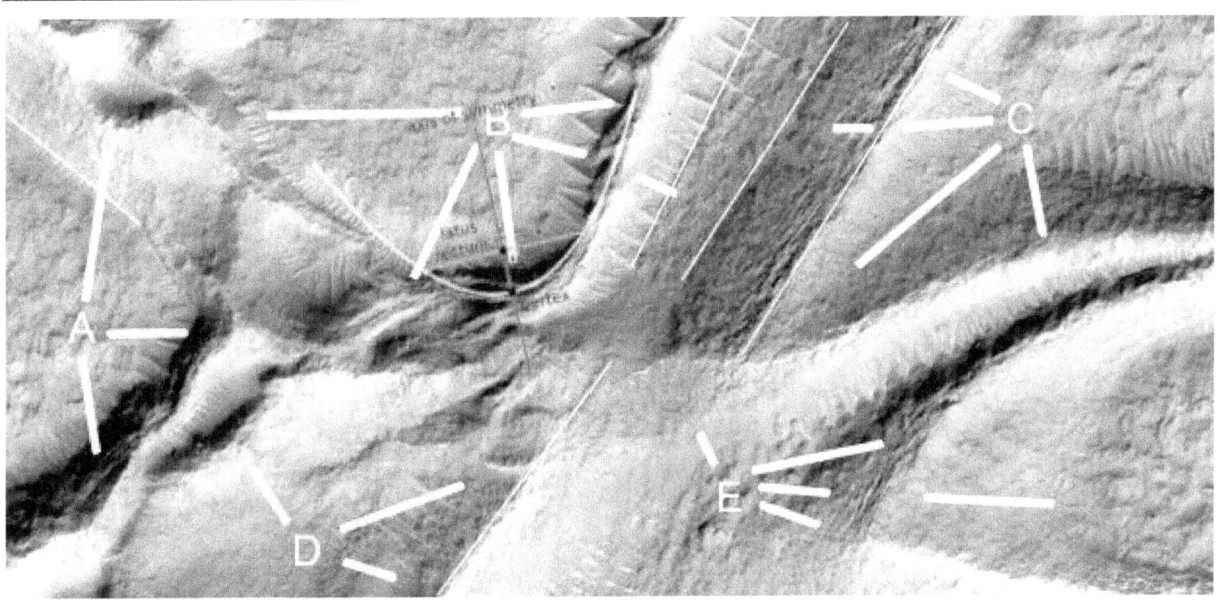

Cymd454h

Hypothesis

A and B show the sides of a water channel, water would have flowed across this at C to another dam. The shape appears so artificial that a natural explanation is hard to sustain.

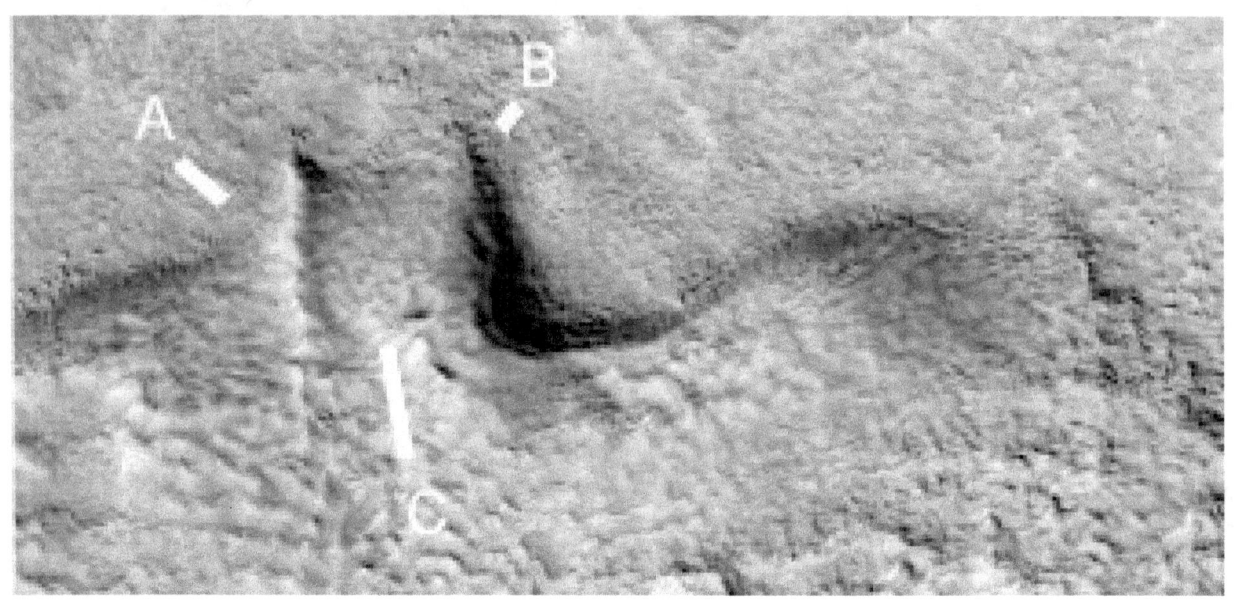

Cymd454h2

A parabola is shown.

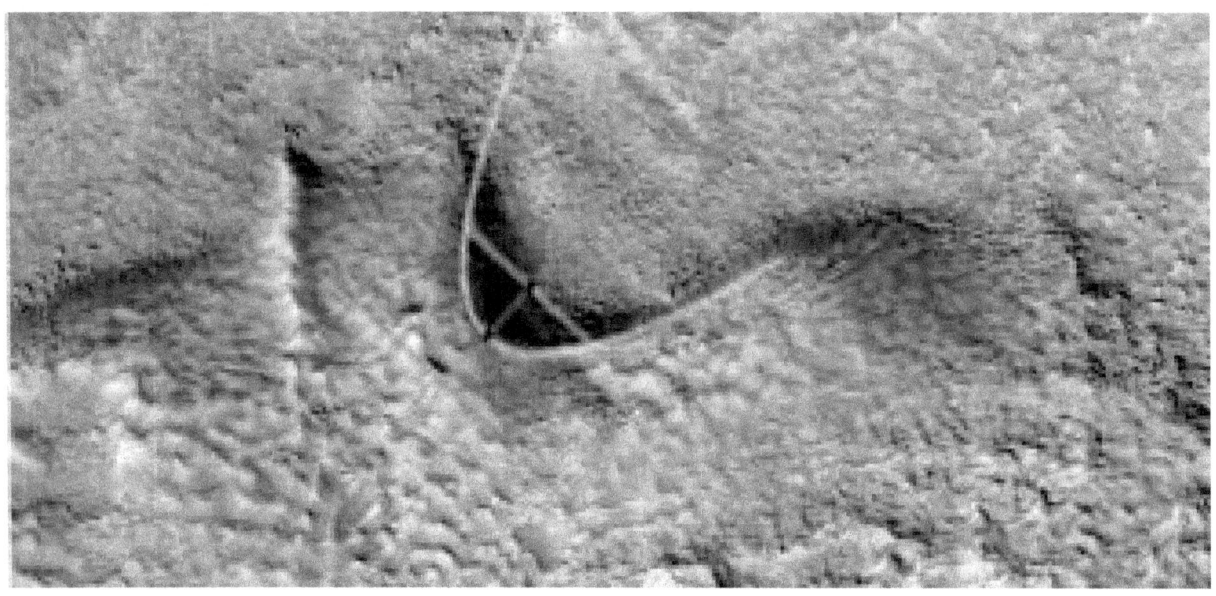

Held1095f

Hypothesis

A shows more dams, turned on its side to fit into the page. B shows a dam wall in good condition at 11 and 3 o'clock, one with cracks at 5 o'clock. C shows more cracks at 5 and 6 o'clock, in good condition at 7 o'clock. D and E also show walls in good condition. F shows more cracks developing.

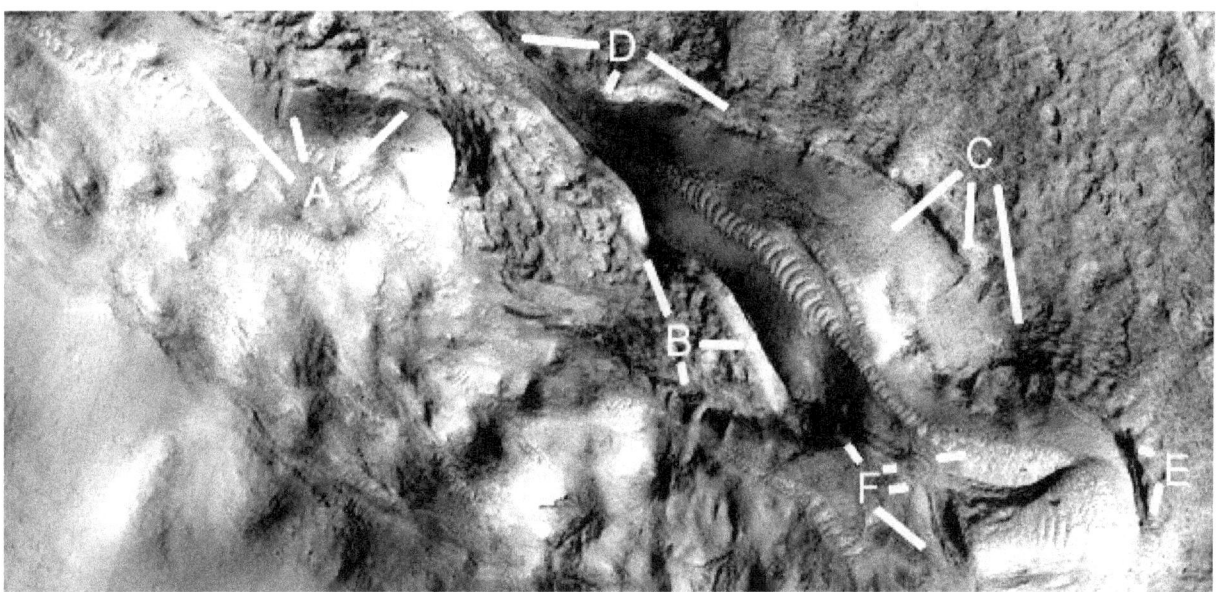

Held1095f2

Hypothesis

At least 5 parabolas occur in the formation.

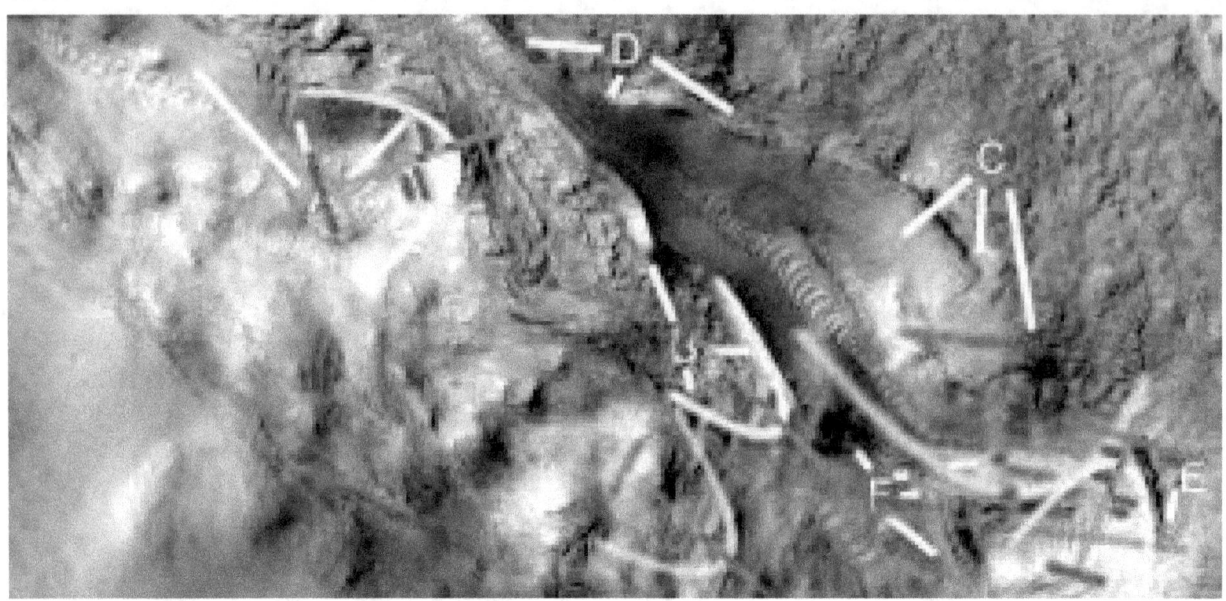

Ect1731k

Hypothesis

A shows a water channel going into a pit dam, B shows another water channel coming from this from 10 to 4 o'clock, also another water channel at 7 o'clock second leg. C shows a water channel coming from the other side of the pit dam to B. D shows a small water channel connecting two pit dams.

Ect1731k2

Hypothesis

Eight parabolas are shown, though there would also be some smaller ones and the water channel at C.

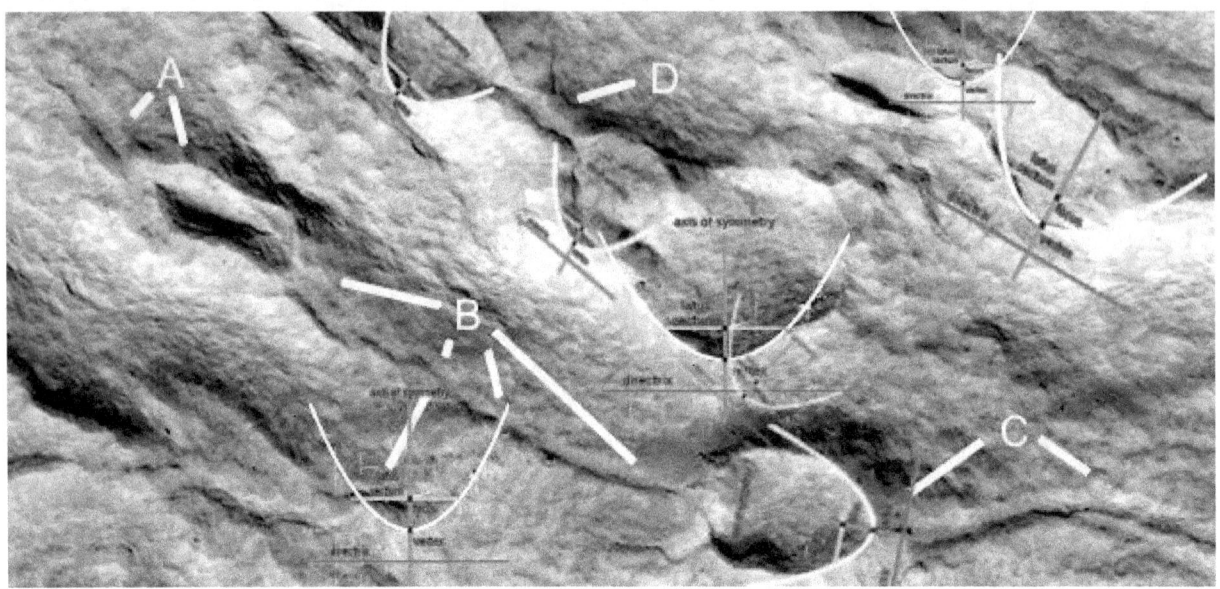

Cities

There are formations that look like cities, these are also clustered around this old Martian equator. Some are also clustered around large extinct volcanoes like Olympus Mons. It adds to the global hypothesis, that these creatures lived together in these buildings in warmer areas.

Cymhh209o

Hypothesis

A shows many rooms, also the walls here appear to be doubled or are collapsed tubes. This is important for the room hypothesis, if someone could go to each room in these tubes then each is accessible. If not then how many could be used is problematic. The thicker ridges also appear hollow at some points elsewhere, B shows a main tube that has some collapsed areas along it. C shows an area that may have eroded to the bare ground, there are faint walls here the same as in the other parts. C at 11 o'clock has very high walls as see from the shadows. Engineers could calculate the height of these walls from the shadow knowing the sun angle from HiRise. The higher the wall the longer the shadow would be inside the room. At C at 8 o'clock the walls are lower as if eroding. D at 5 o'clock shows a rounded formation of rooms like a nexus, at 8 o'clock the walls have collapsed apparently leaving some pillars standing in some cases. E shows a zig zag in this wall or tube, as if the access to it gives straight sections for the entrances. F shows areas where the ceiling appears to have either fallen onto the walls or is still secured above them in parts.

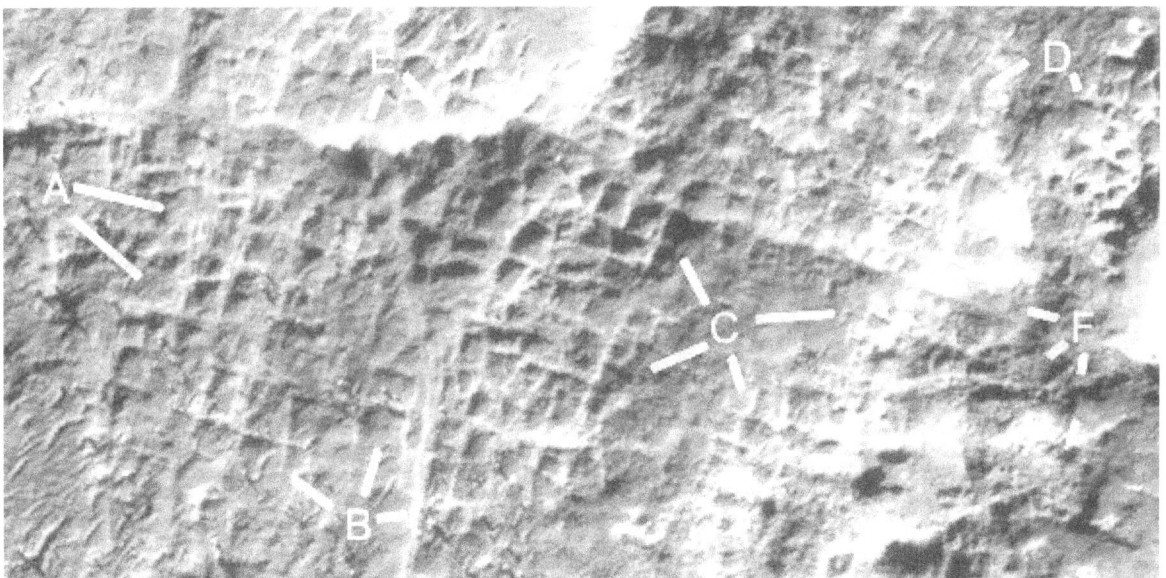

Cymhh361i

Hypothesis

The three dimensional impression is even stronger here, A shows rooms appearing under the smooth ceiling material. B may also be tubes or suspended roads as there is an impression of empty space under them. C at 9 o'clock shows rooms with no ceilings, at 4 o'clock there is still some ceiling or they are full of soil. D at 9 o'clock is like a hill of rooms, at 1 and 2 o'clock there is a road like formation that goes on to 12 and 2 o'clock. The letter E is in a depression surrounded by higher rooms like at 7 and 8 o'clock. F shows more variations in the elevations of the rooms from the shadow. G has many straight walls and may have right angles from directly above it. The rooms at H appear to be partially eroded.

Cymhh469g

Hypothesis

A at 10 o'clock shows a hill with room like shapes on its lower side, at 3 and 5 o'clock are more rooms. B and C show many walled rooms. D shows rooms that may be partially buried by the dark soil, or they ended in this open area. E shows more degraded rooms, F at 10 o'clock shows a nexus where many walls converge to it. At 3 and 4 o'clock there are perhaps rooms under the dark soil. G at 10, 12, and 1 o'clock as well as H at 12 o'clock follow this edge of the rooms, this section may be an intact ceiling with rooms under it.

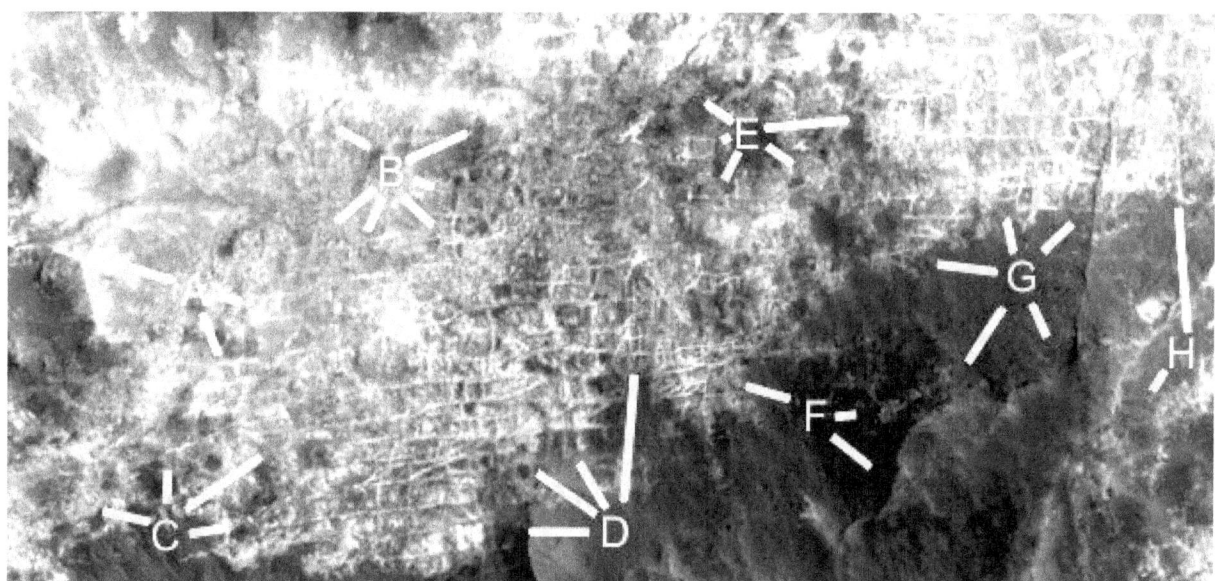

Cymhh469g2

Hypothesis

There are many lines here showing how straight the walls are, but many more could have been drawn as well.

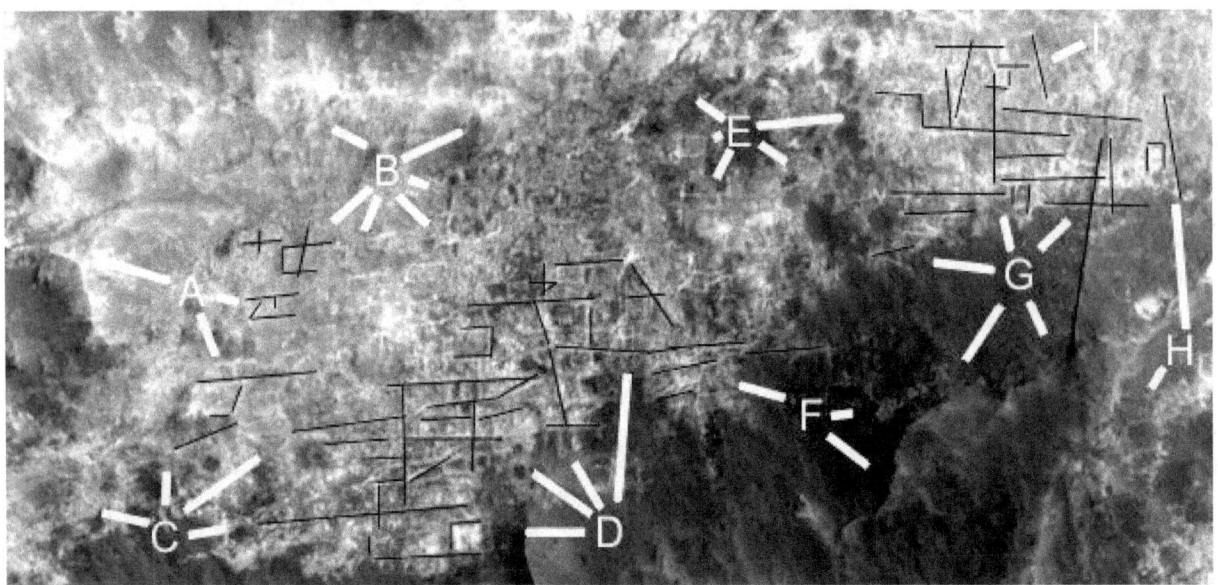

Buildings

Some individual formations look like large buildings, sometimes incorporating parabolas.

Cymhh467

Hypothesis

A may show some collapsed hollow hills. B shows some straight ridges, perhaps interior supports of this larger formation. From C to D is a curved interior support. E may be a collapsed section, F shows some tubes or walls.

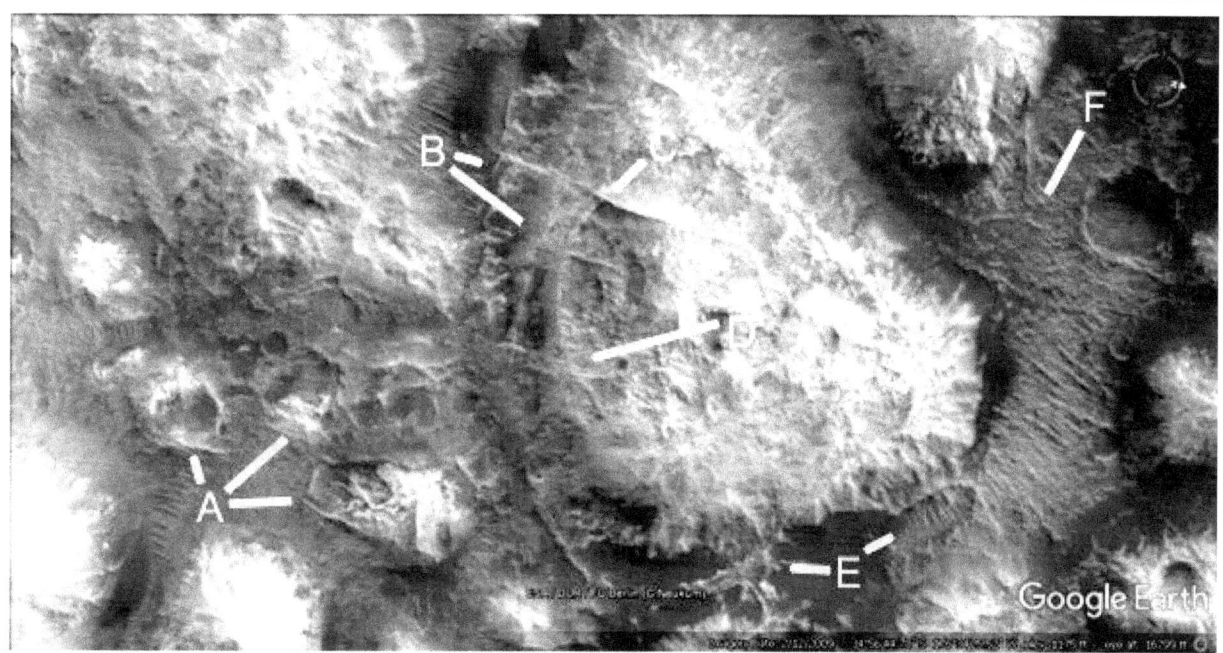

Cymhh467a

Hypothesis

There are two parabolas in this formation, as well as the straight walls.

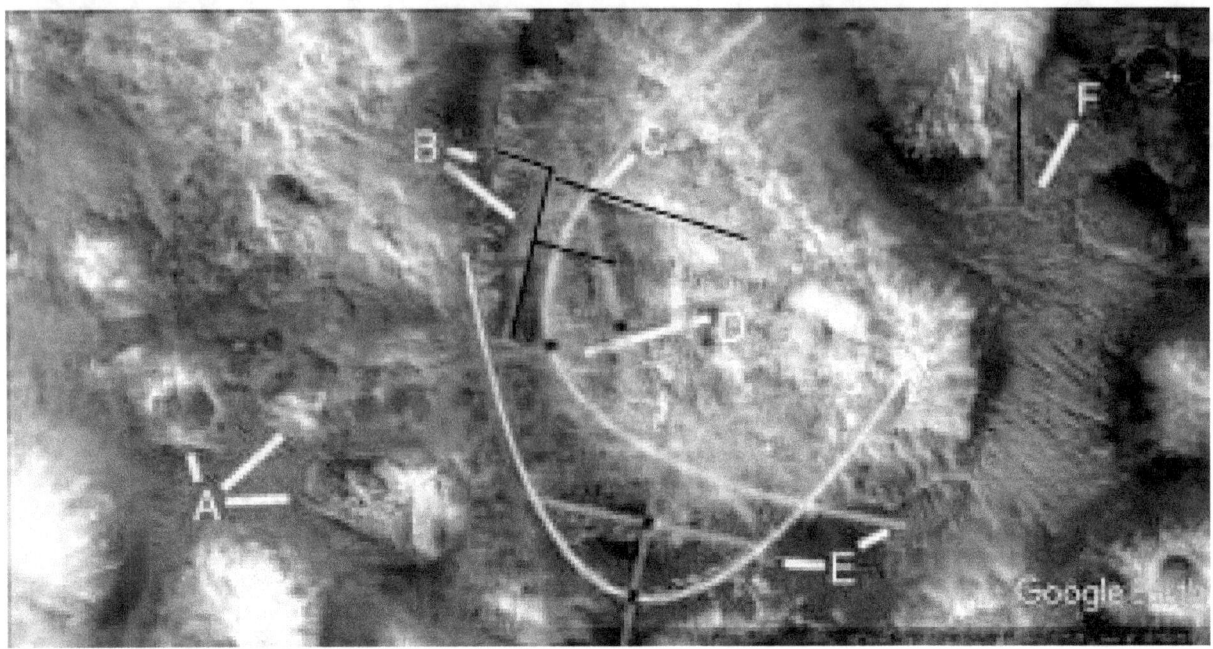

The hills often shows collapsed segments on their roofs so being hollow is implied. That adds to the hypothesis, that they lived in these hollow hills, and travelled between them on these roads.

Prhh944c

Hypothesis

The top of the layer here is shown at A at 12 o'clock, at 10 o'clock is a tube. B shows multiple layers under it, this may be the construction technique. C shows a broken wall segment at 8 o'clock second leg, this may be two thinner layers broken together. At the first leg is a tube. At 9 o'clock second leg is another broken layer. At 6 o'clock the tube appears to come from here, this has a collapsed side and a gap between it and 8 o'clock first leg. At 12 o'clock the texture of the roof is different to the wall layers.

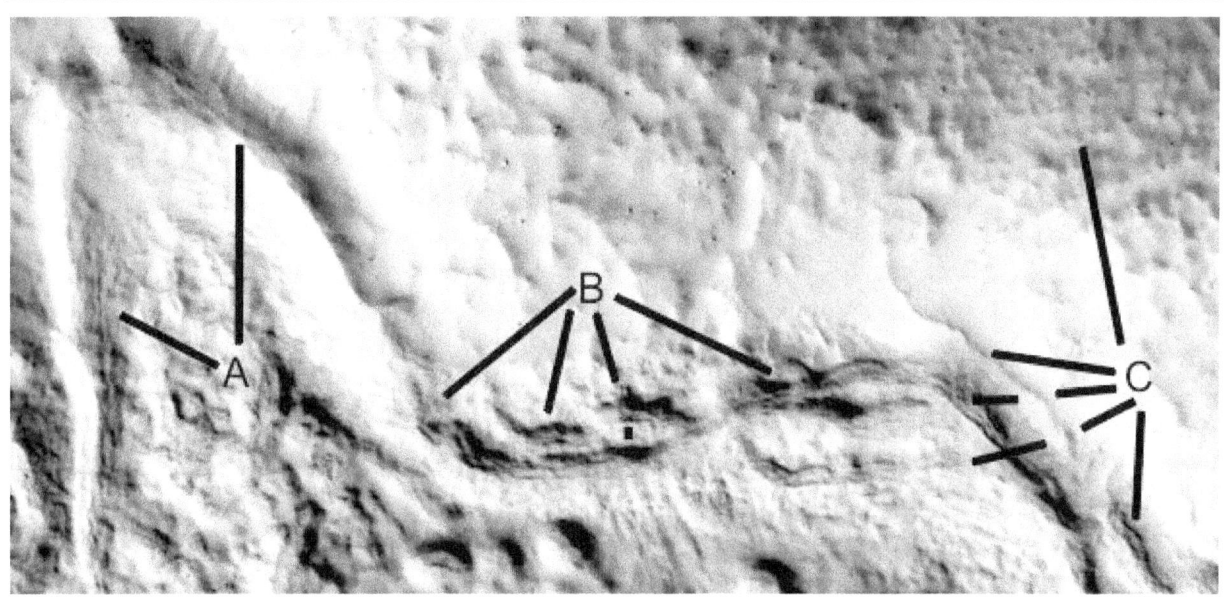

Prhh944c2

Hypothesis

Three parabolas are shown, like a parabolic wave. This can be an approximation to ocean waves which are elliptical.

Prhh944f

Hypothesis

A shows tubes or eroded segments on the roof. B shows contours which may have been used for strengthening the roof. C shows a settled area. D shows many parabolic arcs to strengthen the roof at 9 and 10 o'clock, at 2 o'clock there is an exposed grid perhaps used for reinforcing the roof.

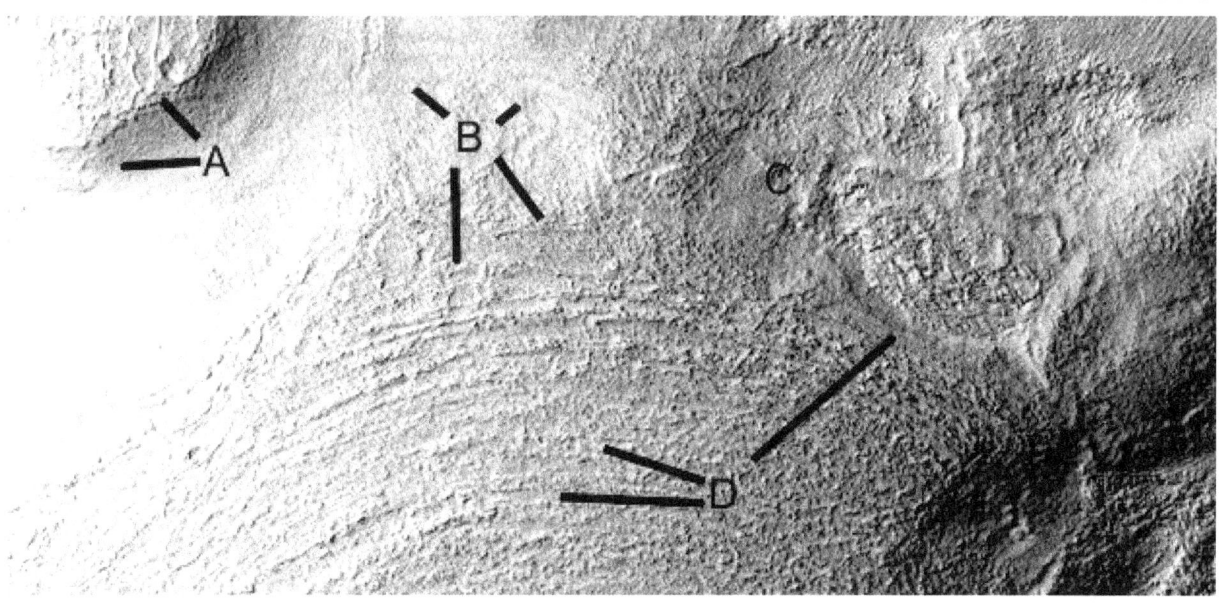

Prhh944f2

Hypothesis

Three parabolas are shown, there are several more but these are the clearest. The axis of symmetry of each is closely aligned but each parabola is smaller than the one surrounding it.

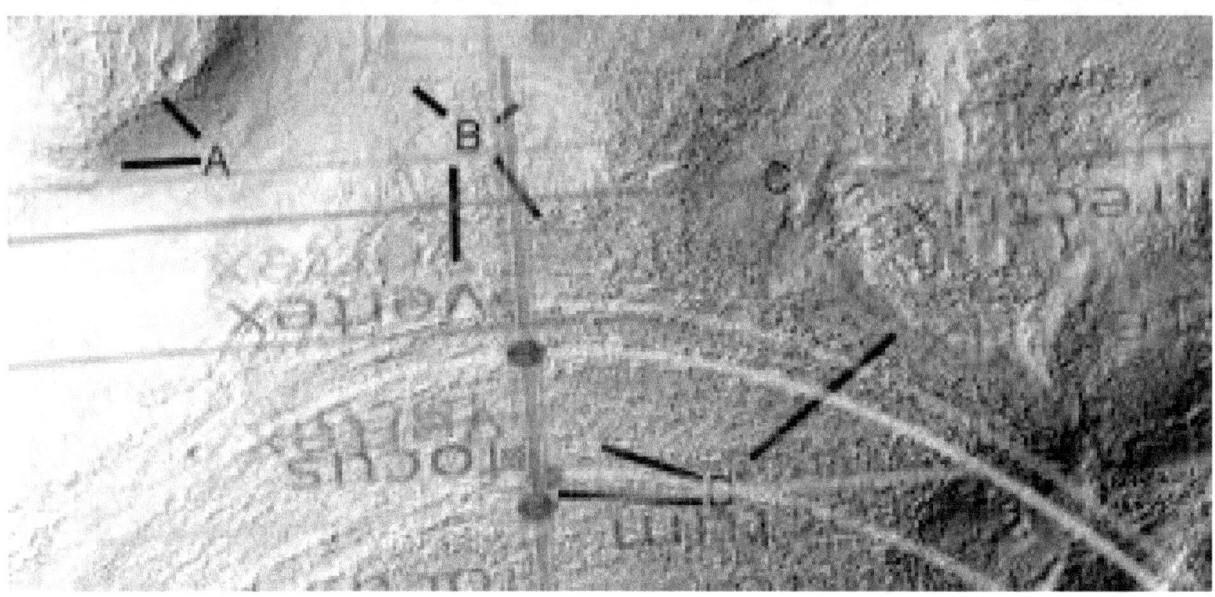

Prhh944j

Hypothesis

This may be a Cobler Dome where the parabolic layers of bricks are exposed. They are less visible at A at 10 o'clock, at 4 o'clock the top of the hill may be peeling off. B shows a smooth skin like cement that may have broken off on the upper side exposing the layers. C shows the parabolic layers, D shows two skins that have eroded away exposing the arcs.

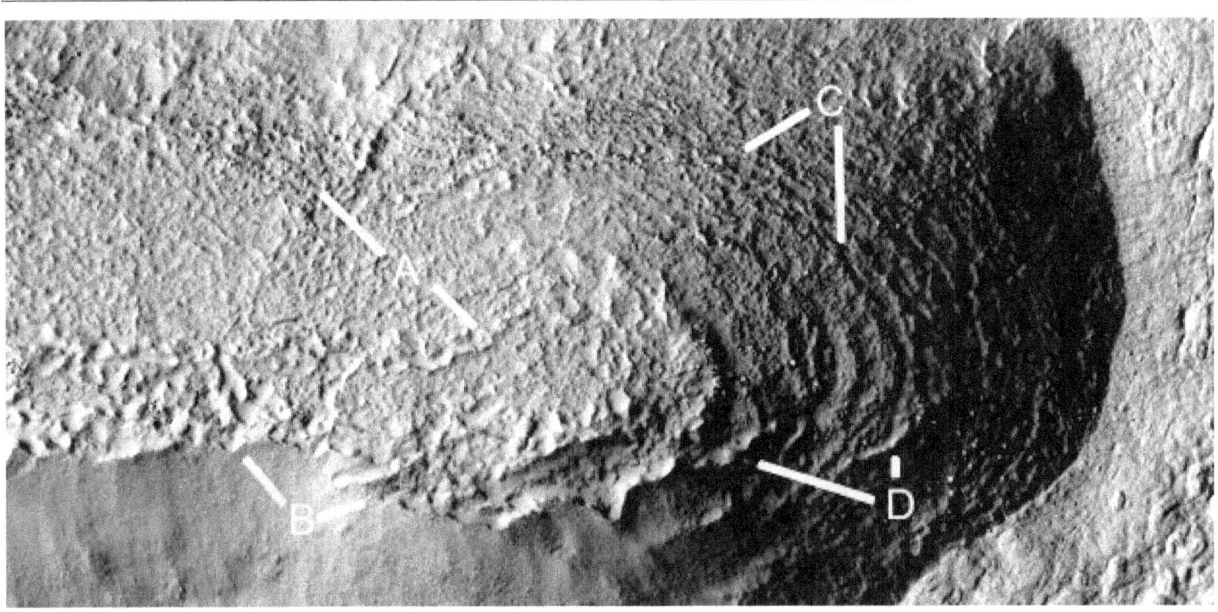

Prhh944j2

Hypothesis

Three parabolas are shown, there are several more which are too faint. Straight ridges are also overlaid by lines.

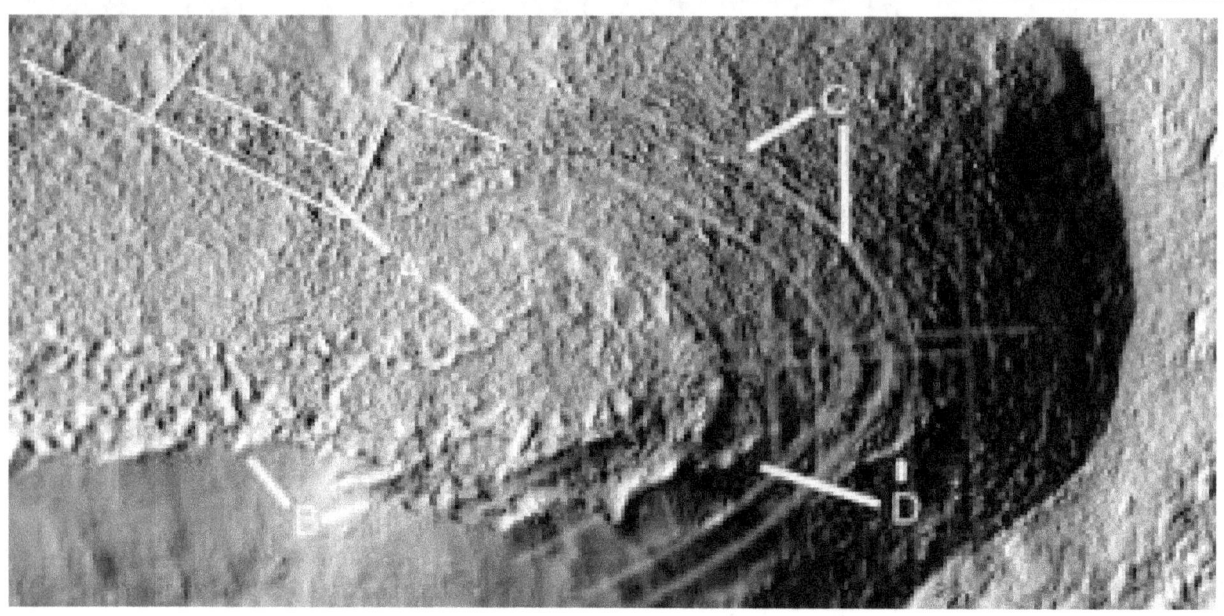

Helhh1117

Hypothesis

A shows the curved segments of the hollow hill roof. B may be a collapsed segment of the roof. C at 2 and 4 o'clock may be a tube, at 5 o'clock an interior support with some settled segments of the roof around it. D at 1 o'clock may show a tunnel going into the hill continuing on at 4 o'clock perhaps as a collapsed tube.

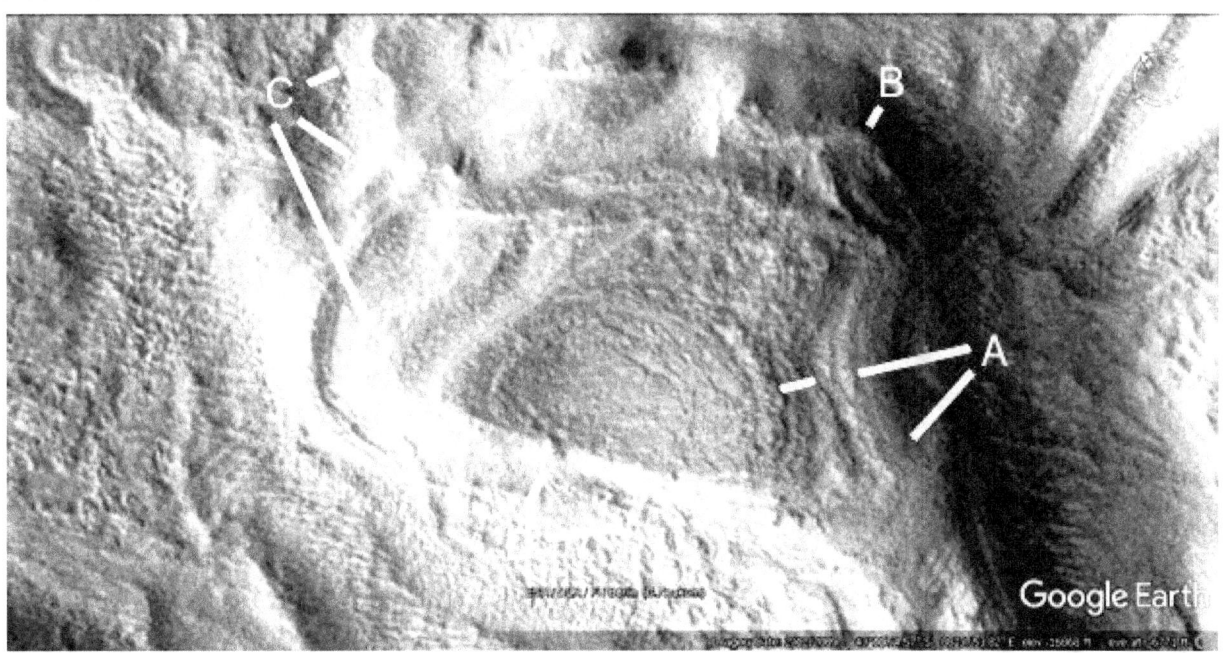

Held117a

Hypothesis

The edge of the rounded segment of the roof forms a parabola, the flat side lines up well with the latis rectum, the name for the line through the focus. The ends of a parabolic formation often deviate from the perfect parabola, shown at E. This may be because the parabola was not used to be a geometric statement to be viewed. Instead it was hypothetically used to make the formations stronger. These edge at E would serve no purpose to continue here as a parabola. This corner may also have been a small parabola to make it stronger.

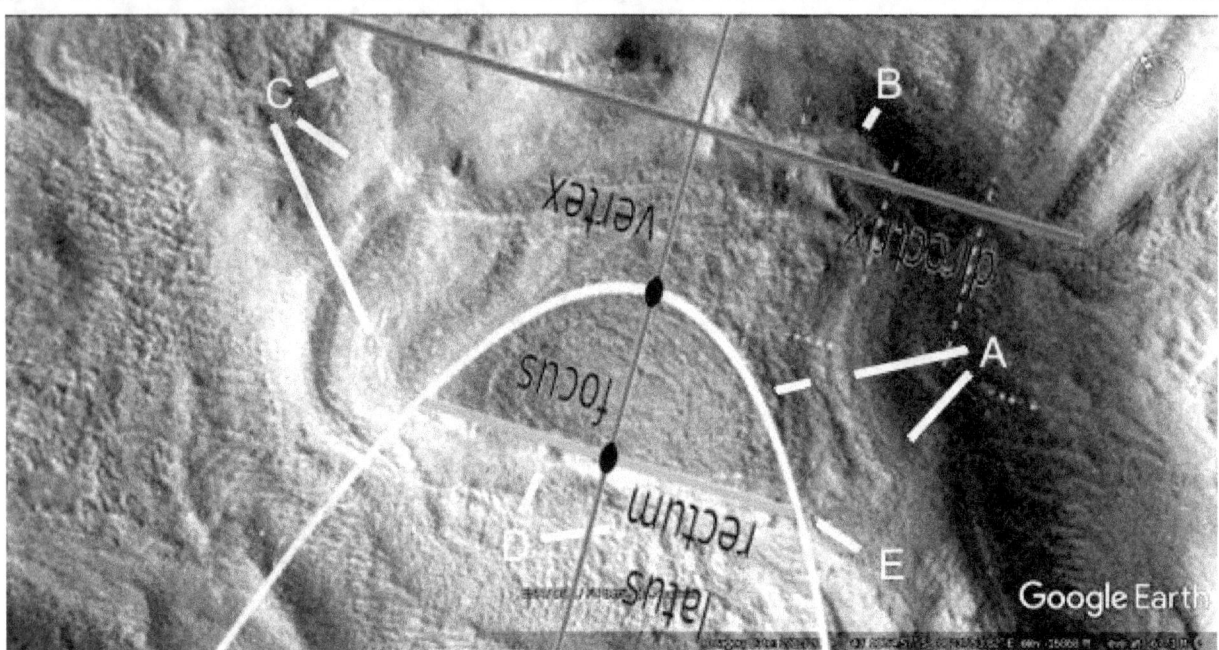

Walled fields

The hypothesis is that these may have been used for farming, or for pools of water containing fish.

Held1186

Hypothesis

These walls are much straighter and with more right angles between them.

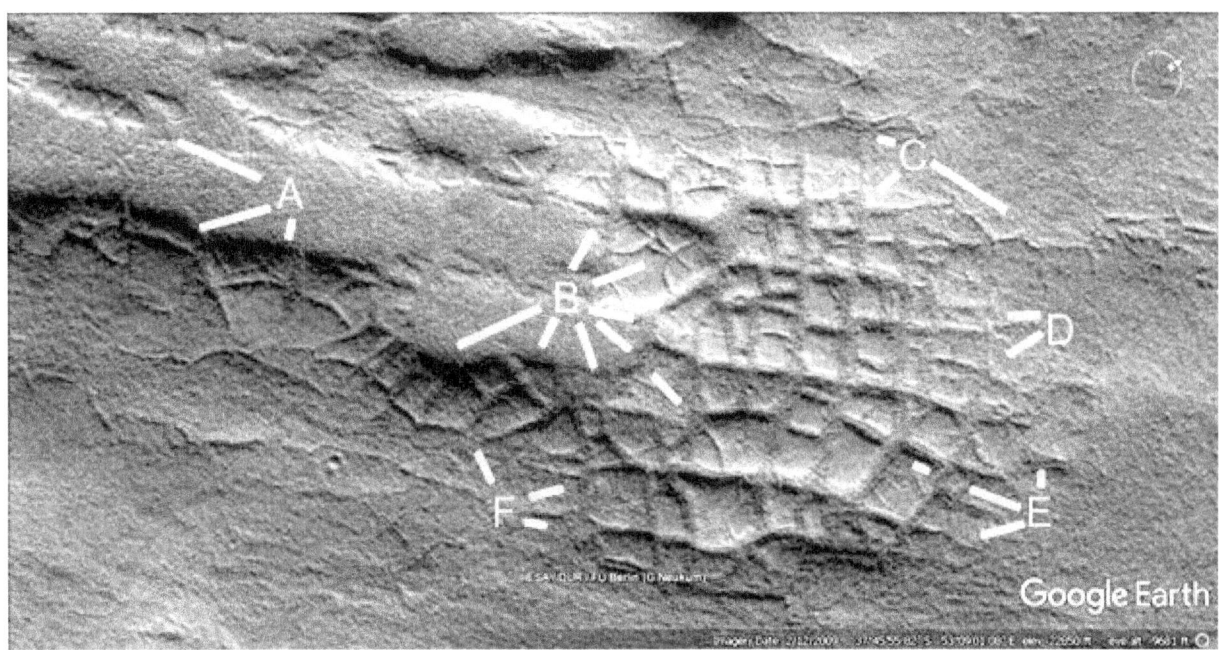

Held1222c

Hypothesis

The walled fields are in better condition here, without gaps. A shows some joins with little erosion, at 8 o'clock however is a much more eroded wall.
B shows an eroded wall at 10 o'clock and where one wall passes over another at 7 o'clock. C shows a much thicker wall between 6 and 10 o'clock, this extends under a wall to a thin wall between 1 and 4 o'clock at D. E shows some wall erosion at 3 and 9 o'clock.

Held1222c2

Hypothesis

The lines indicate how straight the walls are.

Held1222e

Hypothesis

This shows how many walls are hollow. The wall at A at 6 and 7 o'clock has collapsed indicating it was a tube. At 4 and 8 o'clock the walls are intact, it implies these tubes would give a passage in and out of the hills. B shows more collapsed walls, at 3 o'clock one goes into a small hill perhaps a habitat. Above C at 10 o'clock the tube has partially collapsed, the wall forms a side of this hill. At 5, 7, and 8 o'clock the walls have collapsed, at 4 o'clock the wall goes into another hill which may be a habitat. D, F, and G shows more collapsed walls. E shows more narrow walls going through a possible habitat at 2 o'clock.

Held1244

Hypothesis

A shows a possible habitat at 4 o'clock, B shows two others at 8 and 11 o'clock. These may be like the typical hill in this area when the outer skin erodes away. A at 6 o'clock shows many fine walls or tubes going into a nexus at B at 4 o'clock, also with a circle of walls around it. This would be similar to Earth roads where a central meeting place might be bypassed with this ring road. C shows more walls, D shows how they go into a hill at 6 and 9 o'clock. This hill is much flatter, it connects the hypothesis of the other hills in the image being like for example Held1232. It appears as if the roof has collapsed onto the ground. E shows a wider wall coming out of the hill at A.

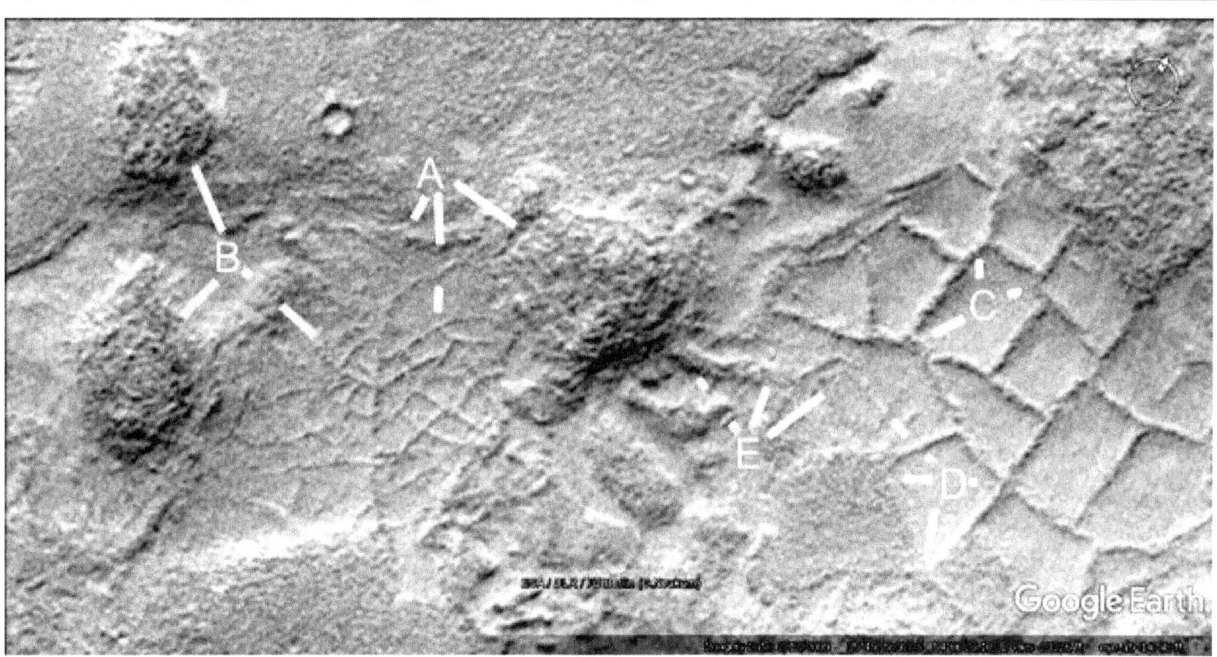

Held1258

Hypothesis

A also implies the hill is artificial, it is approximately parallel to the Latis Rectum of the parabolic wall. B is probably a collapsed hill at 8 o'clock, a wall comes out of it at 7 o'clock. C also shows a network of walls coming out of a hill. The walls at D appear more eroded.

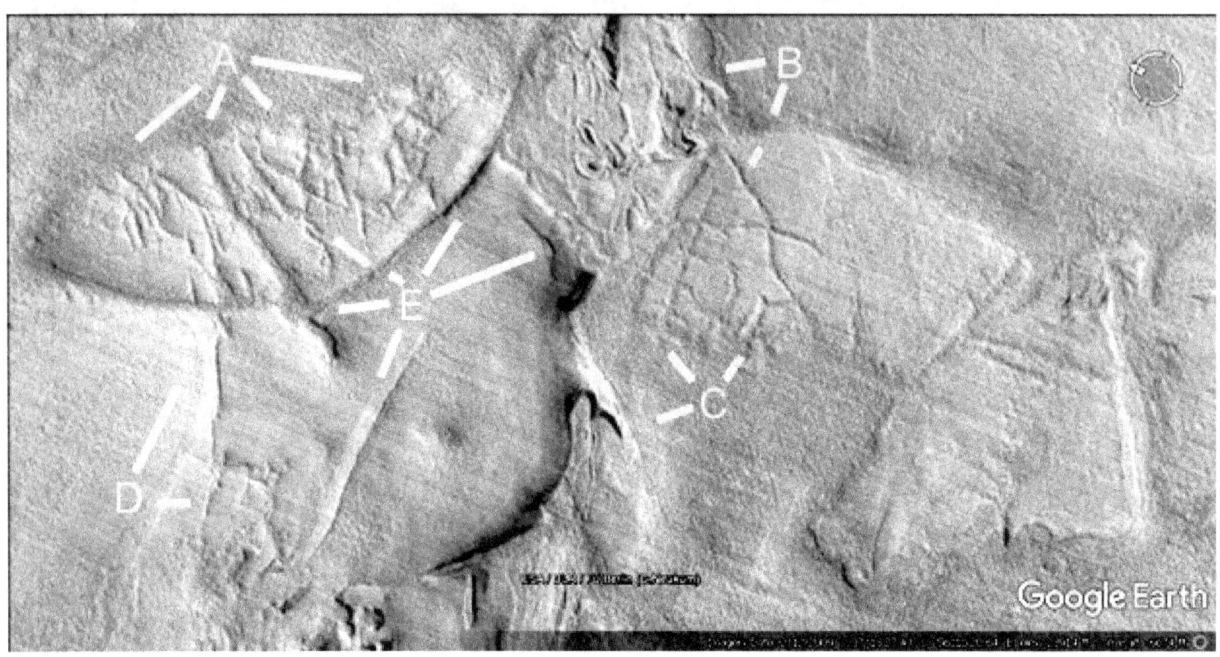

Held1258b

Hypothesis

A parabola is shown, also the lines indicate how straight the walls are.

Held1295b

Hypothesis

A appear to show a water channel or perhaps roadway, perhaps water could come through here and fill some of the walled areas. B shows some of these walls, C shows a parabola. D shows another curved wall, probably a parabola but not long enough to check. Shows many walled fields with smaller walls subdividing them.

Held1295b2

Hypothesis

A parabola is shown, also the lines show how straight the walls are.

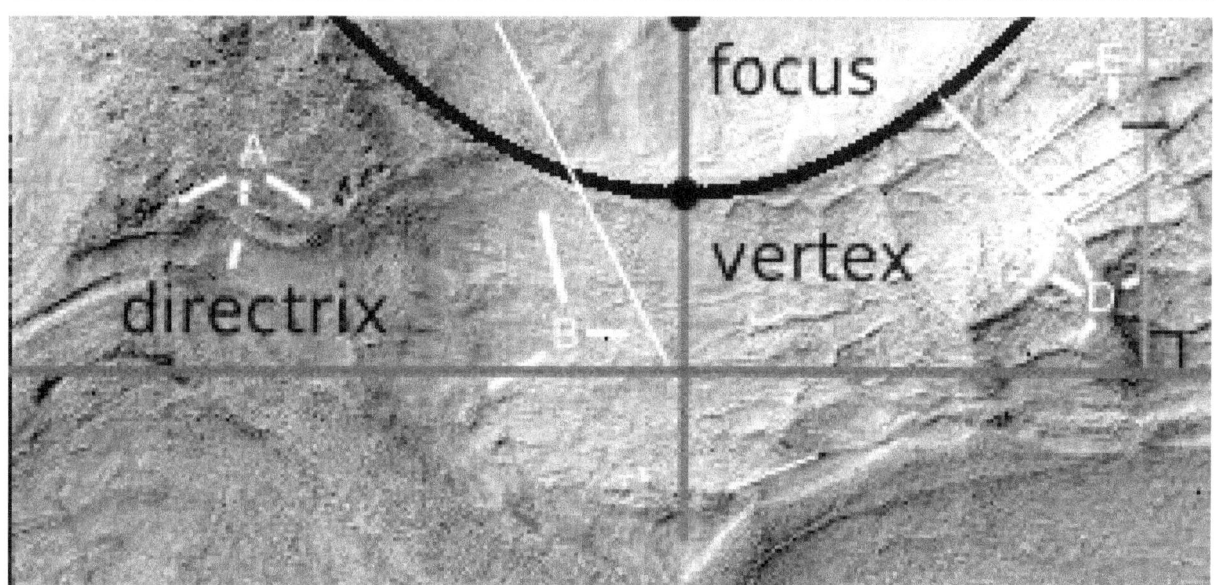

Roads

Some formations also look like roads, they often appear between hills that are hollow. The hypothesis these hills are buildings, either completely constructed or adapted from geological formations. It further ads to the global hypothesis, we use roads and so we might expect Martians to have built them to travel between buildings and cities.

Prhh498

Hypothesis

The hollow hill has collapsed at A, B shows a straight wall still standing. C shows another road going into the hill perhaps with two lanes, this extends to D at 10 and 1 o'clock. There may be another road at 7 o'clock.

Prr499

Hypothesis

This is a closeup of a road, much smoother than the surrounding terrain like cement. It extends past A to B where a tube or raised road intersects it. C shows this tube going down from 10 o'clock, then possibly at 6 and 7 o'clock into the crater.

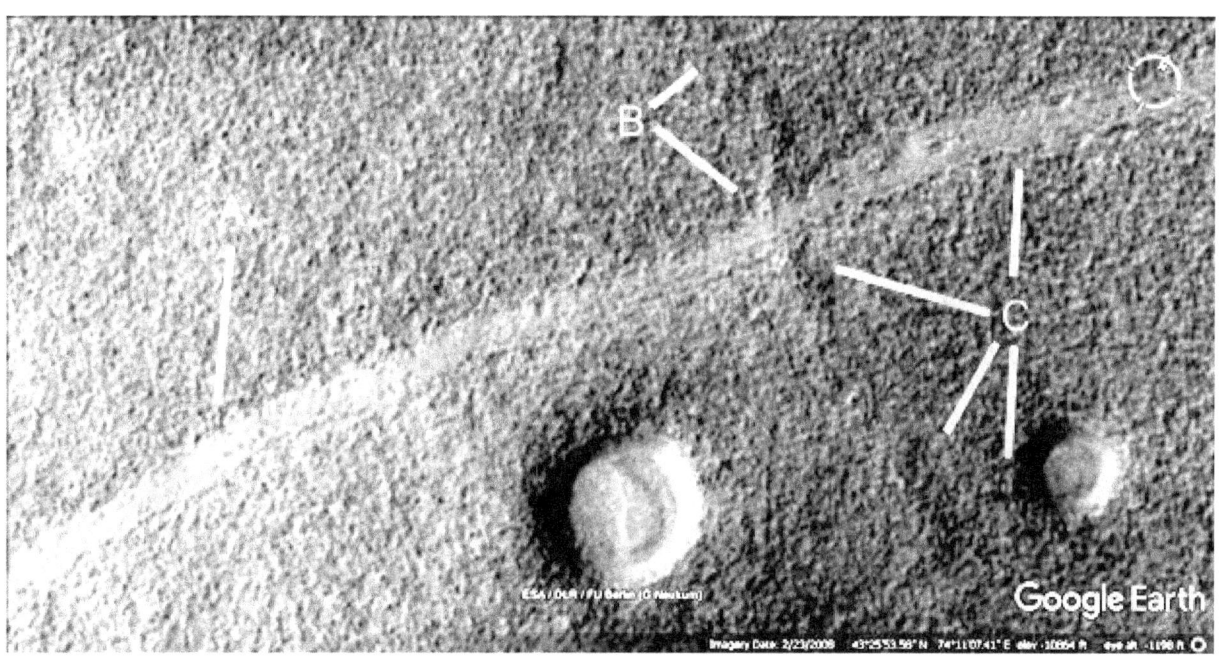

Prr508

Hypothesis

A shows the road continuing on over the pale material, B and C also show pits like altered craters perhaps with the same road material to act as dams.

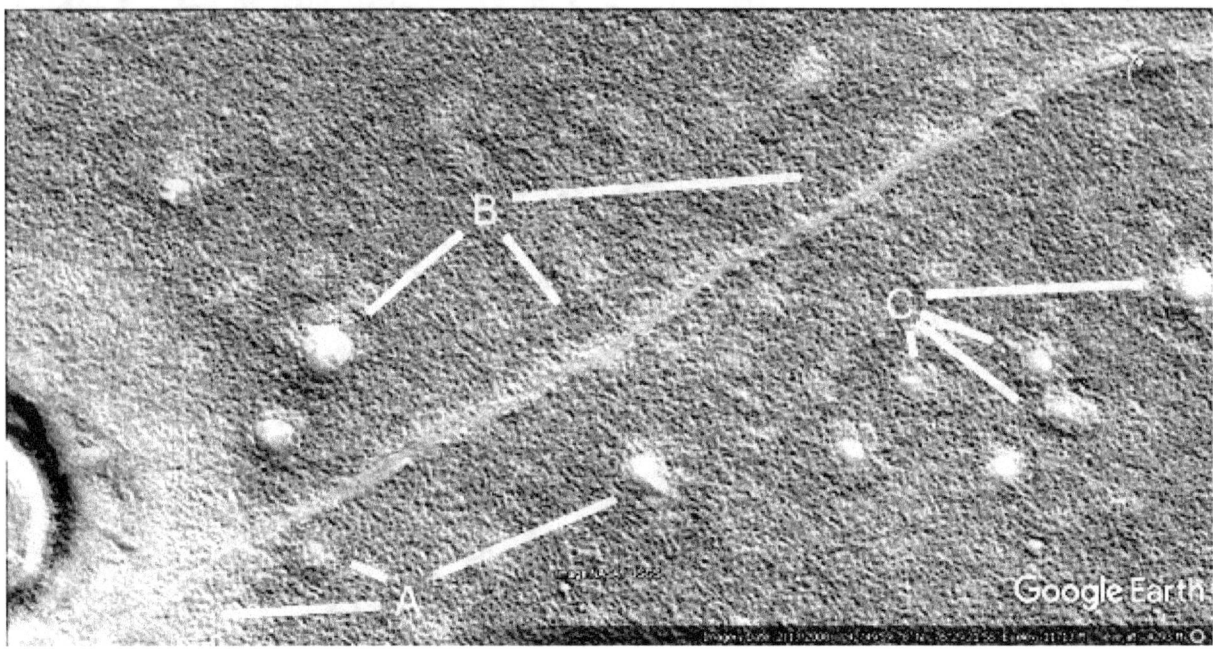

Prr533a

Hypothesis

This closeup of the road shows right angled shapes in it, perhaps like bricks or tiles. This impression continues along the road where it seems to vary in an angular rather than a smooth way. The center is very smooth compared to the surrounding terrain as shown by comparing A at 1 and 5 o'clock. B shows a shape like a gutter along the road's side. C shows a small pit at 10 o'clock that appears to be connected to the road, perhaps a former hollow hill, at 2 o'clock is an angular section on the side of the road.

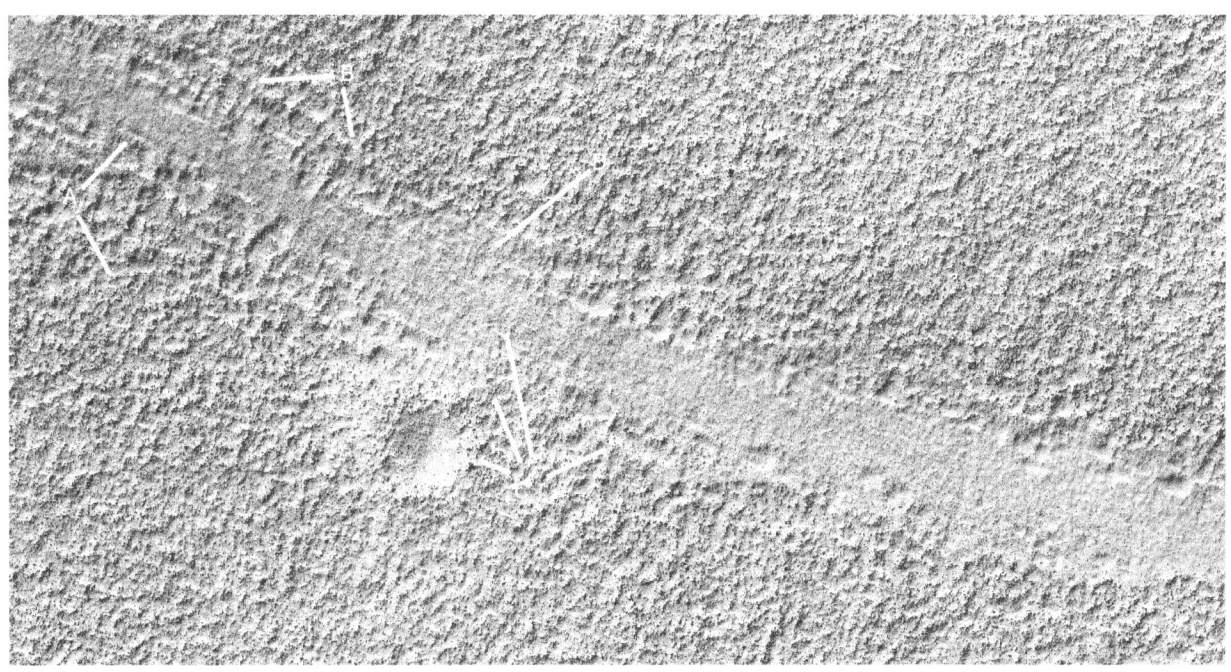

Prhh1821

Hypothesis

A shows more roads, they connect to a crater at 5 o'clock. B shows a road at 6 o'clock going into a small hollow hill, another at 4 o'clock going into a hollow hill. C shows a road connecting to a complex of hollow hills. D and E show many more roads connecting to hollow hills. F and G show roads connecting to the large crater. H shows a major intersection going up the image.

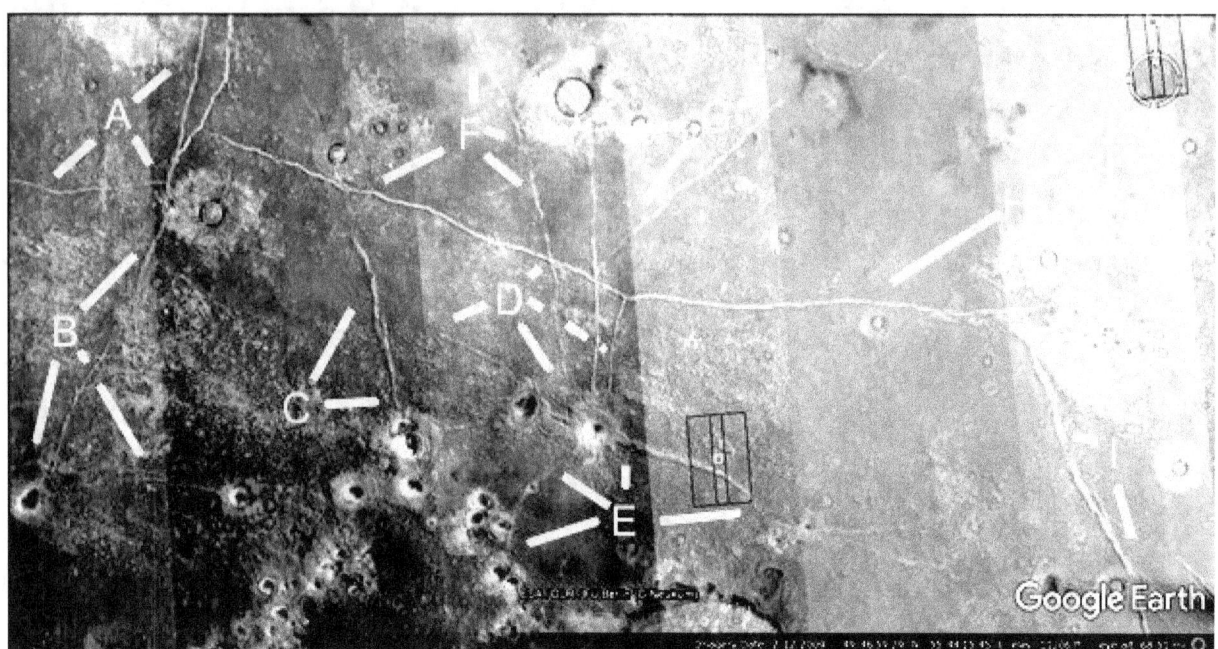

Tubes

A further hypothesis is that some roads were enclosed like tubes. These hypothetical Martians then could have travelled through them to avoid the cold, predators, meteors, etc. Some may also have been raised roads, for example the ground may have been swampy or covered in water. So, much as we do on Earth, they may have built roads raised above this ground to travel on.

Prt641

Hypothesis

A shows a curved tube going from the walled hill at 4 and 5 o'clock to the small crater at 1 o'clock. B at 8'clock shows the walls of the hill, at 7 o'clock a tube comes out of the hill, at 1 and 4 o'clock are two more hollow hills. D shows the curved tube, it connects to another tube shown by B at 8 o'clock. At 9 o'clock is a small tube from the larger one, at 10 o'clock the smaller hill appears to have collapsed. This main tube continues up through E to the right.

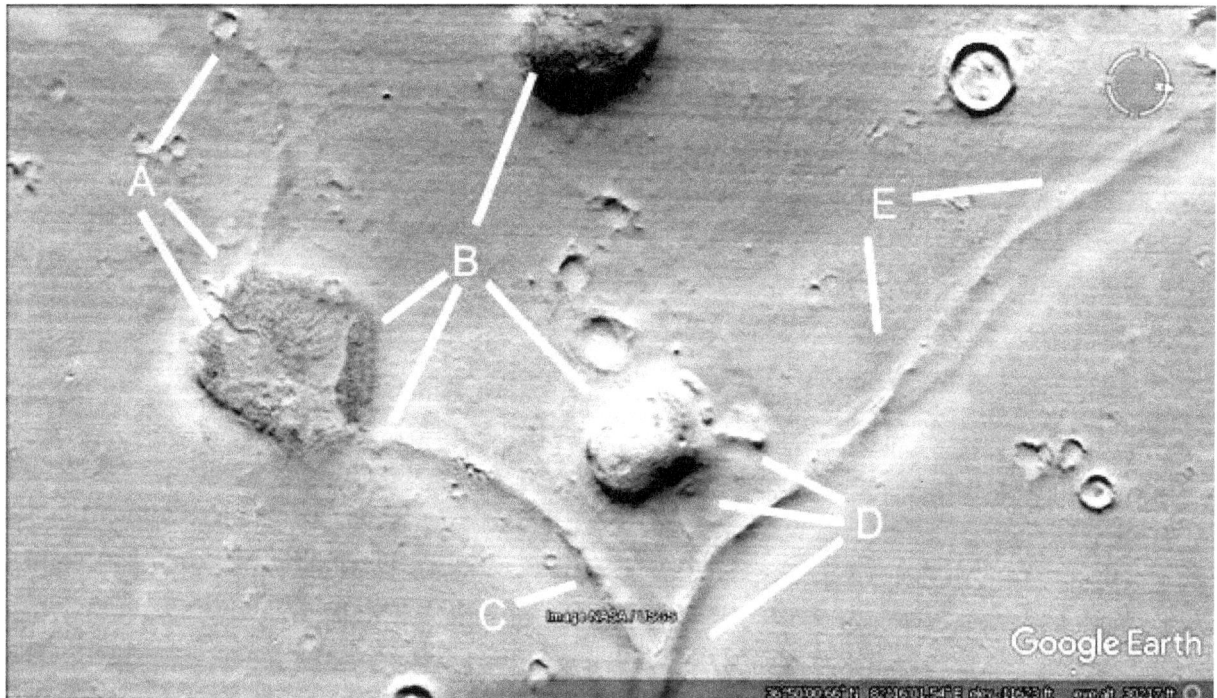

Prt641a

Hypothesis

Two parabolas are shown.

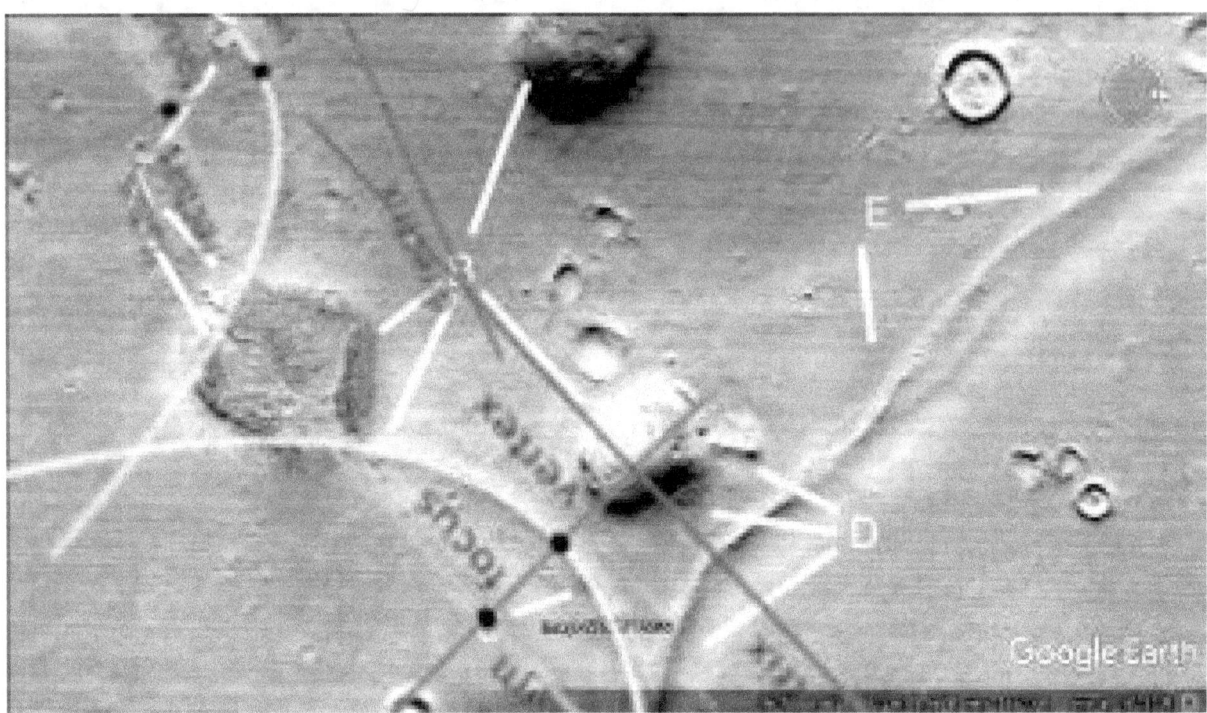

Prt798

Hypothesis

A shows a hollow hill with cavities in the roof, it connects to a wider part of the hill at 6 o'clock. This has a twisted shape like a rope, it continues on through the twisted tube at B to connect to a collapsing hill at 2 o'clock. At 8 o'clock there is another tube. At 3 o'clock the roof has collapsed. D shows another tube going into the hill at 8 o'clock, this connects to the tube at 5 o'clock. This in turn connects to the hill above D with tubes at right angles to it. E shows a collapsed roof at 10 o'clock, at 11 o'clock is a tube. Bat E at 12 o'clock up to F at 6 o'clock is a symmetrical wall.

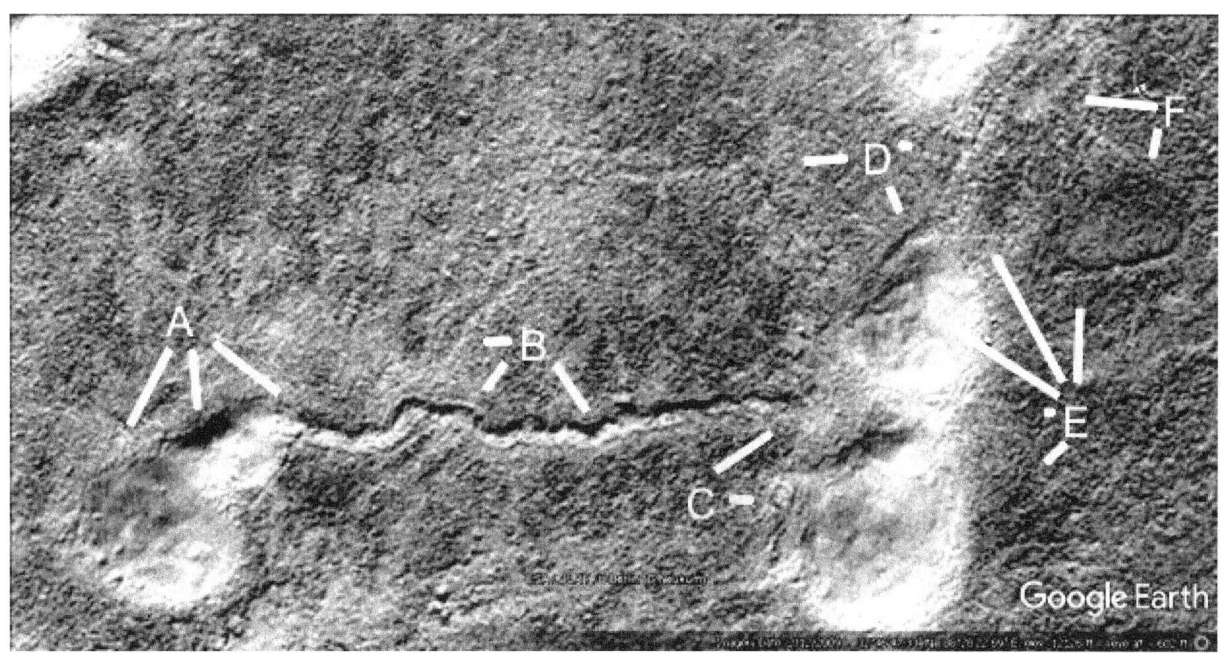

Prt804

Hypothesis

A shows more tubes between collapsed hills. B shows layers in the hill at 2 o'clock like a Cobler Dome. At 11 o'clock the tube from the chain of hills enters the hollow hill. At 3 o'clock is a thicker tube connected to a small hill. C at 8 o'clock shows the circular roof of the hill, it contains two parabolas, at 4 o'clock a tube goes into a small hill with a cavity on the roof. From 11 to 3 o'clock are other tubes. D at 5 o'clock shows the edge of this circular roof, the rest of D shows other tubes. E shows an arc of tubes connected to some collapsing hills.

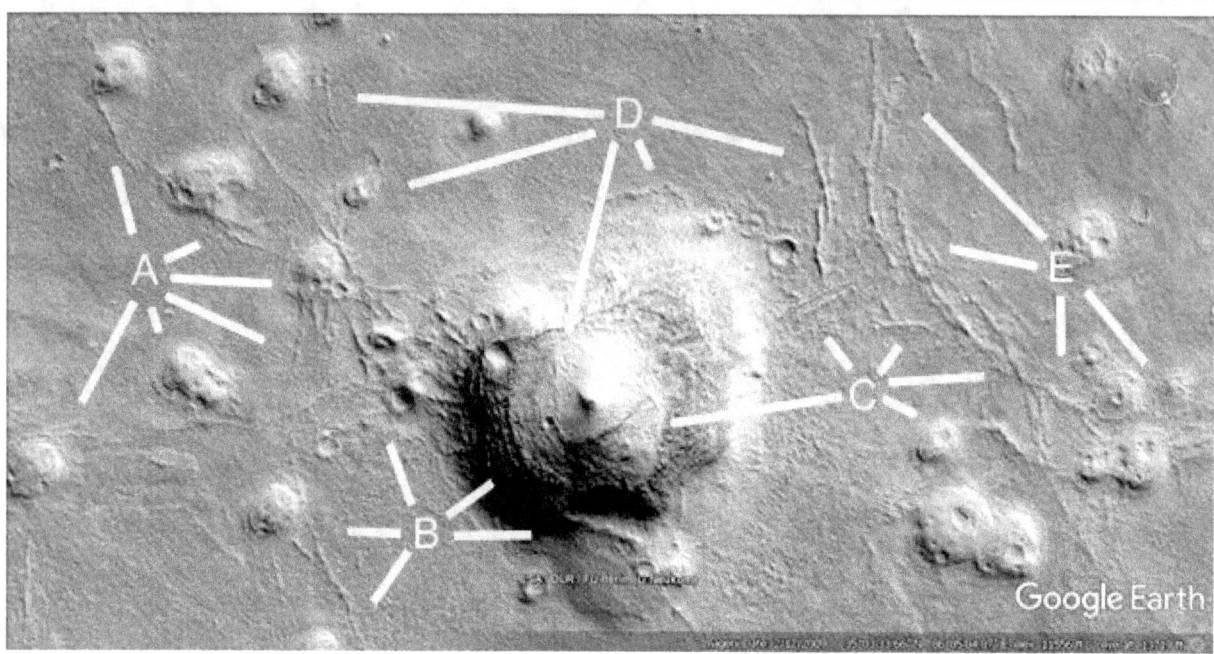

Prt804a

Hypothesis

The roof is close to a circle, here a circle is overlaid onto it. Also two parabolas are drawn onto the dark marks on the roof.

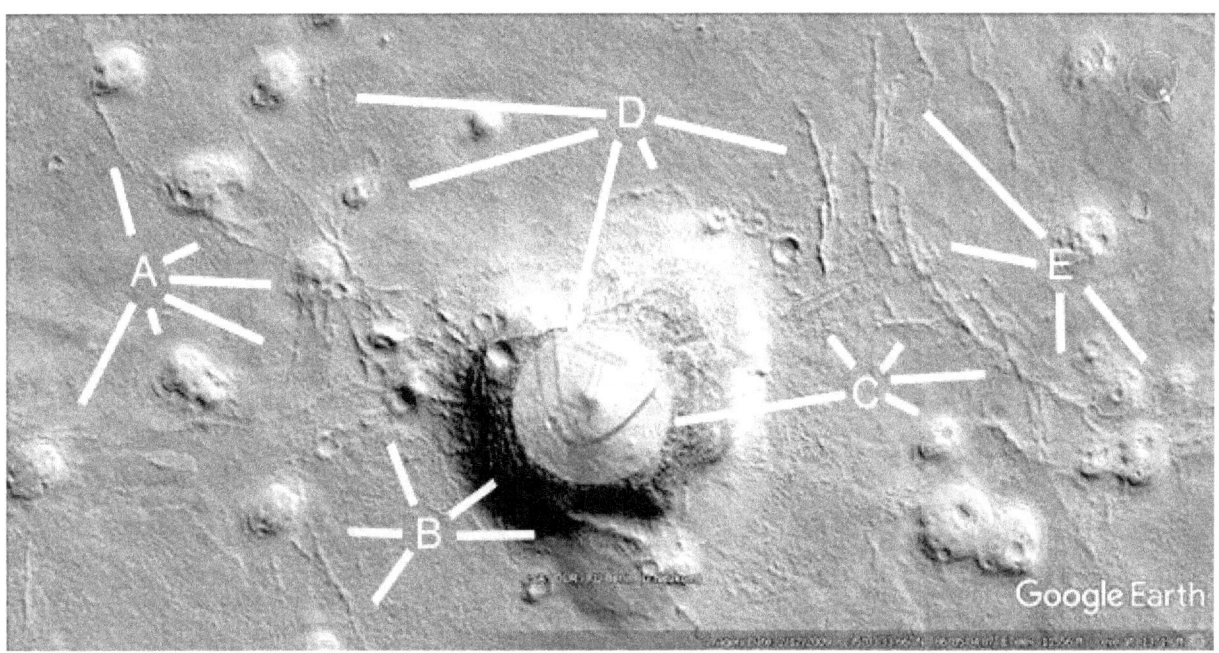

Prt814

Hypothesis

A from 5 to 7 o'clock shows two collapsed hills connected by a tube, the holes in the roof may have been rooms. At 8 o'clock is a tube. B at 10 o'clock shows a collapsed hill connected by a tube to A at 7 o'clock. B from 4 to 7 o'clock shows small hills connected by tubes, also some tubes go to the crater under it. C at 6 o'clock shows many tubes connected to the crater, at 7 o'clock a tube goes through a collapsed hill over to 4 o'clock and then up to the nexus at F at 1 o'clock. At 4 o'clock a forked tube comes out of a collapsed hill. C from 10 to 2 o'clock shows a tube coming out of the collapsed hill continuing over to the nexus. D and E show more tubes connecting to the hills and over to the crater at E at 4 o'clock.

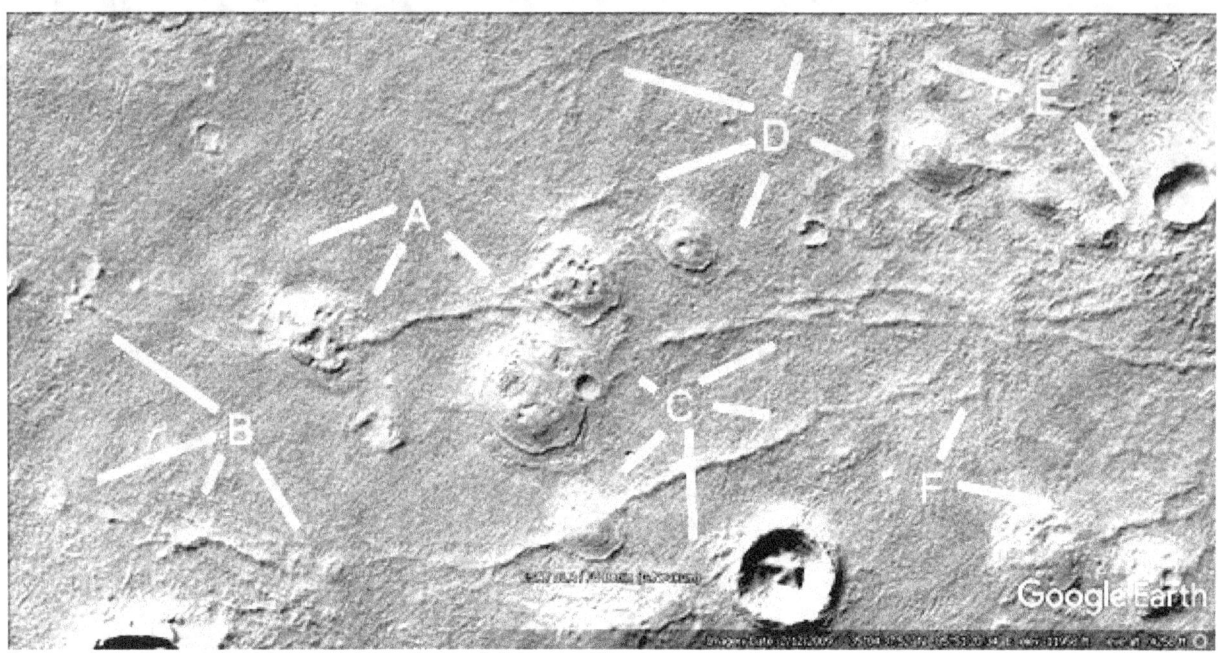

Tube cities

The hypothesis is these large numbers of tubes connected together to form habitats and cities. Some of these may have been underground, others connect to artificial looking hills.

Prt662

Hypothesis

A shows a wavy tube, B shows a clear area surrounded by tubes like a field. C shows tubes going into a crater at 6 and 8 o'clock, at 1 o'clock they go into a rounded area, also shown by F at 10 o'clock, under a nexus. D shows more tubes going into this nexus. E at 6 o'clock shows an intersection of tubes then this goes down, making a right angled turn into a hollow hill at F at 1 o'clock. E at 12 o'clock shows a T intersection, at 4 o'clock there are about four faint parallel tubes going up the image. F at 7 and 8 o'clock shows tubes going into three collapsed hills, also shown by G. H may be a large habitat, at 9 o'clock a tube crosses other tubes at 10 o'clock going up to I at 2,4, and 6 o'clock and a collapsed hill. At 10 and 11 o'clock faint tubes go into the crater. J shows more tubes going into the collapsed hill.

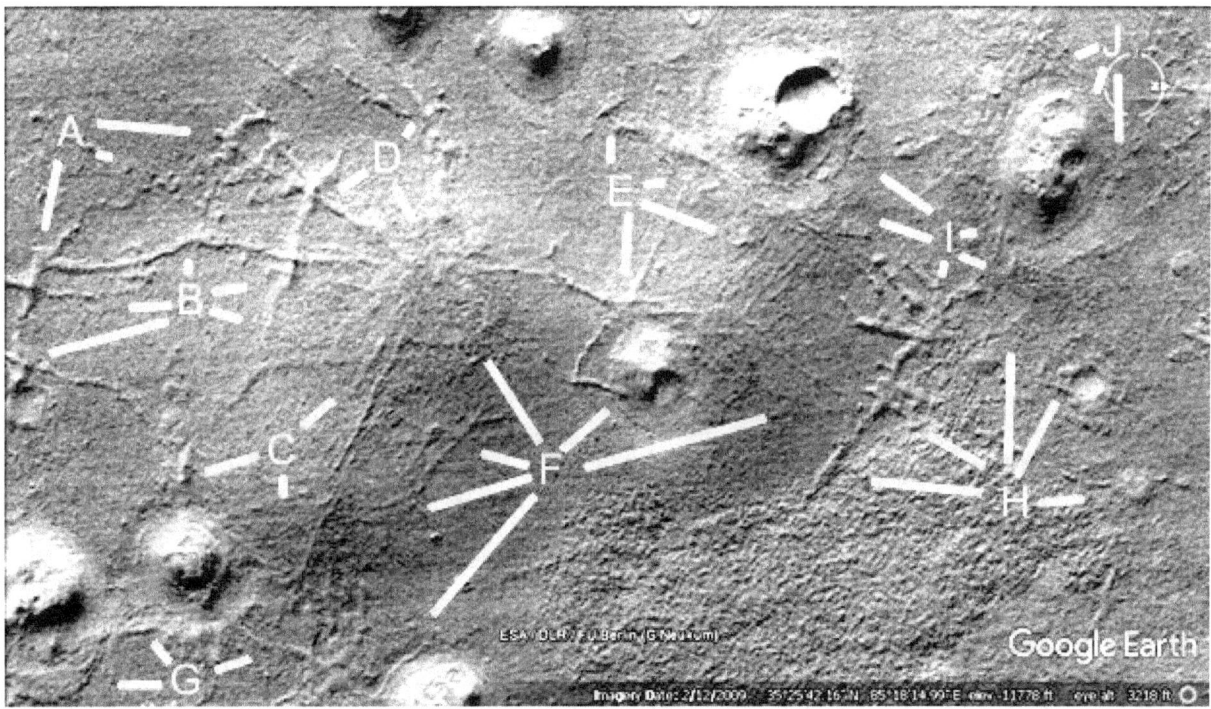

Prt682

Hypothesis

The tubes come together in a large nexus here, there also seems to be flat areas like cement over the tubes. These might act as a roof with rooms under them. A shows a tube crossing another at 2 o'clock, this connects to another tube at 10 o'clock. At 6 o'clock is the edge of the outer circular shape of the nexus. This may have allowed movement around the nexus without going into the centre, like an Earth ring road in many cities. B shows a continuation of the ring road at 3 o'clock, a forked tube at 10 o'clock and at 9 o'clock, and a narrow fork at 8 o'clock. C shows a larger tube at 10 o'clock where it appears to end on top of a small platform. At 1 o'clock the tube is hollow like the roof collapsed. D shows a tube ending at 11 o'clock, some tubes crossing at right angles in a mesh at 2 o'clock. E shows two tubes parallel to each other, further along one tube crosses over the other like a knot. F shows a small hill connecting to the tube at 3 o'clock, a loop of a tube at 5 o'clock with a central tube. From 8 to 10 o'clock is the flattened part of the nexus, whether from erosion or a roof. G shows a small nexus.

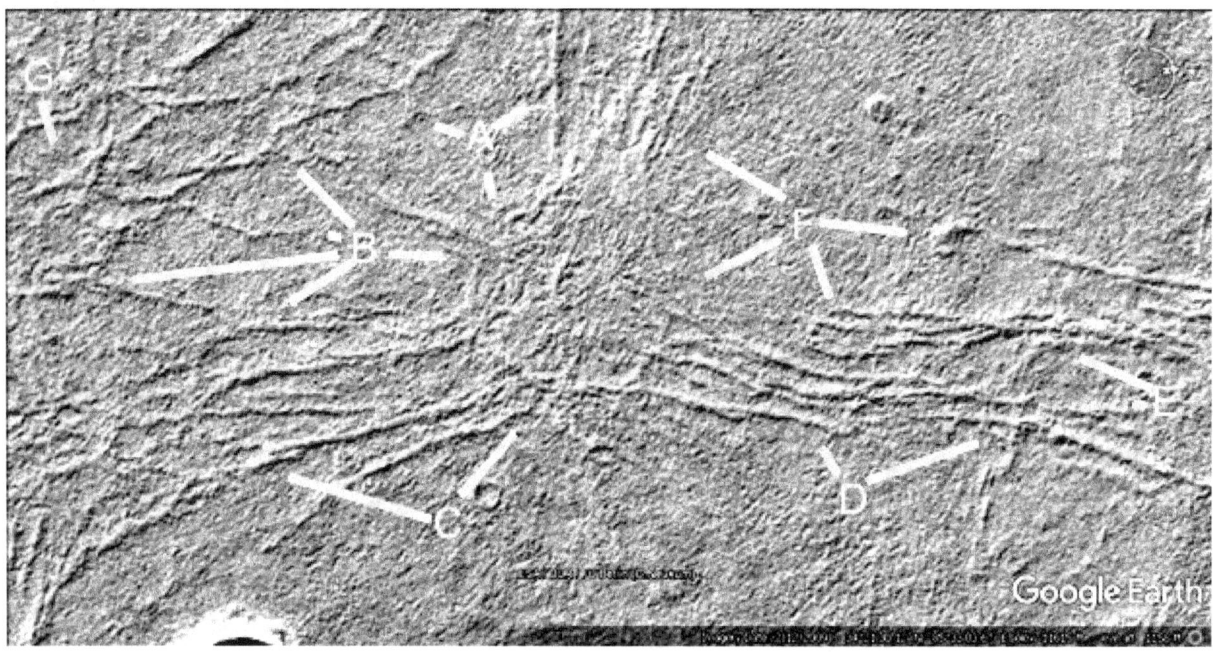

Prt714

Hypothesis

A shows a large nexus at 4 o'clock, it appears to have flat sheets of cement over it so some segments might be rooms. At 1 and 2 o'clock parallel tubes go to the nexus. B shows a squarish area surrounded by tubes, at 7 o'clock there are more like squarish walled segments. At 1 o'clock the crater appears to have been overed over on the right side or this can be an exposed room in the nexus. A wider tube is at 5 o'clock. C shows a T intersection of tubes at 1 o'clock, the tube goes down crossing a long hill at 5 o'clock going into a crater. Another tube crosses the hill from 6 to 7 o'clock. D shows another nexus at 2 o'clock again with flattened segments of a roof. At 4 o'clock this connects to a hill collapsing in many areas. Parallel tubes are shown at 1 o'clock. E shows more tubes, some going into a crater at 4 o'clock. F shows an arc of parallel tubes. G shows tubes exiting under the collapsing hill.

Prt753

Hypothesis

A shows many parallel tubes going through the long hill, continuing as E and E to the large nexus between E and F. This is a flat sheet like a roof in many areas. A at 5 o'clock and D at 7 o'clock show tubes crossing the parallel tubes so someone could have moved from one to another more easily. Above I there are nine parallel tubes going to the nexus, B shows about eight more parallel tubes. Under this is H with a grid or mesh of tubes, this continues on through C with more meshed tubes to the nexus. F shows about six more parallel tubes from 8 to 11 going to the nexus, between E and F there are about twelve more tubes going into the nexus. Between F and G there are about seven more tubes going to the nexus, many more of these form a tube mesh as well.

Some areas appear to be bounded, the hypothesis is they were farmlands or walled off for some reason. Often they have a parabolic boundary.

Farms

The hypothesis is that these large areas were farms, they are often bounded by parabolas with walls. We have something similar on Earth, we build walled fields and larger farms.

Prt857

Hypothesis

A, B, and C show many parallel tubes inside this farming area. Some connect to the craters at A at 7 o'clock. Between A and B there are about six parallel tubes, between B and C there are about four. B from 2 to 4 o'clock shows a tube going into the crater. D shows where many of these tubes converge, there may have been a hollow hill here. E at 7 o'clock shows a small hill and a straight tube extends up the image.

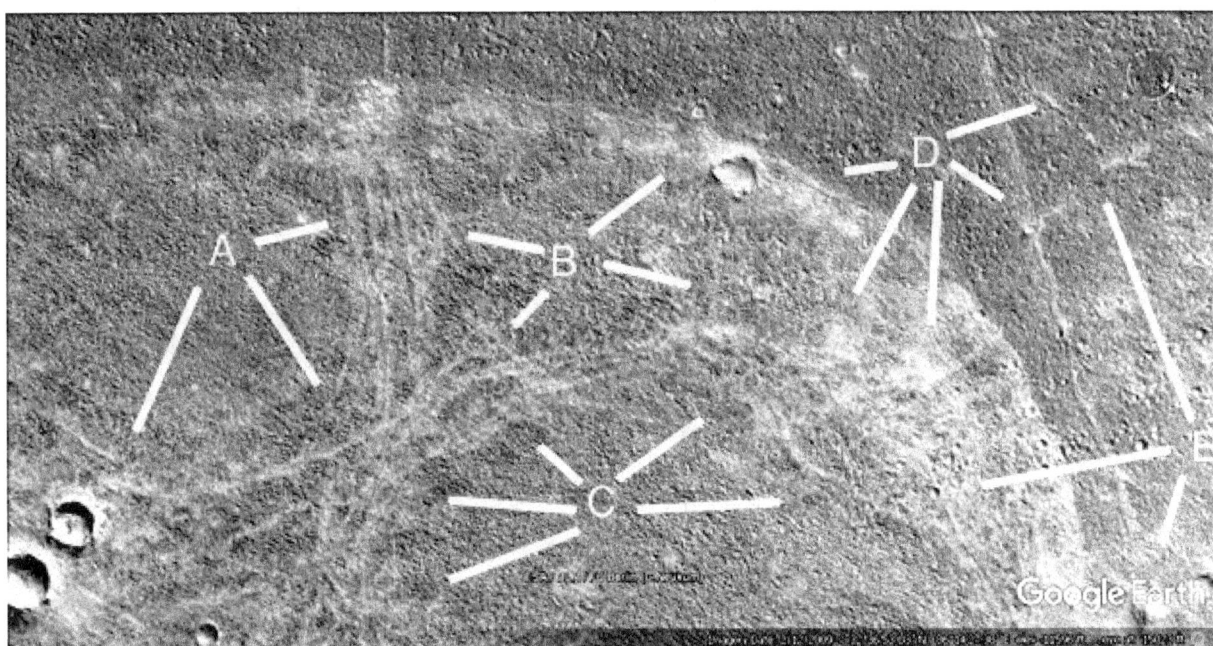

Prt857a

Hypothesis

A parabola is shown. Also the line shows how straight the long tube is.

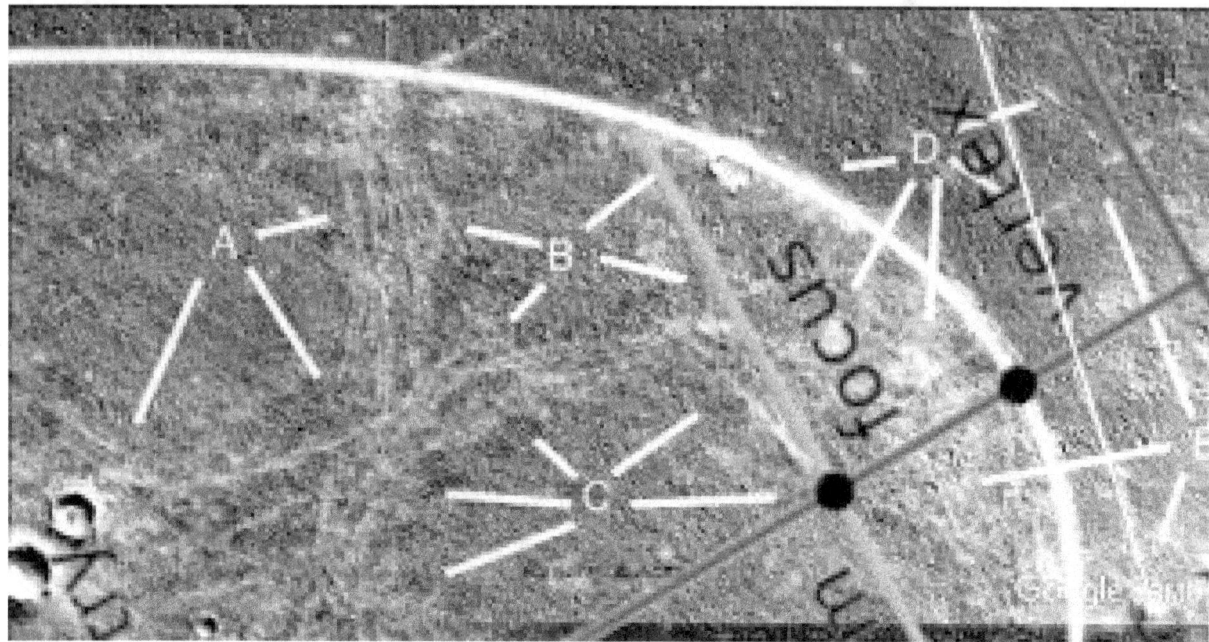

Ecydhh1941

Hypothesis

These curved shapes may have been used for agriculture. Found in many areas of Mars the boundaries are often parabolas. A shows a road or tube going into a crater, B shows the other side of this road and one of the curved pale areas. C shows more of these often shaped as parabolas. At 4 o'clock there is a wall or tube according to the shadows. D shows another tube at 12 o'clock, at 2 o'clock is the other side of the hollow hill. At 7 o'clock is a paler segment of the field. E shows more curved fields and a tube at 3 o'clock going down to a hollow hill at 6 o'clock. F shows another segment of the tube. G shows a tube going to the large crater at 7 o'clock. H, I, J, and K show more tubes and hollow hills.

Ecydhh1941a

Hypothesis

Three parabolas are shown, however the pale curves may all have been parabolas.

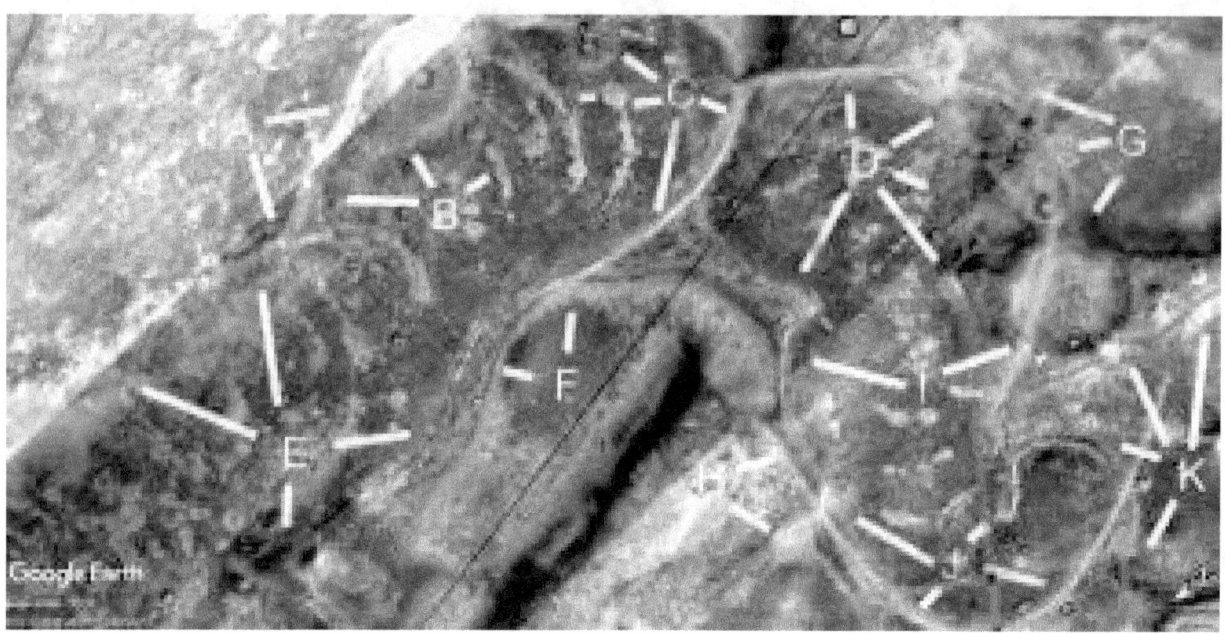

Ecydt1974

Hypothesis

Many walls and pale fields are shown, these may also have been farms.

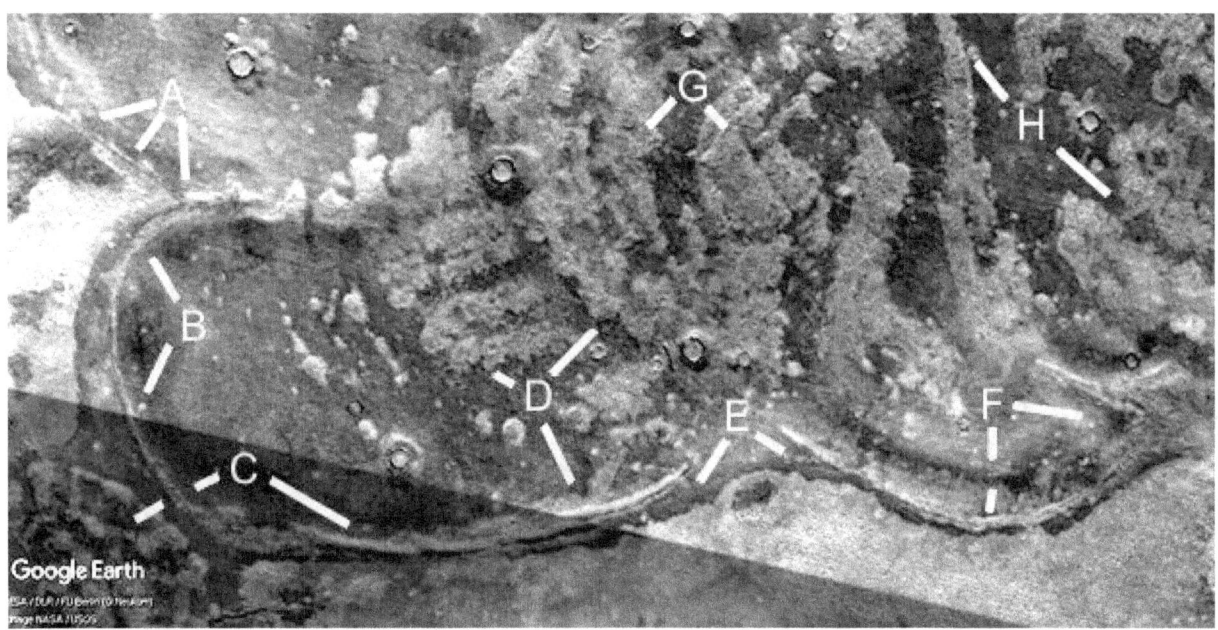

Ecydt1974a

Hypothesis

Three parabolas are shown.

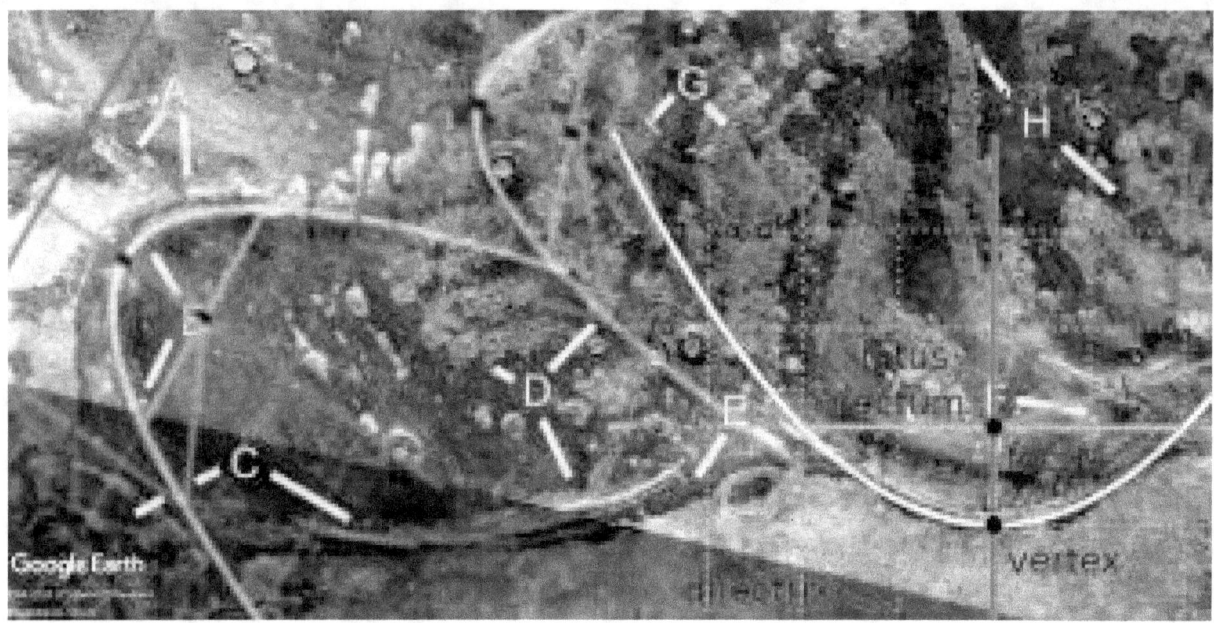

Ishh2306

Hypothesis

These may have been walled fields as often seen near Cydonia. B shows two collapsed hills from 5 to 7 o'clock, C may show tubes or roads in the field. D shows a tube between two craters at 12 o'clock. At 3 and 4 o'clock is a hill connected to a crater.

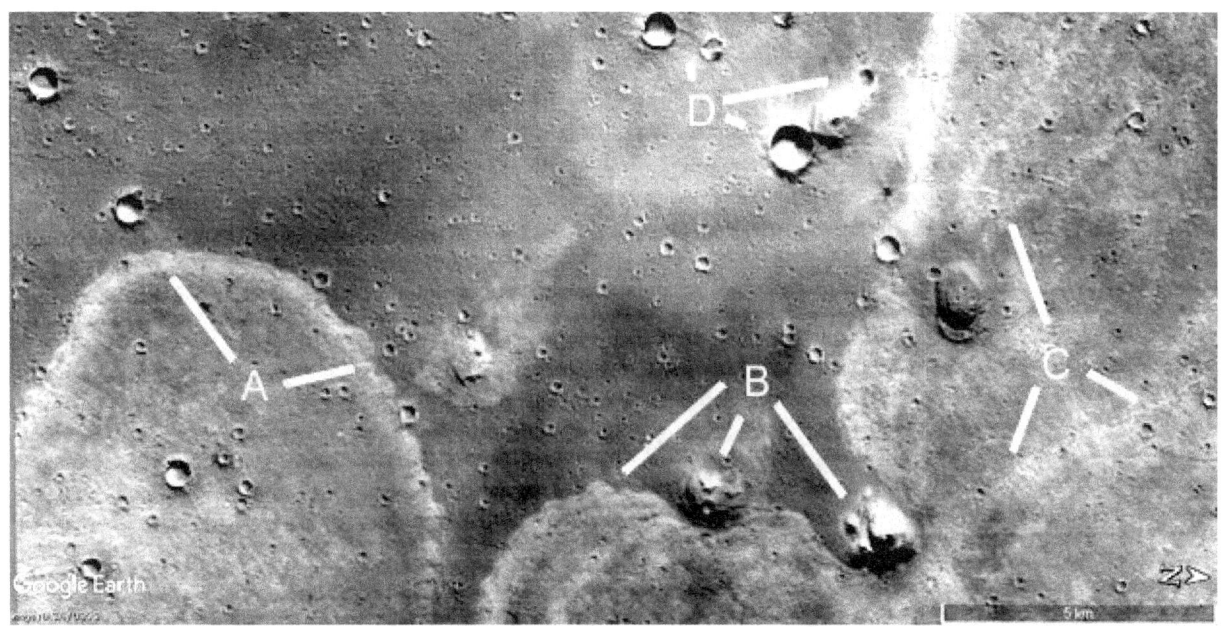

Ishh2306a

Hypothesis

Five parabolas are shown.

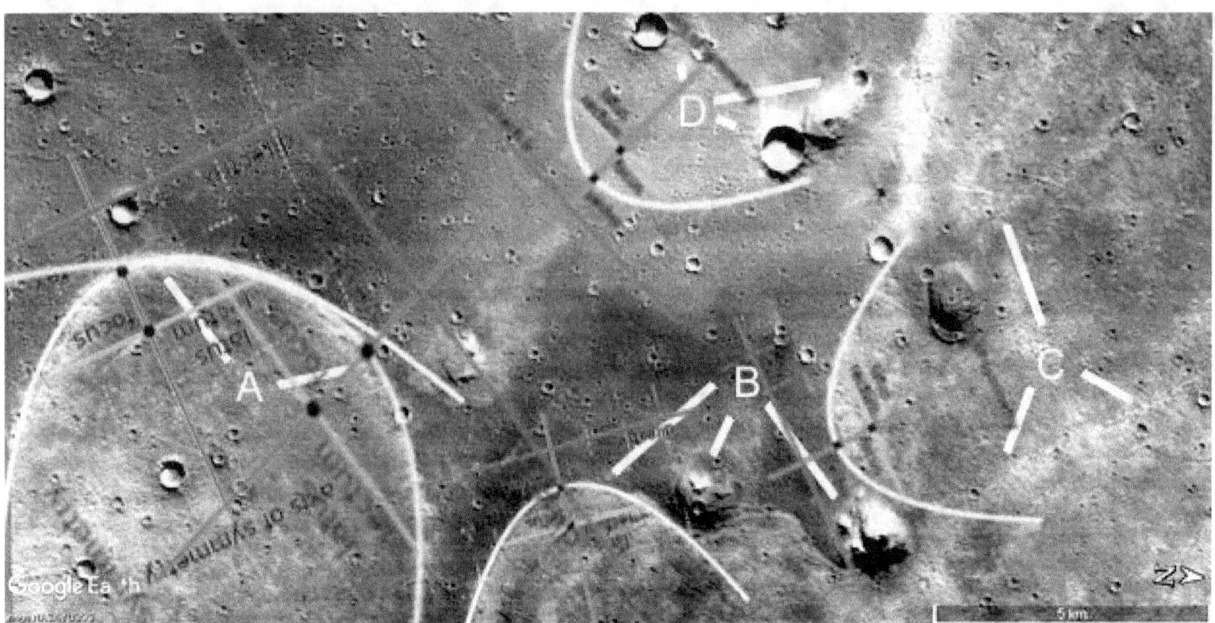

Lakes

The hypothesis is that some water channels and canals connect to larger artificial lakes. This is also something we do on Earth.

Prd886c

Hypothesis

A shows the double walls of this dam at 0 o'clock, also a small cavity in the wall at 8 o'clock. This connects to a star shaped wall from 7 o'clock to 3 o'clock. B shows this dam wall is intact at 10 o'clock, there is a wavy wall like some tubes at 7 o'clock. At 8 o'clock one of the walls is much shorter. C shows this double dam wall continuing at 5 and 9 o'clock, the wall at 12 o'clock has broken up into segments on its end. D shows another walled segment of the dam, below 10 o'clock the wall is more eroded. At 4 o'clock there is a small entrance between the walls.

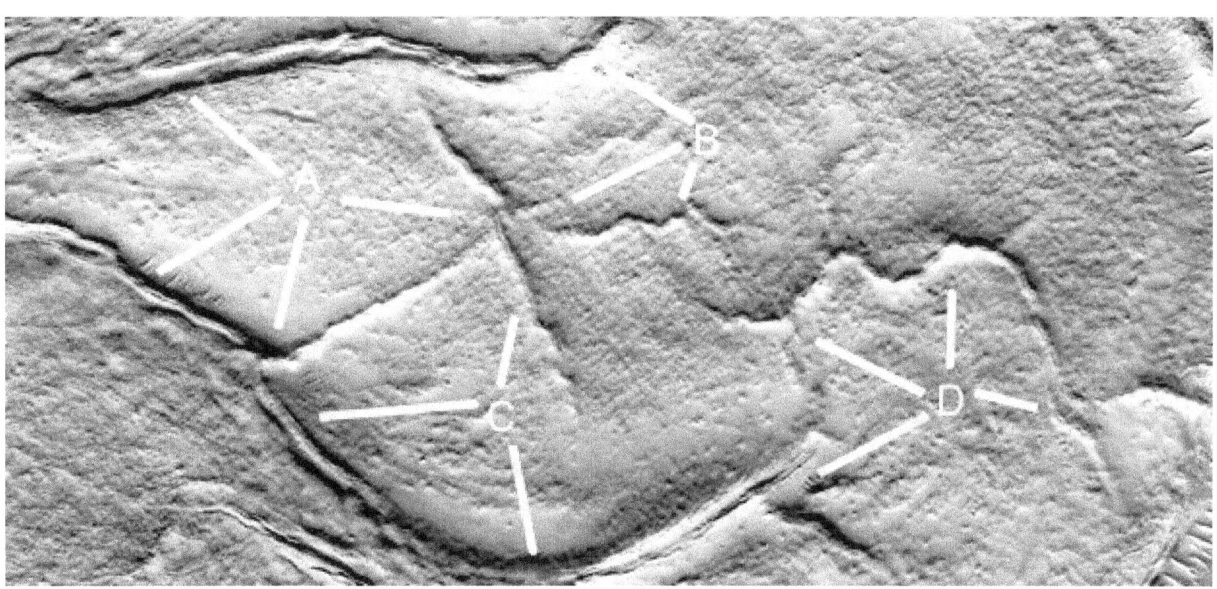

Prd886c2

Hypothesis

A parabola is shown. The axis of symmetry goes approximately through the centre of the star. The focus is also in line with the dam wall between E and F, the latis rectum or line through the focus would then approximately be an extension of this wall. A line is drawn from E to F to illustrate this.

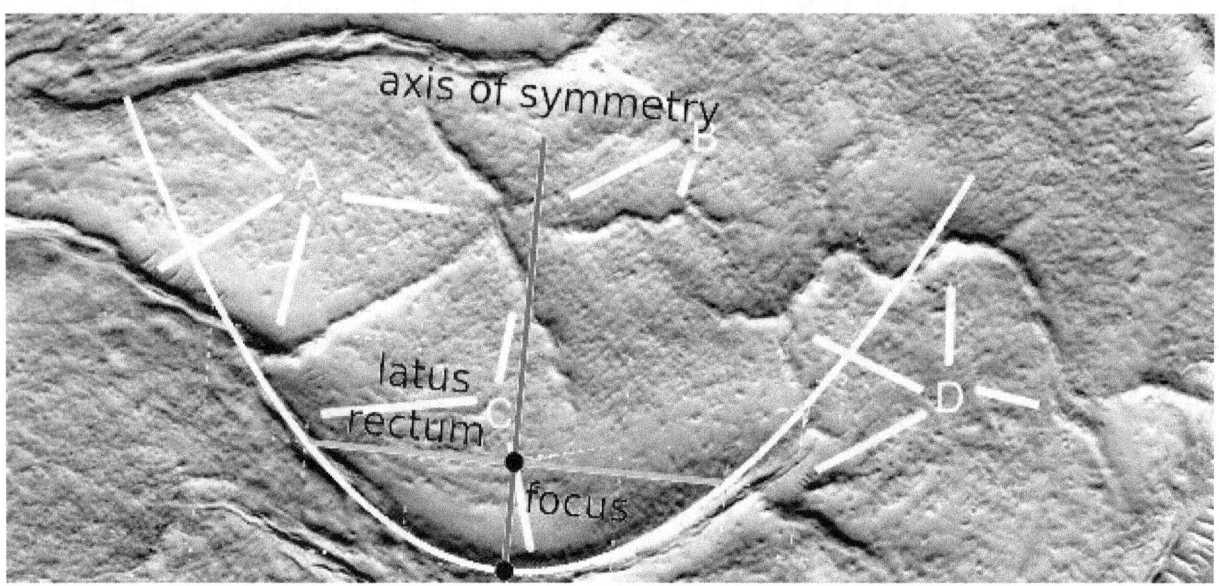

Prd911b2

Hypothesis

Eight parabolas are shown. This is a good example of how natural looking areas in a crater can be looked at more carefully. With a closeup there cold be even six more parabolas here.

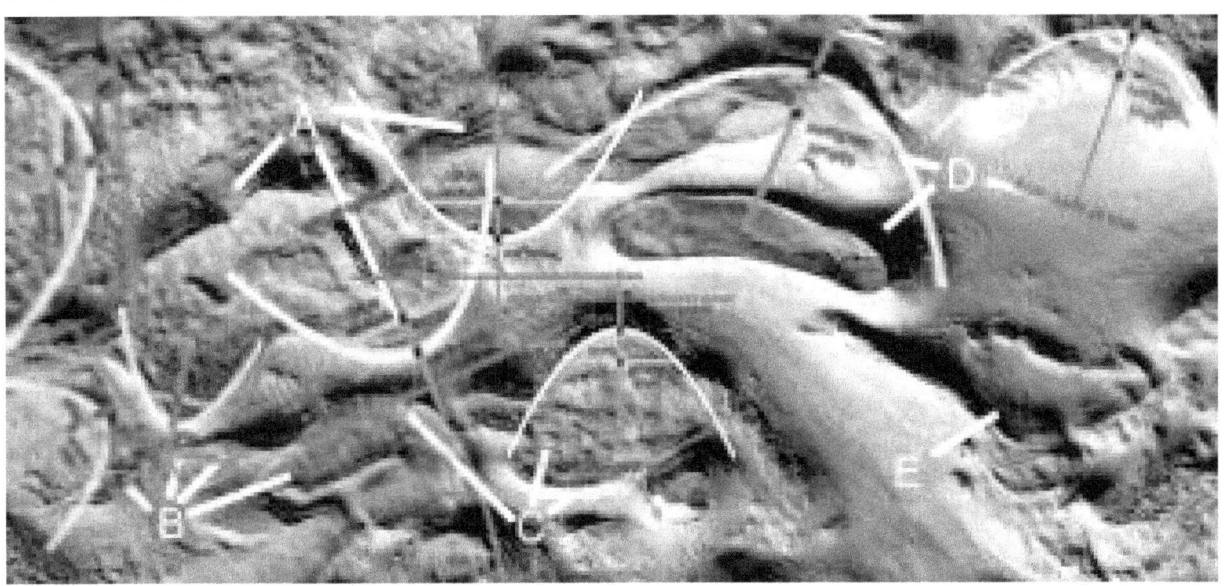

Prhh1018

Hypothesis

Many tubes come out of this formation, A at 8 and 9 o'clock shows a tube intersection. At 3 o'clock is another tube from the pit wall. B shows two more tubes, below the one at 4 o'clock are two small enclosures, also another two between there and C at 8 o'clock. These may all be dams including the large pits. C at 7 o'clock shows many faint tubes coming out of the pit wall. D at 9 o'clock shows the pit wall is doubled with a groove between them. At 5, 6, and 7 o'clock the pit wall is very even and rounded, at 3 o'clock is another tube coming out of the pit wall. E at 12 o'clock shows one of the pale formations inside the pit, these may have been hollow hills and have a similar albedo to parts of the pit walls. At 2 and 9 o'clock the pit wall gets thicker, this part has a roof like a tube but to the right and left it becomes a groove again. It's likely then most of these pit walls are hollow.

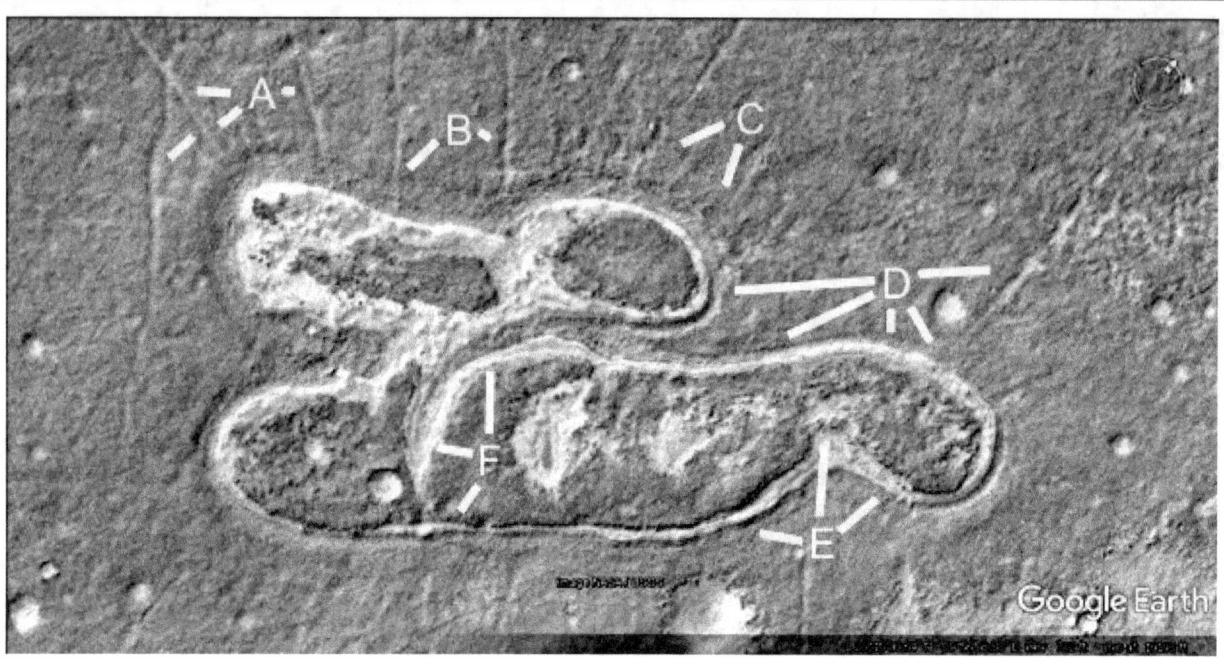

Prhh1018a

Hypothesis

The lines show how straight the tubes are. Also six parabolas are shown to fit onto the edges of the pit dams.

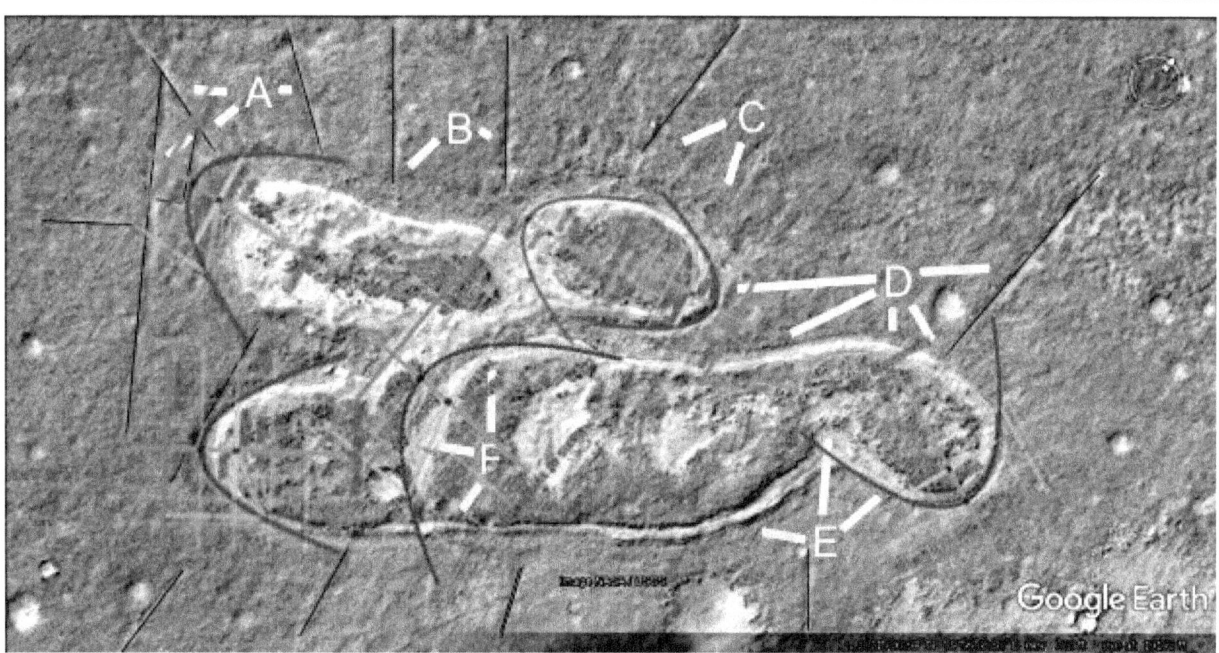

Geometry

The hypothesis is that two hyperbolas were constructed, the one shown here is close to the old Martian equator.

Prt1055

Hypothesis

This shows a nearly perfect hyperbola forming a tangent to the large crater, and to a smaller crater on the left.

Prt1055a

Hypothesis

This shows a hyperbola overlaid onto the formation, it shows it is nearly a perfect hyperbola. It deviates a small amount to the left at A as if affected by the gravity of passing near a planet or moon. B at the top of the image shows two other walls, C shows a road like shape connecting to the crater. B in the crater shows concentric circles which might indicate orbits around the sun, or the surface of a planet with the outer circle being the atmosphere. D is a line or chord drawn as a tangent to the smaller crater, it is at right angles to the vertical transverse axis, the dark line which nearly bisects the large crater. With the inaccuracies inherent from the age of this formation, also in fitting the hyperbola, this may have been intended to go through the center of the crater.

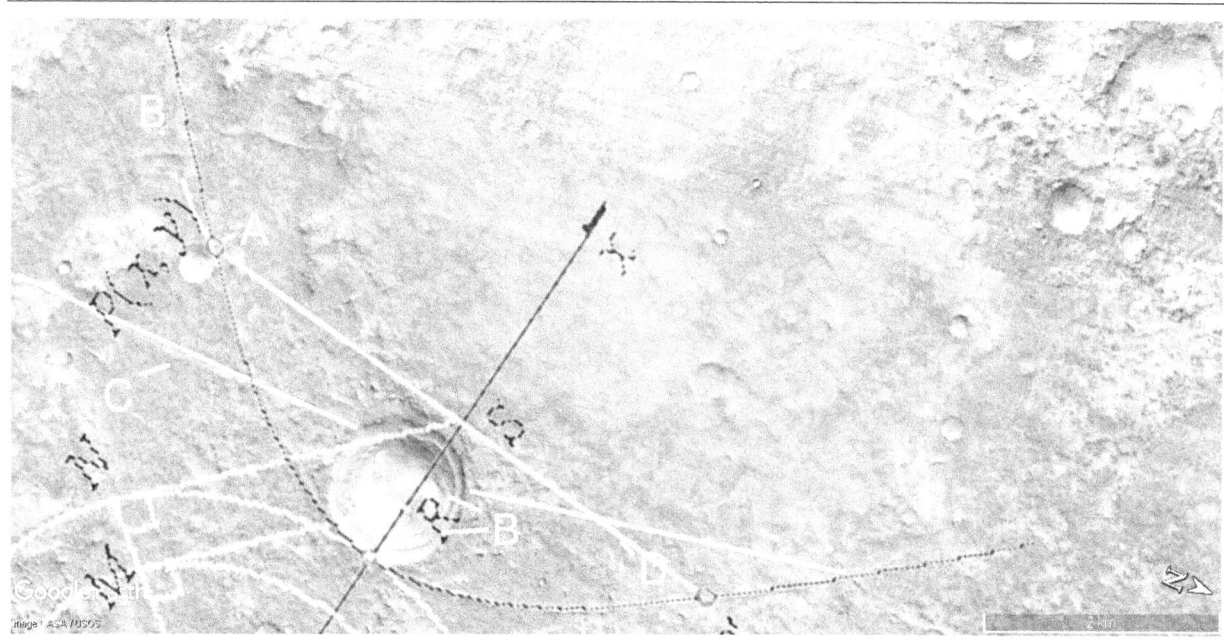

Conclusions

This introduction is intended to show an outline of the global hypothesis, explained in more details throughout the Martian Hypotheses books. There are hypothetical dams in it to collect water, also canals, water channels and lakes. There are two kinds of hypothetical cities, one based on more conventional rooms and walls. The other appears to be based on interconnected tubes. Hypothetical buildings are shown with collapsed areas like rooms. These are often connected with roads and tubes to each other and to farms, canals, dams, craters, and the oceans. With this overview the additional images in these books shows how these hypotheses repeat in many areas and extend into a more detailed global hypothesis. If these are natural then they are highly unusual, the parabolic formations do not appear to occur naturally.

Index

50, 60, and 70 degrees, 8
6 tubes in parallel going into the hill, 118
a wall or tube, 325
agriculture, 325

am floor, 273
amphitheater, 58
angled hill, 100
angled turn, 103

angles, 8
angular, 47, 48, 49, 310
angular walled segment, 214
approximately straight tube, 109
arc of parallel tubes, 87, 321
arch like bends in it, 109
arches, 62, 66, 125
arches and pillars, 123
arches or pillars, 77
arches or pillars in it are exposed, 84
arcs in the roof, 168
arcs on the roof, 204
area free of tubes, 70
Argd1444a, 270
as if it has been repaired, 201
asteroid belt, 241
atmosphere, 239, 241, 337
axis of symmetries, 209
axis of symmetry, 136, 161, 293, 332
bank, 39, 272
bend in the tube, 100
bisects the crater, 240
bisects the large crater, 239, 337
branched tubes, 68
branches, 60
branchings, 147
brick, 49
bricks, 47, 49, 50, 310
bridge, 9
broken layer, 159, 290

broken wall fragments, 157
broken wall segment, 159, 290
building blocks in the wall, 139
buried, 33
buried area, 33
buried walls, 33
canal, 183, 277
canal embankment, 275
canal wall, 275
canal wall like a single segment, 275
canals, 183, 247, 277
cavity, 9, 14, 25, 51, 193, 217, 219, 221, 228
cavity in the hill, 80
cavity in the tube segment, 149
cavity in the wall, 131
cavity on its roof, 88
cavity on the roof, 105, 315
cavity under it, 79
ceiling, 27, 35, 285, 286
ceiling degradation, 35
ceiling material, 31
ceilings, 10, 25
cement, 6, 15, 43, 45, 48, 61, 232, 245, 247, 275, 308
cement flaking off the roof, 201
cement roof, 77
cement skin is peeling off, 172
central meeting place, 302
centre of the nexus, 98
centre of the star, 136, 332
change directions in the tubes, 74

channel, 183, 277

chaotic exterior, 139

chord, 209, 239, 337

circle, 9, 210, 211

circle and hyperbola, 242

circle is overlaid, 106, 316

circle of walls, 302

circular roof of the hill, 105, 315

clean edge like cement, 273

Cobler Dome, 58, 105, 119, 162, 169, 294, 315

Cobler Dome but with three layers, 172

Cobler Dome or Amphitheatre, 168

collapse, 228, 285

collapsed, 39, 42, 52, 61, 64, 69, 193, 272, 296, 301, 307, 313

collapsed hill, 70, 73, 101, 103, 112, 115, 171, 195, 303

collapsed hill connected by a tube, 111, 317

collapsed hill with a central wall, 123

collapsed hill with exposed rooms, 148

collapsed hill with the holes as rooms, 114

collapsed hill., 191

collapsed hills, 74, 103, 119, 172, 329

collapsed hills connected by a tube, 111, 317

collapsed hills connected by tubes, 117

collapsed hollow hill, 53, 65, 66, 212, 217, 219, 221, 230, 273

collapsed hollow hills, 19, 125, 288

collapsed nexus, 74

collapsed on its roof, 137

collapsed roof, 6, 64, 104, 179, 214, 314

collapsed segment, 193

collapsed segment of the hill, 83

collapsed tile segment, 275

collapsed tube, 103, 174, 214, 236, 273, 296

collapsed tube goes to a crater, 102

collapsed tube or parallel tubes, 202

collapsed tube or tunnel going to the crater, 102

collapsed tunnel, 199

collapsed wall, 273

collapsed walls, 301

collapsing hill, 84, 89, 90, 104, 314

collapsing hills, 105, 315

collapsing tube breaking into segments like pillars, 149

complex of hollow hills, 311

concave, 230

concentric circles, 239, 337

concentric rings, 119

connect to the crater, 118

connected to a tube, 97

connections between these hill like tubes, 152

connections to hills, 108

constructed, 139

constructing, 141

construction technique, 127, 159, 290

continuation of the ring road, 75, 320
converge to the center, 28
crack, 129, 263
cracked top of the wall, 181
cracked wall, 181
cracks, 275, 281
cracks developing in the dam floor, 141
cracks have been sealed, 141
crater, 41, 51, 56, 60, 66, 67, 69, 199, 214, 233, 235, 263, 313
crater rim, 54, 60
craters, 54, 66
creep in the dam, 268
Crowned Face, 244
curved fields and a tube, 325
curved interior support, 19, 288
curved pale areas, 325
curved road, 226
curved shapes, 325
curved tube, 64, 68, 69, 214, 217, 313
curved tube going into the crater, 84
curved tubes, 218
curved wall, 27, 228, 234, 305
curved walls, 221, 222, 229
Cydonia Face, 244
Cymd259c, 262
Cymd259c2, 263
Cymd280a, 264
Cymd280a2, 265
Cymd280i, 266
Cymd280i2, 267

Cymd408a, 268
Cymd408a2, 269
Cymd454h, 279
Cymd454h2, 280
Cymd460a, 4
Cymd460a2, 5
Cymdhh470b, 33
Cymdhh470b2, 34
Cymhh209o, 284
Cymhh361i, 285
Cymhh464g, 6
Cymhh464g2, 7
Cymhh464i, 8
Cymhh464i2, 9
Cymhh465e, 10
Cymhh465e2, 11
Cymhh465g, 12
Cymhh465g2, 13
Cymhh465h, 14
Cymhh465h2, 14
Cymhh466d, 15
Cymhh466d2, 16
Cymhh466k, 17
Cymhh466k2, 18
Cymhh467, 19, 288
Cymhh467a, 20, 289
Cymhh469g, 21, 286
Cymhh469g2, 22, 287
Cymhh469h, 23
Cymhh469h2, 24

Cymhh469i, 25
Cymhh469i2, 26
Cymhh469j, 27
Cymhh469j2, 28
Cymhh469l, 29
Cymhh469l2, 30
Cymhh469m, 31
Cymhh469m2, 32
Cymhh471f, 35
Cymhh471f2, 36
Cymhh471g, 37
Cymhh471g2, 38
Cymmh464d, 6
dam, 143, 183, 193, 263, 264, 268, 277, 279
dam floor may be cement, 145
dam not a crater, 169
dam wall, 127, 139, 145, 263, 264, 266, 281
dam wall extends above this surface, 145
dam wall has layers, 141
dam wall is breaking, 129
dam wall is flat like cement, 139
dam wall is intact, 135, 332
dams, 18, 44, 127, 206, 211, 230, 281, 309, 334
dark line like a collapsed tunnel, 79
dark soil, 29
darker area may be a repair, 202
darker ceiling, 37
darker soil, 12
degraded hill with a tube, 70
degraded hollow hill, 64

degraded or buried, 6
degraded rooms, 21, 287
degraded skin on the roof, 99
degraded tube, 62
degraded tube going into the crater, 113
degraded tube going to a smaller hill, 202
degraded tubes, 65, 66
degraded wall, 141, 145, 221
degrading floor, 6
degrading into two walls, 127
degrading layer, 156
disconnected tube, 68
disintegrated roof, 29
dome, 31, 35, 56
double dam wall, 135, 332
double parabola, 209
double parabolas, 209
double wall, 131, 132, 138, 143, 183, 185, 273, 277
double wall as the roof collapsed, 149
double wall as the tube collapses, 137
double wall forming, 132
double wall of a pit dam, 181
double wall or collapsed tube, 148, 149
double walls, 131, 137
double walls of this dam, 135, 332
Ect1619, 273
Ect1619a, 274
Ect1643, 275
Ect1643a, 276
Ect1731k, 282

Ect1731k2, 283
Ecydhh1941, 325
Ecydhh1941a, 326
Ecydt1974, 327
Ecydt1974a, 328
edge has a raised lip, 157
edge of a patched roof, 174
edge of the smooth cement skin, 174
edge of this circular roof, 105, 315
ellipse, 210, 211
ellipses, 170
elliptical, 160, 211, 291
elliptical hill, 123
enclosure, 61, 233
enclosures, 206, 334
Engineers, 266, 285
entrance, 233, 285
entrance to the next walled area, 137
entrances, 230
eroded hollow hill, 68
eroded nexus or collapsed hill, 70
eroded parallel tubes, 78
eroded road or tube, 79
eroded rooms, 15
eroded tube, 77, 118, 148
eroded tube segments, 67, 98
eroded tubes, 67, 112, 212
eroded wall material, 33
eroded wall., 299
exposed grid, 160, 292

exposed parabolic layers, 166
exposed roof of a hollow hill, 179
exposed room in the nexus, 87, 321
exposing the layers, 162, 294
extended tube, 68
external walls, 249
eye like shape, 186
faint rooms, 14
faint tube going into the crater, 102
faint tubes, 206, 334
faint tubes go into the crater, 72, 319
faint tubes go into the hill, 79
faint wall parallel to the larger wall, 148
faint walls, 31
fainter ridges, 7
farm, 251
farming, 101
farming area, 121, 323
farms, 327
ferns, 244, 251
Ferns, 208
Fibonacci branches, 244
fine walls or tubes, 302
finer wall, 27
fit of a hyperbola, 240
flat areas like cement over the tubes, 75, 320
flat cement areas, 80
flat sheet like a roof, 96, 322
flat sheets of cement, 87, 321
flat smooth area like cement, 80

flat tube, 66
flattened part of the nexus, 75, 320
flattened segments of a roof, 87, 321
floors, 31
foci, 209
focus, 240, 297
fork, 77
fork and a small hill, 125
forked tube, 73, 75, 76, 97, 101, 111, 112, 125, 317, 320
forked tube ends, 77
foundations, 18
four faint parallel tubes, 72, 319
four parallel tubes, 70
furniture, 6, 10, 29, 31
gap growing, 275
geometric statement, 209, 210
geometry, 242
gravity, 239, 337
grid of tubes, 114
grid or mesh of tubes, 96, 322
groove, 67, 183, 185, 206, 208, 210, 214, 241, 277, 334
grooves, 66, 208, 210, 219, 223
grout, 275
growing crops, 211
gutter, 47, 310
habitat, 273, 301, 302
have broken off, 58
Held1095f, 281
Held1095f2, 281

Held117a, 297
Held1186, 298
Held1222c, 299
Held1222c2, 300
Held1222e, 301
Held1232, 302
Held1244, 302
Held1258, 303
Held1258b, 304
Held1295b, 305
Held1295b2, 306
Helhh1117, 296
Hellas, 273
highly eroded, 33
highly eroded dam, 4
highly eroded tubes, 112
hill composed of rooms, 37
hill connected to a crater, 82, 329
hill connecting to this mesh, 88
hill is artificial, 303
hill is collapsing, 82
hill is smooth like cement, 169
hill roof is settling, 174
hill wall, 217
hill with tubes going into it, 91
hills have collapsed, 172
hole in the tube, 137
holes as rooms, 115
holes in the roof, 100, 111, 317
hollow, 10, 12, 61, 127, 206, 214, 285, 334

hollow hill, 39, 41, 42, 45, 47, 52, 53, 54, 56, 61, 62, 64, 88, 105, 116, 121, 133, 150, 177, 189, 191, 193, 197, 199, 205, 232, 247, 264, 272, 307, 310, 311, 315, 323, 325

hollow hill roof, 296

hollow hill with cavities in the roof, 104, 314

hollow hills, 52, 56, 58, 69, 133, 313

hollow like a collapsed tube, 118

hollow of eroded rooms, 31

hollow walls, 273

horizontal tube, 123

hyperbola, 238, 239, 244, 336, 337

hyperbola is a tangent, 240

hyperbolic orbit, 242

hyperbolic wall, 241

Images, main section, 4

imprint of a completely eroded hill and tube, 112

Index, 338

inner wall, 211

intact ceiling, 21, 25, 35, 287

intact ceilings, 14, 25, 27, 33

intact flat roof like in the nexus, 99

intact pale ceilings, 29

intact roof, 27, 217

intact roofed segment, 214

intact rooms, 15, 35

interior support, 204, 212, 273, 296

interior support exposed, 83, 107

interior supports, 19, 53, 66, 189, 195, 203, 219, 288

internal arches or pillars, 108

intersecting at right angles, 228

intersecting tube, 74

intersection, 113

intersection of tubes, 72, 319

intersection with the other tube, 99

intersections, 74

intersects another tangent, 210

irregular rooms, 31

Ishh2306, 329

Ishh2306a, 330

knot, 244

knotted tube, 123

lanes, 42, 307

large habitat, 72, 319

large nexus, 75, 76, 87, 96, 320, 321, 322

large tubes, 119

large walled field, 78

larger amphitheatre, 119

larger hollow wall, 273

larger nexus, 74

larger rooms, 12, 14

larger tubes coming out of the nexus, 76

latis rectum, 297

Latis Rectum, 303

layer is shown also peeling off, 156

layer like a step in a dome, 172

layers, 15, 58

layers going into the crater, 169

layers in the collapsing hill, 102

layers in the hill, 105, 169, 315

layers in the side of the hill, 166

layers of skin breaking off, 137

leaf, 208, 209

lighter walls, 15

line down the centre, perhaps a tube, 118

line of hills, 82

lip, 264

lip dams, 133

long entrance, 247

long tube going into a hill, 123

longer wall, 233

loop of a tube, 75, 320

main nexus, 97

major and minor axis, 210

major intersection, 311

many tubes cross each other, 97

mathematical pattern, 242

meeting place, 9, 62, 65, 66, 67

mesh of tubes, 76, 88

meshed tubes, 96, 322

more easily eroded, 15

more tubes, 108

movement around the nexus, 75, 320

multiple layers under it, 159, 290

multiple parallel tubes, 78

narrow fork, 75, 320

narrow tube, 148

narrow wall, 273

narrow walls, 301

Nefertiti, 244

nest, 12

network of walls, 303

nexus, 9, 14, 21, 65, 87, 89, 96, 111, 114, 115, 285, 287, 302, 317, 321, 322

nexus at ground level, 76

nexus hill, 76

nexus of walls or tubes, 17

nine parallel tubes, 70

nine parallel tubes going to the nexus, 96, 322

old equator, 244

one tube crosses over the other like a knot, 75, 320

one wall passes over another, 299

open rooms, 12

optical illusion, 211

orbits, 239, 337

outer circular shape of the nexus, 75, 320

outer skin, 302

pair of parallel tubes, 103

pair of tubes, 116

pale curves, 326

pale fields, 327

pale wall, 7

parabola, 5, 16, 40, 55, 57, 59, 63, 85, 92, 94, 101, 120, 122, 126, 130, 136, 140, 142, 157, 158, 165, 173, 175, 178, 180, 184, 190, 194, 198, 200, 204, 213, 215, 218, 224, 225, 231, 234, 265, 267, 272, 276, 278, 280, 297, 304, 305, 324, 332

parabolas, 20, 69, 81, 106, 124, 128, 134, 143, 144, 146, 151, 160, 161, 163, 167, 170,

182, 188, 192, 204, 207, 209, 220, 222, 229, 231, 246, 248, 253, 269, 274, 281, 283, 289, 291, 293, 295, 313, 316, 325, 326, 328, 330, 333, 335

parabolic, 15, 133, 263

parabolic arch, 263

parabolic arches on its roof, 150

parabolic arcs to strengthen the roof, 160, 292

parabolic dams, 270

parabolic dome, 153

parabolic domes, 155

parabolic domes from ground level, 154

parabolic groove, 18

parabolic layers, 162, 294

parabolic layers of bricks are exposed, 162, 294

parabolic segment, 156

parabolic shape, 209

parabolic tube, 123

parabolic wall, 247, 303

parabolic wave, 160, 291

parallel latis rectums, 231

parallel lines, 208, 209

parallel ridges on the roof, 164

parallel to go into the nexus, 98

parallel tubes, 62, 73, 77, 96, 98, 101, 109, 113, 117, 121, 147, 322, 323

Parallel tubes, 87, 321

parallel tubes go to the nexus, 87, 321

parallel tubes going into the crater, 73

parallel tubes going to the crater, 107

parallel tubes going to this crater, 77

parallel tubes or walls, 78

partially buried, 6, 9, 10, 21, 31, 287

partially collapsed, 301

passage in and out of the hills, 301

patch, 39, 45, 58, 264, 272

patched, 45

peeled, 197, 264

peeled off, 53

peeled skin, 197

peeling, 264

pillar, 263, 285

pillars or arches, 112

pillars used in its construction, 129

pit, 15, 47, 51, 62, 206, 219, 226, 230, 231, 232, 233, 264, 310, 334

pit dam, 133, 138, 186, 245, 247, 282

pit dam wall, 182

pit dam walls are cracking, 143

pit dams, 145, 183, 187, 207, 249, 277, 282, 335

pit wall, 206, 212, 213, 214, 219, 223, 225, 228, 334

pit walls, 206, 219, 334

pits, 44, 49, 52, 206, 230, 309, 334

planet or moon, 239, 337

posts, 223

Prca480, 39, 271

Prca480a, 40, 272

Prd1033, 230

Prd1050e, 232

Prd1059, 245
Prd1062, 247
Prd1062a, 248
Prd1065, 249
Prd1065a, 250
Prd879c, 131
Prd879d, 132
Prd882, 133
Prd886c, 135, 331
Prd886c2, 136, 332
Prd886d, 137
Prd886e, 138
Prd891b, 139
Prd891b2, 140
Prd891c, 141
Prd891c2, 142
Prd901a, 143
Prd901a2, 144
Prd911b2, 146, 333
Prd965b, 181
Prd965b2, 182
Prd965c, 183, 277
Prd965c2, 184, 278
Prd965d, 185
Prd965g, 186
Prd965g2, 187
Prhh1000, 199
Prhh1000a, 200
Prhh1001, 201
Prhh1006, 202

Prhh1014, 203
Prhh1014a, 204
Prhh1015, 205
Prhh1018, 206, 334
Prhh1018a, 207, 335
Prhh1019, 208
Prhh1019a, 209
Prhh1021, 210
Prhh1021a, 211
Prhh1030, 212
Prhh1030a, 213
Prhh1031a, 214
Prhh1031a2, 215
Prhh1031c, 216
Prhh1031d, 217
Prhh1031d2, 218
Prhh1031e, 219
Prhh1031e2, 220
Prhh1031f, 221
Prhh1031f2, 222
Prhh1031k, 223
Prhh1031k1, 224
Prhh1031k2, 225
Prhh1031l, 226
Prhh1031l2, 227
Prhh1031l3, 228
Prhh1032, 228
Prhh1032a, 229
Prhh1033a1, 231
Prhh1033a2, 231

Prhh1068c, 252
Prhh1068c2, 253
Prhh1821, 311
Prhh498, 42, 307
Prhh533, 46
Prhh563, 53
Prhh569, 54
Prhh569a, 55
Prhh570, 56
Prhh570a, 57
Prhh578, 58
Prhh578a, 59
Prhh580, 60
Prhh581a, 61
Prhh751, 91
Prhh751a, 92
Prhh752, 93
Prhh752a, 94
Prhh789, 102
Prhh936, 150
Prhh936a, 151
Prhh939, 152
Prhh939a, 153
Prhh940, 154
Prhh944a, 156
Prhh944b, 157
Prhh944ba, 158
Prhh944c, 159, 290
Prhh944c2, 159, 291
Prhh944f, 160, 292

Prhh944f2, 161, 293
Prhh944j, 162, 294
Prhh944j2, 163, 295
Prhh945, 164
Prhh945a, 165
Prhh946, 166
Prhh946a, 167
Prhh949, 168
Prhh951, 169
Prhh951a, 170
Prhh952, 171
Prhh956, 172
Prhh956a, 173
Prhh958b, 174
Prhh958b2, 175
Prhh958c, 176
Prhh960, 177
Prhh960a, 178
Prhh961, 179
Prhh961a, 180
Prhh994, 195
Prhh994a, 196
protruding walls, 27
Prr493, 41
Prr499, 43, 308
Prr508, 44, 309
Prr509, 45
Prr533a, 47, 310
Prr533b, 48
Prr533c, 49

Prr533d, 50
Prr546, 51
Prr561, 52
Prt1051a, 233
Prt1051c2, 234
Prt1053, 235
Prt1054, 236
Prt1054a, 237
Prt1055, 238, 336
Prt1055a, 239, 337
Prt1055b, 240
Prt1055c, 241
Prt1055d, 242
Prt1055e, 243
Prt1056, 244
Prt1059a, 246
Prt1066, 251
Prt586, 62
Prt586a, 63
Prt592, 64
Prt593, 65
Prt594, 65
Prt605, 66
Prt615, 67
Prt641, 68, 312
Prt641a, 69, 313
Prt657, 70
Prt662, 71, 318
Prt671, 72
Prt681, 73

Prt682, 74, 319
Prt684, 76
Prt686, 77
Prt687, 78
Prt697, 79
Prt704, 80
Prt704a, 81
Prt708, 82
Prt712, 83
Prt713, 84
Prt713a, 85
Prt714, 86, 321
Prt719, 87
Prt738, 88
Prt742, 89
Prt746, 90
Prt753, 95, 322
Prt756, 96
Prt762, 97
Prt762a, 98
Prt772, 99
Prt774, 100
Prt774a, 101
Prt792, 103
Prt798, 104, 314
Prt804, 105, 315
Prt804a, 106, 316
Prt805, 107
Prt812, 108
Prt813, 109

Prt813a, 110
Prt814, 111, 317
Prt819, 112
Prt822, 113
Prt827, 114
Prt838, 115
Prt841, 116
Prt842, 117
Prt844, 118
Prt849, 119
Prt849a, 120
Prt857, 121, 323
Prt857a, 122, 324
Prt860, 123
Prt860a, 124
Prt863, 125
Prt863a, 126
Prt870d, 127
Prt870d2, 128
Prt870f, 129
Prt870f2, 130
Prt912, 147
Prt912a, 148
Prt912b, 149
Prt978, 187
Prt978a, 188
Prt986, 189
Prt986a, 190
Prt987, 191
Prt987a, 192

Prt989, 193
Prt989a, 194
Prt998, 197
Prt998a, 198
raised road, 43, 308
ratio 3
    4
       5, 187
rectangle, 164
rectangular cavity, 164
rectangular dam, 172
rectangular enclosure, 235
rectangular mesh of tubes, 77
rectangular meshes of tubes, 78
rectilinear rooms, 9
regular bulges like the tube is collapsing, 77
regular shapes like tiles, 129
regular spacing like tiles, 275
regular spacings, 223
regular tile spacings, 275
regular undulations, 129
regular vertical pillars, 139
reinforcing the roof, 160, 292
repaired, 54, 132
ridge like grout, 275
ridges like grout, 275
right angled, 47, 310
right angled intersection, 77
right angled segment, 201
right angled triangle, 187

right angled tube, 61, 216, 223

right angled turn into a hollow hill, 72, 319

right angled walled formation, 172

right angles, 18, 54, 208, 218, 228, 235, 286

right angles in a mesh, 75, 320

right angles to another tube, 98

ring of tubes, 101

ring road, 302

ring roads, 78

road, 41, 42, 43, 44, 45, 46, 47, 48, 49, 50, 52, 53, 56, 210, 216, 226, 236, 239, 286, 307, 308, 309, 310, 337

road connects to a crater, 116

road or tube, 325

road went through the hill, 118

roads, 10, 52, 56, 210, 211, 219, 226, 311

roads connecting to hollow hills, 311

roads connecting to the large crater, 311

roof, 9, 27, 39, 58, 62, 64, 176, 197, 199, 206, 217, 272, 334

roof close to an ellipse, 168

roof has collapsed, 99, 104, 302, 314

roof has collapsed onto the floor, 99

roof intact, 172

roof is close to a circle, 106, 316

roof is settling, 168, 201

roof material, 12

roof partially exposed, 202

roof skin, 202

roof there is a tube, 83

roof with rooms under them, 75, 320

roofs, 27, 52

room, 27, 285

room hypothesis, 285

room like shapes, 21, 287

rooms, 6, 9, 10, 12, 14, 21, 25, 29, 31, 35, 37, 111, 286, 287, 317

rooms in the hollow hill, 107

root pattern, 244

round nexus, 9

rounded dome, 27

sealed cracks, 141

sealed rooms, 9

secondary wall, 138

segment of the tube, 325

segments might be rooms, 87, 321

semicircular shape, 28

settled, 53, 60, 176

settled area on its roof, 98

settled area on the roof, 91

settled part of the roof, 179

settled roof, 53, 107, 174

settled roof segments, 205

settled roofs, 179

settled segments of the roof, 296

several parallel layers, 139

shallow pits, 211

shallow wall, 211

sharp turn in the tube, 79

sharp wall, 131

sharp wall intersection, 131

short wall, 233
signs of the cement breaking off, 156
six parallel tubes, 121, 323
six parallel tubes go into the nexus, 97
skin, 53, 58, 264
skin is breaking off, 99
skin peeled, 62
skins that have eroded away, 162, 294
slope, 241
small cavity in the wall, 135, 332
small collapsed hill, 80, 84, 89
small crater connected to a hill, 109
small crater has many tubes, 97
small domes, 31
small entrance between the walls, 135, 332
small forked tube, 98
small habitat connected by tubes, 80
small hill connected to the tube, 147
small hill is connected to a tube, 80
small hills connected to tubes, 97
small hollow hill, 311
small nexus, 73, 75, 76, 320
small nexus connecting the tubes, 107
small partially collapsed hills, 80
small regular mounds, 141
small rooms, 7, 12
small tube, 149
small tube coming out of its apex, 156
small tube goes into the hill, 79
small tube going into a hill, 82

small tube going into the crater, 84, 102
small tubes connecting to craters, 91
smaller hills connected by tubes, 166
smaller tube, 91
smaller tube segments remaining intact, 97
smooth, 47, 310
smooth cement side, 174
smooth dam floor, 268
smooth dam floor like cement, 143
smooth edge to this layer like cement, 172
smooth interior, 247
smooth like cement, 145
smooth sides of the hill like cement, 156
smooth skin like cement, 162, 294
smooth slope, 214, 230
smoother ceiling material, 37
smoother ground, 58
smoother hollow like an arch, 139
squarish area surrounded by tubes, 87, 321
squarish array of parallel tubes, 78
squarish array of walls or tubes, 82
squarish cavity, 176
squarish segment, 176
squarish tiles, 275
squarish walled segments, 87, 321
standard parabola, 81
standard shaped parabola, 158
star shaped wall, 135, 332

straight, 28, 171

straight grooves, 18, 208

Straight lines, 10

straight ridges, 19, 288

straight sections, 182

straight segments, 189

straight tube, 89, 103, 115, 121, 171, 323

straight tube from the hill to the crater, 113

straight tube going to a collapsing hill, 107

straight wall, 6, 10, 177

straight walls, 12, 20, 186, 199, 289

strengthening the roof, 160, 292

supports, 223

surface of a planet, 239, 337

suspended roads, 286

symmetrical formation, 114

symmetrical wall, 104, 314

symmetrical walls, 114

symmetry, 226

T intersection, 72, 77, 97, 319

T intersection of tubes, 87, 321

T intersection with a small tube, 113

T junction, 226

tangent, 210, 239, 337

tangent to the large crater, 238, 336

tangents, 210

ten parallel tubes, 88

the roof skin is peeling, 99

thicker tube, 105, 315

thicker wall, 273

thicker wall or tube, 114

thickness of the wall, 157

thinner layers broken together, 159, 290

three cavities in the roof, 112

three dimensional, 286

three parallel tubes, 108

three parallel tubes come from hills, 82

three tubes going into a crater, 90

three tubes interleave, 97

tiles, 47, 275, 310

tiles or patches, 168

top of the dam wall is degrading, 138

top of the layer, 159, 290

trapezoid, 205

trapezoidal, 195

trapezoids, 14

tree branches, 147

triangles, 8, 14

triangular, 7, 14

triangular array of tubes, 77

triangular hill, 102

triangular walls, 14

tube, 39, 40, 43, 52, 56, 58, 60, 61, 64, 65, 66, 68, 69, 138, 171, 191, 193, 197, 199, 206, 208, 209, 216, 217, 219, 228, 232, 233, 235, 236, 247, 272, 285, 296, 301, 308, 313, 334

tube and layers, 137

tube appears to be degrading, 102

tube between two craters, 113, 329

tube coming out of the hill, 114

tube connected to a small crater, 112
tube connecting the parallel tubes, 78
tube connects to the hill, 115
tube crosses each one, 78
tube crosses other tubes, 72, 319
tube crosses the hill, 87, 321
tube crossing another, 75, 320
tube ending in a small hill, 107, 116
tube ends in a triangular hill, 149
tube exiting the hill, 90
tube fragments, 125
tube free area, 101
tube from the chain of hills, 105, 315
tube from two small hills, 114
tube goes into the hill, 79
tube goes through a collapsed hill, 111, 317
tube going from the hill, 168
tube going into a collapsing hill, 88
tube going into a crater, 107, 114
tube going into a hill, 82
tube going into the crater, 70, 77, 79, 80, 115, 121, 323
tube going into the hill, 83, 100, 104, 314
tube has collapsed, 79
tube intersection, 65, 67, 206, 334
tube intersections, 65, 66, 74
tube is breaking up into segments, 125
tube is hollow like the roof, 75, 320
tube is in good condition, 149
tube is more eroded, 137

tube mesh, 89, 96, 112, 168, 322
tube mesh around a hill, 112
tube network, 67
tube nexus, 62, 65, 66, 68, 73, 101
tube on the roof, 100, 107
tube on the roof of the hill, 84
tube or straight collapsed part of the roof, 202
tube parallel to the side of the hill, 98
tube running up the hill side, 154
tube segments, 67, 68
tube terminates, 88
tube that may have collapsed, 109
tube to the crater, 80
tube turns sharply to the left, 82
tubes, 19, 54, 56, 58, 61, 62, 64, 65, 66, 67, 68, 74, 75, 206, 207, 208, 210, 219, 221, 230, 235, 286, 288, 320, 334, 335
tubes and a settled area, 201
tubes and hollow hills, 325
tubes are collapsing, 99
tubes are much thicker, 78
tubes at right angles, 76
tubes at right angles to it, 104, 314
tubes between collapsed hills, 105, 315
tubes branching, 68
tubes came after the crater, 78
tubes climbing the hill, 76
tubes come down the hill parallel, 76
tubes come out of the hill, 76, 98
tubes coming off this road, 116

tubes coming out of a hill, 147
tubes coming out of the hill, 101
tubes coming to a nexus, 101
tubes connect to a collapsed hill, 115
tubes connect to a larger segment like a room, 99
tubes connect to a pair of craters, 116
tubes connect to the nexus, 76
tubes connect to this road, 117
tubes connected to the crater, 111, 317
tubes connecting hills, 109, 112
tubes connecting into a longer tube, 84
tubes connecting to a long hill, 91
tubes connecting to each other, 89
tubes connecting to hills, 89
tubes connecting to hollow hills, 108
tubes connecting to the hill, 82
tubes connecting to the hills, 111, 317
tubes converge, 121, 323
tubes cross each other, 76
tubes crossing the parallel tubes, 96, 322
tubes fork many times, 76
tubes forking like a tree, 116
tubes from the crater, 70
tubes from the crater and collapsed hill, 119
tubes go into the collapsing hills, 89
tubes go into the crater, 117
tubes go to the crater, 111, 317
tubes going down the side of the hill, 168
tubes going into a collapsed hill, 102

tubes going into a crater, 72, 76, 80, 319
tubes going into the collapsed hill, 72, 319
tubes going into the collapsing hill, 93
tubes going into the crater, 82
tubes going into the hill, 123
tubes going into the hill with three cavities, 112
tubes going into the hollow hill, 118
tubes going into the nexus, 73
tubes going into this nexus, 72, 319
tubes going into three collapsed hills, 72, 319
tubes going to craters, 73
tubes in varying condition, 125
tubes like a field, 72, 319
tubes making a walled area, 74
tubes must have come after the crater, 117
tubes on the roof, 115
tubes or eroded segments on the roof, 160, 292
tubes or roads in the field, 329
tunnel, 232, 296
twenty more tubes going into the nexus, 97
twisted shape like a rope, 104, 314
twisted tube, 104, 314
two hills linked by tubes, 73
two parabolas, 105, 209, 315
two straight tubes at right angles, 113
two straight walls, 165

two tubes parallel, 75, 320

unnatural shape for a crater, 114

unusual shaped curved roof, 93

upper layer, 172

volcanic ash, 15

wall, 42, 54, 183, 193, 195, 211, 212, 214, 216, 219, 223, 228, 230, 233, 236, 244, 247, 263, 266, 275, 277, 285, 301, 303, 307

wall and crater, 240

wall bulges outwards in an arc, 138

wall erosion, 299

wall in a dam, 133

wall intersection, 228

wall is eroded or breaking, 268

wall is in better condition, 127

wall like an interior support, 174

wall like some tubes, 135, 332

wall of a collapsed tube, 148

wall of the tube has split, 99

wall on the edge of this roof, 202

wall or sealed crack, 141

wall segment, 230

walled, 54

walled area with a collapsed hill, 113

walled areas, 305

walled dam, 193

walled enclosure, 233

walled field, 90

walled fields, 76, 107, 299, 305, 329

walled hill, 69, 313

walled or tubed field, 74

walled room, 21, 287

walled segment of the dam, 135, 332

walled segments around a hill, 202

walled structures, 6

walls, 8, 9, 10, 11, 14, 15, 17, 19, 25, 27, 28, 29, 31, 33, 35, 54, 69, 185, 199, 211, 220, 228, 230, 233, 235, 239, 245, 249, 251, 281, 286, 288, 305, 313, 337

walls are hollow, 301

walls converge, 21, 287

walls have collapsed, 301

walls have dark spots, 138

walls of rooms, 10

walls or tubes, 137

water channel, 141, 143, 145, 279, 282, 305

water table, 187

wavy, 233

wavy tube, 67, 72, 88, 109, 123, 319

wider tube, 87, 321

wider tube perhaps like a room, 76

worn down walls, 7

zig zag, 110

www.ingramcontent.com/pod-product-compliance
Lightning Source LLC
Chambersburg PA
CBHW062212220526
45471CB00009B/3166